ASYMPTOTIC THEORY OF ꓩꓯPARATED FLOWS

Boundary-layer separation from a rigid body surface is one of the fundamental problems of classical and modern fluid dynamics. The major successes achieved since the late 1960s in the development of the theory of separated flows at high Reynolds numbers are in many ways associated with the use of asymptotic methods. The most fruitful of these has proved to be the method of matched asymptotic expansions, which has been widely used in mechanics and mathematical physics. There have been many papers devoted to different problems in the asymptotic theory of separated flows, and we can confidently speak of the appearance of a new and very productive direction in the development of theoretical hydrodynamics. This book will be the first to present this theory in a systematic account.

The book will serve as a useful introduction to the theory, and will draw attention to the new possibilities that application of the asymptotic approach provides.

Asymptotic Theory of Separated Flows

Vladimir V. Sychev
*Central Aero-Hydrodynamic
Institute, Zhukovskii, Russia*

Anatoly I. Ruban
*University of Manchester,
England*

Victor V. Sychev
*Central Aero-Hydrodynamic
Institute, Zhukovskii, Russia*

Georgi L. Korolev
*Central Aero-Hydrodynamic
Institute, Zhukovskii, Russia*

Translated by
Elena V. Maroko
*Moscow Physical-Technical
Institute, Zhukowskii, Russia*

English edition edited by

A.F. Messiter
University of Michigan

Milton Van Dyke
Stanford University

CAMBRIDGE
UNIVERSITY PRESS

CAMBRIDGE UNIVERSITY PRESS
Cambridge, New York, Melbourne, Madrid, Cape Town, Singapore, São Paulo

Cambridge University Press
The Edinburgh Building, Cambridge CB2 8RU, UK

Published in the United States of America by Cambridge University Press, New York

www.cambridge.org
Information on this title: www.cambridge.org/9780521455305

© Cambridge University Press 1998

First published 1998
This digitally printed version 2008

A catalogue record for this publication is available from the British Library

Library of Congress Cataloguing in Publication data

Asimptoticheskaia teoriia otryvnykh techenii. English
Asymptotic theory of separated flows / Vladimir V. Sychecv ... [et
al.].
p. cm.
Translation of: Asimptoticheskaia teoriia otryvnykh techenii.
Includes bibliographical references.
ISBN 0-521-45530-8
1. Fluid dynamics. 2. Boundary value problems – Asymptotic theory.
I. Sychev, V. V. (Vladimir Vasil'evich)   II. Title.
TA357.A8213 1997   97-31078
532'.0533–dc21   CIP

ISBN 978-0-521-45530-5 hardback
ISBN 978-0-521-06587-0 paperback

Contents

Preface

The major successes achieved since the late 1960s in the development of the theory of separated flows of liquids and gases at high Reynolds numbers are in many respects associated with the use of asymptotic methods of investigation. The most fruitful of these has proved to be the method of matched asymptotic expansions, which has also become widely used in many other fields of mechanics and mathematical physics (Van Dyke, 1964, 1975; Cole, 1968; Lagerstrom and Casten, 1972). By means of this method, important problems have been solved concerning boundary-layer interaction and separation in subsonic and supersonic flows, the nature of these phenomena has been made clear, and basic laws and controlling parameters have been ascertained. The number of original papers devoted to different problems in the asymptotic theory of separated flows now amounts to many dozens. We can confidently speak of the appearance of a new and very productive direction in the development of theoretical hydrodynamics. However, it has not yet received an adequate systematic account, and the objective of this book is to make an attempt to partially fill this gap.

Of necessity, the authors have restricted themselves to consideration of a range of problems most familiar to them, associated with two-dimensional separated flows of an incompressible fluid, assumed to be laminar. Thus, the book excludes the results of many investigations of separated gas flows at transonic, supersonic, and hypersonic speeds, although chronologically the problem of boundary-layer separation in supersonic flow was solved earlier than the others (see Chapter 1).

Here we leave completely untouched the study of turbulent boundary-layer interaction and separation, as well as the large class of problems of internal flows in channels and tubes. Detailed reviews have been devoted to many of the above-mentioned problems, among which we should point out especially those of Brown and Stewartson (1969), Neiland (1974, 1981), Stewartson (1974, 1981), Lagerstrom (1975), Messiter (1979, 1983), Adamson and Messiter (1980), and Smith (1982a).

Work in the field of the asymptotic theory of separated flows is still in a stage of intensive development, and the authors are well aware of the fact that while this book is being prepared for publication many new results will appear. Nevertheless, we hope that the book may serve as a useful introduction to the theory, and may draw attention to the new possibilities that application of the asymptotic approach provides us in the study of many problems. In the authors' opinion, the main value of such an approach does not consist in the fact that it provides an approximate method of solving a problem (whose accuracy may then prove very low if the actual value of the small parameter in the problem is not very small compared with unity), but rather that it reveals the physical mechanism of a phenomenon and establishes its controlling dimensionless parameters, which play the role of similarity parameters. This circumstance quite makes up for the known difficulty of constructing a set of matched asymptotic expansions for a large number of flow subregions. The results of asymptotic analysis should always be taken into account (or even introduced directly) for the construction of reliable numerical methods of computation. Only in such a way can one take into account all the local singularities of a flow and create an adequate mathematical model.

Finally, one of the important consequences of the asymptotic theory of viscous separated flows of an incompressible fluid is its deep relationship with the classical theory of ideal fluid flows with free streamlines. This refers not only to the problem of establishing the asymptotic behavior of the global flow field past a blunt body of finite dimensions (Chapter 6), but also to the local singularities of the solutions of Kirchhoff's theory at the points of flow separation from a body surface, whose occurrence is impossible to explain within the limits of inviscid-flow theory, since in reality they are associated with the process of interaction between the boundary layer and the external potential flow (Chapters 1, 2, 5).

Several authors have contributed to the writing of this book, so it is possible that (despite the editor's efforts) the reader will find that the style of narration throughout the book is not quite uniform. Chapters 1 and 6 were written by V. V. Sychev, Chapters 2, 4, and Section 2 of Chapter 3 by A. I. Ruban, Chapter 5 and Sections 1 and 3 of Chapter 3 by Vic. V. Sychev, and Chapter 7 by G. L. Korolev and A. I. Ruban. The authors are grateful to S. N. Timoshin, who wrote Section 7 of Chapter 6 at their request. The authors also consider it a pleasant duty to acknowledge the great help of K. E. Ivashkina, G. E. Tsarkova, and E. B. Ivanova in preparing illustrations and graphs.

The authors would like to express special gratitude to Professor G. Yu. Stepanov, who took pains to read the manuscript and made some valuable comments.

V. V. Sychev

Preface to the English Edition

The authors are honored by the publication of the English edition by Cambridge University Press, edited by two colleagues who are well known for their work in fluid mechanics – Professor A. F. Messiter of the University of Michigan and Professor M. D. Van Dyke of Stanford University. The authors are deeply indebted to them for their interest in this book, the initiative for the translation into English, and for their considerable effort in the editing, which has no doubt improved the original text. Our work rests heavily on the present state of development of the asymptotic analysis of problems in fluid mechanics, which in many respects is due to Professor Van Dyke and Professor Messiter.

Since the publication of the Russian edition of this book in 1987, there have appeared numerous papers devoted to the asymptotic theory of separated flows. Therefore, in the preparation of the English edition the authors have attempted to bring the book more up to date. In particular, Chapters 2 and 6 have been expanded by the addition of two new sections, (Section 4 and Section 6, respectively) and Chapter 7 has been completely revised. Elsewhere we have confined ourselves to improvements and short additions in the text and to expanding the list of references.

V. V. Sychev

1

The Theory of Separation
from a Smooth Surface

The problem of studying flow separation from the surface of a solid body, and the flow that is developed as a result of this separation, is among the fundamental and most difficult problems of theoretical hydrodynamics. The first attempts at describing separated flow past blunt bodies are known to have been made by Helmholtz (1868) and Kirchhoff (1869) as early as the middle of the nineteenth century, within the framework of the classical theory of ideal fluid flows. Although in the work of Helmholtz one can find some indication of the viscous nature of mixing layers, which are represented in the limit as tangential discontinuities, nonetheless before Prandtl (1904) developed the boundary-layer theory at the beginning of the twentieth century there were no adequate methods for investigating such flows. Prandtl was the first to explain the physical nature of flow separation at high Reynolds number as separation of the boundary layer. In essence, Prandtl's boundary-layer theory proved to be the foundation for all further studies of the asymptotic behavior of either liquid or gas flows with extremely low viscosity, i.e., flows of media similar to air or water, with which one is most commonly concerned in nature and engineering. Therefore we begin with a brief description of the main ideas of this theory.

1.1 Boundary-Layer Theory

The Navier–Stokes equations with corresponding boundary and initial conditions are usually used as the basic system of relations describing a viscous fluid flow. These equations reflect sufficiently well the behavior of real liquids and gases.

For an incompressible fluid in the absence of body forces, the Navier–

1

Stokes equations can be written in the form

$$\frac{\partial \mathbf{V}}{\partial t} + (\mathbf{V} \cdot \nabla)\mathbf{V} + \frac{1}{\rho}\nabla p = \nu \nabla^2 \mathbf{V}, \quad \nabla \cdot \mathbf{V} = 0. \tag{1.1.1}$$

The symbols adopted here are standard: $\mathbf{V}$ is the flow velocity vector, p the hydrodynamic pressure, ρ the density, and ν the kinematic viscosity of the fluid.

Let us consider the problem of flow around a body in a uniform stream. The boundary conditions for this problem will be the condition that the fluid adhere to the solid surface and also the conditions defining the values of the flow variables at infinity.

Henceforth, we shall always write both the equations and boundary conditions in dimensionless form. For this purpose we shall refer the independent variables $\mathbf{r}$ to the characteristic body length L, the components of the velocity vector $\mathbf{V}$ to a characteristic speed U_0 (e.g., to an unperturbed flow speed), and the pressure to twice the dynamic pressure ρU_0^2; and we take L/U_0 as the characteristic time.

The main dimensionless parameter of the problem will be the Reynolds number

$$\mathrm{Re} = \frac{U_0 L}{\nu}.$$

The Navier–Stokes equations written in dimensionless form differ from (1.1.1) by the fact that on the right-hand side the quantity Re^{-1} appears instead of ν as the coefficient of the highest derivatives. For the study of two-dimensional (plane) fluid flows, it is convenient to write the equations in orthogonal curvilinear coordinates x, y referred to the body surface $y = 0$ (see, e.g., Rosenhead, 1963):

$$\frac{\partial u}{\partial t} + \frac{u}{h}\frac{\partial u}{\partial x} + v\frac{\partial u}{\partial y} + \frac{\kappa v u}{h} + \frac{1}{h}\frac{\partial p}{\partial x} = \frac{1}{\mathrm{Re}}\frac{\partial}{\partial y}\left\{\frac{1}{h}\left[\frac{\partial(hu)}{\partial y} - \frac{\partial v}{\partial x}\right]\right\},$$

$$\frac{\partial v}{\partial t} + \frac{u}{h}\frac{\partial v}{\partial x} + v\frac{\partial v}{\partial y} - \frac{\kappa u^2}{h} + \frac{\partial p}{\partial y} = -\frac{1}{\mathrm{Re}}\frac{1}{h}\frac{\partial}{\partial x}\left\{\frac{1}{h}\left[\frac{\partial(hu)}{\partial y} - \frac{\partial v}{\partial x}\right]\right\},$$

$$\frac{\partial u}{\partial x} + \frac{\partial(hv)}{\partial y} = 0, \quad h = 1 + \kappa(x)y. \tag{1.1.2}$$

Here κ is the dimensionless curvature of the body surface, which is defined as positive for a convex wall, as shown in Figure 1.1, when the center of curvature is located on the side $y < 0$. The coordinate x is measured from some point O on the body surface, y is measured along a normal to the surface, and u and v are the corresponding dimensionless velocity components tangential and normal to the wall.

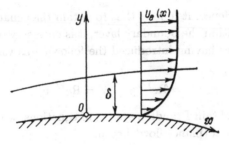

Fig. 1.1 Velocity profile for the flow in the boundary layer.

On the basis of the continuity equation, we may introduce the stream function ψ, defined by the relations

$$\frac{\partial \psi}{\partial y} = u, \qquad \frac{\partial \psi}{\partial x} = -(1 + \kappa y)v. \qquad (1.1.3)$$

Its use as a dependent variable is often found to give a more convenient way of writing the momentum equations.

From a mathematical point of view, the investigation of a flow at high Reynolds number is equivalent to consideration of the asymptotic behavior of a solution of the Navier–Stokes equations as Re → ∞. If we take the limit as Re → ∞, maintaining the orders of magnitude of the dimensionless independent and dependent variables introduced above, then the viscous terms in (1.1.2), which are inversely proportional to the Reynolds number, will disappear and we obtain the Euler system of equations. Since there are no terms containing the highest derivatives of the dependent variables, the order of this system is lower than the order of the original system of Navier–Stokes equations. This circumstance does not allow us to satisfy all the boundary conditions of the problem. One has to abandon the boundary condition requiring that the fluid adhere to the solid surface, and to allow an arbitrary slip of the fluid along the wall, which contradicts the real physical situation. The way to resolve this question, indicated by Prandtl (1904), is known to follow from the fact that at arbitrarily high Reynolds numbers there always exists a viscous layer along the wall, across which the tangential velocity component u varies from zero to the value prescribed by the theory of ideal fluid flows (Figure 1.1). The relative thickness δ of this boundary layer decreases as the Reynolds number increases, and can be determined from the condition of a balance of inertia and viscous forces. It is easy to find that $\delta = O(\mathrm{Re}^{-1/2})$. The transverse (normal to the wall) component v of the velocity vector inside the boundary layer also turns out to be of order $\mathrm{Re}^{-1/2}$. Hence, it follows that to obtain the equations describing the fluid flow within the boundary layer, it is necessary to take the limit as $\mathrm{Re} \to \infty$ after having introduced the following as variables of order unity:

$$Y = \mathrm{Re}^{1/2}y, \quad V = \mathrm{Re}^{1/2}v. \qquad (1.1.4)$$

As a result, we obtain Prandtl's system of equations, which for the case of two-dimensional (plane) flow[1] become

$$\frac{\partial u}{\partial t} + u\frac{\partial u}{\partial x} + V\frac{\partial u}{\partial Y} + \frac{\partial p}{\partial x} = \frac{\partial^2 u}{\partial Y^2}, \quad \frac{\partial p}{\partial Y} = 0, \quad \frac{\partial u}{\partial x} + \frac{\partial V}{\partial Y} = 0. \quad (1.1.5)$$

Thus, if one starts with the Navier–Stokes equations (1.1.2), considering them to be the exact equations of viscous fluid dynamics, then the solution for the external inviscid flow field (described by the Euler system of equations) and that for the boundary layer (described by the Prandtl equations) should be considered as the leading terms of two asymptotic representations, outer and inner, which hold as $\mathrm{Re} \to \infty$. It is this approach to boundary-layer theory that emphasizes the completely equivalent status of the Euler and Prandtl equations, in the sense that (contrary to a still widespread notion) the former are not to be regarded as "exact" equations, while the latter are "approximate."

The boundary conditions for the system (1.1.5) in the case of a steady flow will be: first, the no-slip condition for the fluid at the body surface

$$u(x,0) = V(x,0) = 0; \qquad (1.1.6)$$

second, the condition of matching the solutions for the boundary layer and the outer inviscid flow field, which can be written as

$$\lim_{Y\to\infty}\{u(x,Y)\} = \lim_{y\to 0}\{U(x,y)\} = U_e(x); \qquad (1.1.7)$$

and finally, the condition defining the boundary-layer velocity profile at some cross section $x = x_0$, which is usually referred to as an initial condition:

$$u(x_0, Y) = u_0(Y). \qquad (1.1.8)$$

In the case of a steady flow, the conditions (1.1.6)–(1.1.8) determine the solution of the system (1.1.5) (if it exists) uniquely for all $x \geq x_0$ under the conditions that $U_e(x_0) > 0$ and the function $u_0(Y)$ is continuous (Nickel, 1958).

---

[1] A more detailed derivation and analysis of the boundary-layer equations can be found, for instance, in the book by Loitsianskii (1962).

As we shall see, the existence of a solution of the boundary-layer equations depends mainly upon the character of the pressure distribution along the wall, that is, upon the longitudinal pressure gradient, determined by the solution of the problem of inviscid flow over the body:

$$\left.\frac{\partial p}{\partial x}\right|_{y=0} = p'_e(x) = -U_e(x)U'_e(x). \qquad (1.1.9)$$

Thus, if $p'_e(x) \leq 0$ for $x > x_0$, then the solution of the problem (1.1.5)–(1.1.8) exists for all $x \geq x_0$; if this condition is not met, there exists a point $x_1 > x_0$ such that the problem has at least one solution in the range $x_0 \leq x \leq x_1$, $0 \leq Y < \infty$ (Oleinik, 1963). This case will be considered in more detail below.

For the solution of many problems, it is convenient to use a dimensionless stream function in the boundary layer, which on the basis of the continuity equation (1.1.5) is defined by relations equivalent to (1.1.3):

$$\frac{\partial \Psi}{\partial Y} = u, \quad \frac{\partial \Psi}{\partial x} = -V; \quad \Psi = \mathrm{Re}^{1/2}\psi. \qquad (1.1.10)$$

Then the momentum equation (1.1.5) and boundary conditions (1.1.6)–(1.1.8) take the form

$$\frac{\partial \Psi}{\partial Y}\frac{\partial^2 \Psi}{\partial x \partial Y} - \frac{\partial \Psi}{\partial x}\frac{\partial^2 \Psi}{\partial Y^2} + p'_e(x) = \frac{\partial^3 \Psi}{\partial Y^3}; \qquad (1.1.11)$$

$$\Psi = \frac{\partial \Psi}{\partial Y} = 0, \quad Y = 0;$$

$$\frac{\partial \Psi}{\partial Y} \to U_e(x), \quad Y \to \infty; \qquad (1.1.12)$$

$$\frac{\partial \Psi}{\partial Y} = u_0(Y), \quad x = x_0.$$

Henceforth, while studying steady flows we shall usually apply the equations in this form.

Our consideration of Prandtl's boundary-layer theory corresponds to the so-called hierarchical principle of constructing asymptotic expansions, according to which these expansions in each of two (or more) regions are built up alternately, term by term (see Van Dyke, 1969). Thus, for instance, as we have observed in the case of unseparated flow around a body, first of all we determine the leading term of the outer asymptotic expansion, which corresponds to the solution of the problem of ideal fluid flow over a given body. Next, we find the leading term of the inner asymptotic expansion, corresponding to the solution of the boundary-layer problem with a given pressure distribution (i.e., known from solving the first problem); then it becomes possible to determine the second term of the outer asymptotic expansion, which takes into consideration the displacement effect of the boundary layer and the wake upon the pressure distribution; then we can find the second term of the inner expansion, which recognizes this change in the pressure distribution, etc.

The hierarchical principle of constructing asymptotic expansions corresponds to an iterative procedure of successively refining the solutions for the external flow field and the boundary layer, as pointed out by Prandtl (1935) himself.

However, as we shall see later, the hierarchical procedure described above faces certain difficulties, or even proves to be entirely inapplicable, for the analysis of separated flows.

1.2 The Problem of Boundary-Layer Separation

The existence of a solution of the boundary-layer equations under the conditions mentioned above means that everywhere in the boundary layer the dimensionless velocity component v normal to the body surface is of order $\mathrm{Re}^{-1/2}$, so in the limit as $\mathrm{Re} \to \infty$ the boundary layer becomes a surface of tangential discontinuity which coincides with the body surface. Such flows are unseparated. They can be realized, for example, near a plate placed along the flow or at the surface of a thin airfoil at a small angle of attack.

In general, for realization of unseparated flow, rather severe restrictions are required, which in the majority of cases are not fulfilled in practice; this leads to the formation of flows of another kind, with flow separation from the surface.

The physical nature of the flow separation phenomenon was first explained by Prandtl (1904) on the basis of his boundary-layer theory. The appearance of flow separation is always due to the action of an adverse (positive) pressure gradient. Under this influence, fluid elements moving inside the boundary layer and having a smaller amount of kinetic energy than elements at its outer edge are more strongly decelerated. The velocity profile in the boundary layer is gradually changed, as shown in Figure 1.2, so that at some cross section $x = x_S$ the dimensionless skin friction $\tau = u_Y|_{Y=0}$ becomes zero for the first time, and for $x > x_S$ there appears a region of slow reverse flow. This region expands rapidly, moving fluid elements from the boundary layer to finite distances from

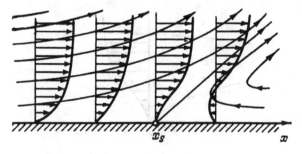

Fig. 1.2 Schematic representation of the flow in a boundary layer near the separation point $x = x_S$.

the body surface. In the limit as Re $\to \infty$, the separated flow therefore looks as if the surface of tangential discontinuity, to which the boundary layer reduces, coincides with the body surface only up to a certain point, but then moves away and becomes a "free" streamline, representing in the limit the so-called mixing layer.

Determination of the ideal fluid flow field corresponding to this limiting state is one of the fundamental and most complicated problems of hydrodynamics. We shall investigate it in one of the later chapters of this book.

To elucidate the conditions for flow separation, it is necessary to examine the mechanism of boundary-layer separation under the action of an adverse pressure gradient to study the local flow structure close to the point of detachment of the zero streamline from the solid wall. Proceeding from the picture of the flow shown in Figure 1.2, this point is usually identified with the point $x = x_S$ where the skin friction first reaches zero. Therefore, from a mathematical point of view the study of boundary-layer separation from a smooth surface is related to the analysis of the behavior of the solution of the boundary-layer equations in the vicinity of the point of zero friction. This problem has been the subject of numerous studies. As a rule, these have been based on the assumption that the pressure distribution along the body surface is regular at the separation point, i.e., that the positive pressure gradient here is finite and distributed along a finite section of the surface. Analyzing the solution of the boundary-layer equations for a linearly decreasing external-flow velocity, Howarth (1938) and Hartree (1939) were the first to discover singular behavior of the solution around the zero-friction point. Later, Landau and Lifshitz (1944) studied the character of this singularity and found that as this singular point is approached, the velocity component

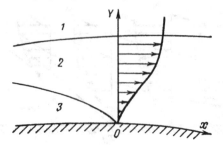

Fig. 1.3 Structure of the flow ahead of the point of zero skin friction $x = 0$.

normal to the wall increases as $(x_S - x)^{-1/2}$. This gave a formal justification for identifying the zero-friction point with the boundary-layer
separation point. Finally, Goldstein (1948) carried out the most detailed
and rigorous mathematical analysis of the situation that arises here.

Goldstein showed that immediately ahead of the separation point
(which for the sake of convenience we will choose as the origin of coordinates, i.e., we will take $x_S = 0$), the solution of the boundary-layer
equations has a "two-layer" structure (Figure 1.3).[2] In the inner (near
the wall) region 3, the thickness of which decreases as $x \to -0$ according to the law $Y \sim (-x)^{1/4}$, the solution of the equation (1.1.11) for the
stream function has the form[3]

$$\Psi = (-x)^{3/4} f_0(\eta) + (-x) f_1(\eta) + O\left[(-x)^{5/4}\right], \quad \eta = Y(-x)^{-1/4}. \quad (1.2.1)$$

The first term of this expansion, for which

$$f_0 = \tfrac{1}{6}\lambda_0\eta^3, \quad \lambda_0 = p_e'(0),$$

corresponds to the known asymptotic velocity profile as $Y \to 0$ at the
point of zero skin friction: $u = \partial\Psi/\partial Y \sim \lambda_0 Y^2/2$. The solution of the
equation for the function $f_1(\eta)$ satisfying the conditions of adherence to
the surface and absence of exponential growth as $\eta \to \infty$ is

$$f_1 = \alpha_1\eta^2.$$

Here, α_1 is an arbitrary constant which generally should be considered
different from zero and positive, since the shear stress at the wall $\tau = \left.\dfrac{\partial^2\Psi}{\partial Y^2}\right|_{Y=0} = 2\alpha_1(-x)^{1/2} + \cdots$ is greater than zero when $x < 0$.[4] In

---

[2] In this figure, region 1 is the external potential flow.

[3] A more detailed description of Goldstein's solution is given in Chapter 4.

[4] Solutions corresponding to the case $\alpha_1 = 0$ are analyzed in detail in Chapter 4.

the outer (main) part 2 of the boundary layer, where $Y = O(1)$, the solution as $x \to -0$ is represented in the form of the following asymptotic expansion:

$$\Psi = \Psi_0(Y) + (-x)^{1/2}(2\alpha_1/\lambda_0)\Psi_0'(Y) + O\left[(-x)^{3/4}\right],$$

where $\Psi_0(Y)$ defines the profile of the longitudinal velocity component at the separation point: $u(0, Y) = \Psi_0'(Y)$.

For the transverse velocity component, we obtain the following expression:

$$V = -\frac{\partial \Psi}{\partial x} = (-x)^{-1/2}\frac{\alpha_1}{\lambda_0}\Psi_0'(Y) + O\left[(-x)^{-1/4}\right].$$

Hence, it follows that if $\alpha_1 \neq 0$, then the value of $V(x, Y)$ increases indefinitely as the point of zero skin friction is approached, and the boundary-layer theory, according to which $V = O(1)$, no longer holds. Another very important result of Goldstein's (1948) investigation was the proof that in the general case it is impossible to continue the solution through the point $x = 0$; that is, it is impossible to construct an asymptotic solution compatible with (1.2.1) as $x \to +0$. It turns out that if $\alpha_1 \neq 0$, then as $x \to +0$ the second term, proportional to x, in the expansion analogous to (1.2.1) for the stream function $\Psi(x, Y)$ in the inner region contains an imaginary constant.

Thus, the results of Goldstein's analysis cannot be considered as the description of the real situation in the vicinity of the boundary-layer separation point. At the same time, they are quite important since the appearance of the singularity studied by Goldstein in solving the problem of a flow in the boundary layer with a given adverse pressure gradient always attests to the impossibility of unseparated flow.

1.3 Self-Induced Separation

The alternative to the situation considered above is to give up the assumption of regularity of the pressure distribution at the boundary-layer separation point as Re $\to \infty$. Instead, one should assume that in real flows near the separation point there is always an interaction of the boundary layer with the external potential flow such that the pressure gradient leading to separation is self-induced. The results of the numerical analysis by Catherall and Mangler (1966) give support to this assumption, since they were the first to show that if the pressure gradient is not given in advance, but is determined in some vicinity of the separation point by the boundary layer itself, then Goldstein's singularity can be absent and the solution can easily be continued through the zero-friction point.

Rational solutions of this kind corresponding to boundary-layer separation in a supersonic flow were first obtained by Neiland (1969) and by Stewartson and Williams (1969). In this case the relationship between the pressure and the slope of the streamlines is local, and the singularity in the pressure distribution is due to the formation of a compression wave in the external flow, a wave which in the limit transforms into a shock wave.

The case of an incompressible fluid is more complicated. If the singular character of the pressure distribution in the vicinity of the separation point still results from the interaction between the boundary layer and the external flow, the appropriate singular solution must be contained in the class of possible solutions for an ideal fluid that might represent the flow outside the boundary layer. At first sight this seems impossible, because of the elliptic character of the problem of incompressible fluid flow around a body. However, if one refers to the theory of flows of an ideal fluid with free streamlines, one finds that it contains a local solution which has all the necessary properties. Let us examine this in more detail.

The solution to the problem of flow around a smooth obstacle, for instance a circular cylinder, in Kirchhoff's scheme is, generally speaking, not unique. Solutions can be constructed with different positions for the point of separation of the zero streamline from the body surface (Figure 1.4). If the origin of the curvilinear orthogonal coordinate system Oxy is placed at the point of detachment of the zero streamline from the body surface, the curvature of the free streamline will be defined by the expression

$$\kappa = -kx^{-1/2} + \kappa_0 + O(x^{1/2}), \quad x \to +0 \qquad (1.3.1)$$

(κ_0 is the dimensionless curvature of the body surface at the separation point, taken to be finite), and the magnitude of the pressure gradient along the body surface in the vicinity of this point will be equal to

$$p_e'(x) = k(-x)^{-1/2} + \frac{16}{3}k^2 + O\left[(-x)^{1/2}\right], \quad x \to -0;$$
$$p_e'(x) = 0, \quad x > 0, \qquad (1.3.2)$$

where k is an arbitrary constant, related to the position of the separation point.

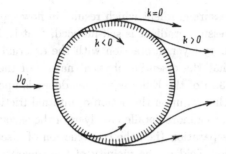

Fig. 1.4 The form of the free streamlines for Kirchhoff flows with various positions of the separation point.

A detailed derivation of the relations that are the basis for the formulas (1.3.1), (1.3.2) can be found in the book by Birkhoff and Zarantonello (1957) or in the work of Imai (1953) and of Ackerberg (1970).

It should be noted that for $k < 0$ the flow is physically impossible: in this case the free streamline intersects the body surface (Figure 1.4), since its curvature according to (1.3.1) acquires an infinitely large positive value. The case $k > 0$ is usually rejected on the basis that the corresponding infinitely large positive pressure gradient (1.3.2) would lead to earlier separation of the flow as it approaches the point $x = 0$. Therefore the requirement $k = 0$, which is referred to as the condition of "smooth separation" of Brillouin (1911) and Villat (1914),[5] is usually used as the condition defining a unique solution of the problem. But in this case the positive pressure gradient along the surface of the body is totally absent, and the appearance of separation cannot be explained from a physical standpoint.

The way out of this paradox was indicated for the first time in the work of Sychev (1972). The main idea of this work consists in the fact that the constant k is taken to be positive and depends upon the Reynolds number such that

$$k(\mathrm{Re}) \to 0 \quad \text{as} \quad \mathrm{Re} \to \infty. \qquad (1.3.3)$$

This assumption implies that the limiting flow state as $\mathrm{Re} \to \infty$ will be an inviscid flow satisfying the Brillouin–Villat condition. However, such a limit process is singular: at any arbitrarily high Reynolds number there will be a vanishingly small vicinity of the point $|x| = 0$ with a

---

[5] For example, for a circular cylinder the separation point, according to this condition, is located at $\theta = 55°$, where θ is the polar angle measured from the forward stagnation point (Brodetsky, 1923).

large adverse pressure gradient which results in flow separation. As we shall see, this pressure gradient is self-induced; that is, it is due to the process of boundary-layer interaction with the external potential flow. This indicates that the essential physical nature of the singular solutions (1.3.1), (1.3.2) of the Euler equations depends upon the viscosity of the medium, that is, upon the action of internal friction forces.

Before we turn to a more detailed analysis of the asymptotic theory of boundary-layer separation (i.e., the construction of a set of expansions describing the flow field in the vicinity of the separation point), it is useful to give some simpler reasoning about the orders of magnitude of the flow variables, which allows us to elucidate the flow structure around the separation point and to understand the physical processes that take place there.

We consider a small vicinity of the separation point with a characteristic longitudinal dimension $\Delta x \ll 1$, in which an adverse pressure gradient causes flow separation. According to (1.3.2) and (1.3.3) the order of magnitude of the longitudinal pressure difference here is[6]

$$\Delta p \sim k(\Delta x)^{1/2}; \qquad (1.3.4)$$

that is, it is a quantity much less than unity. But for a vanishingly small pressure difference to cause flow separation, it is necessary that the pressure gradient $\Delta p/\Delta x$ be of an order of magnitude much greater than unity,[7] and this implies that the main part of the boundary layer with thickness δ of the order of $\mathrm{Re}^{-1/2}$, where $Y = O(1)$ and the tangential velocity component $u = O(1)$, will be locally inviscid, since the viscous term $\partial^2 u/\partial Y^2$ on the right-hand side of the momentum equation (1.1.5) remains of order unity here. From the balance of the inertia terms and the pressure gradient it follows that the change of velocity Δu across the region considered will be of order Δp, and then from the continuity equation it follows that the relative change of the displacement thickness of the main part of the boundary layer also has the same magnitude: $\Delta \delta/\delta \sim \Delta p$.

Now it is necessary to consider a thinner region (a sublayer) next to the wall. There are two reasons for doing so. First, the inviscid character of the flow in the main part of the boundary layer does not allow us

---

[6] The symbol $\sim$ will be used to mean that in the limit the ratio of the functions indicated does not tend to zero or infinity; that is, these functions have values of the same order of smallness.

[7] Landau and Lifshitz (1944) were the first to point out the possibility that the ratio $\Delta p/\Delta x$ must also be proportional to $(\Delta x)^{-1/3}$.

to satisfy the no-slip condition at the body surface. Second, next to the wall, where $u \ll 1$, the effect of nonlinearity in the longitudinal component of the momentum equation appears when $u \sim \Delta u$. If the action of viscosity turns out to be stronger than the nonlinear effect, then the velocity profile to leading order will remain unchanged across the sublayer next to the wall. Consequently, such a situation cannot be consistent with separation of the flow. On the other hand, if the action of friction forces is found to be weaker than the nonlinear effect, then the Bernoulli equation will hold for the flow under consideration. However, this is also impossible, because for the flow quite close to the wall, the square of the velocity would become negative as a result of the pressure increase. Thus, the sublayer next to the wall must be both nonlinear and viscous at the same time. Then from the momentum equation (1.1.5) it follows that in this sublayer with thickness $\mathrm{Re}^{-1/2}\Delta Y$

$$u\frac{\Delta u}{\Delta x} \sim \frac{\Delta p}{\Delta x} \sim \frac{\Delta u}{(\Delta Y)^2}. \tag{1.3.5}$$

From the condition of matching the solutions in the region considered and in the flow region ahead of it, where the skin friction is finite, we have

$$u \sim \Delta Y. \tag{1.3.6}$$

Further, since $u \sim \Delta u$, the first of the relations (1.3.5) gives

$$u \sim \Delta u \sim \sqrt{\Delta p}. \tag{1.3.7}$$

Moreover, from the condition $u \sim \Delta u$ and the continuity equation, it follows that the change in displacement thickness of the viscous sublayer has the same order as its thickness ΔY. Then, comparing (1.3.6) with (1.3.7), we find that the effect of the viscous sublayer upon the change of the thickness δ of the entire boundary layer will be

$$\Delta\delta \sim \mathrm{Re}^{-1/2}\Delta Y \sim \mathrm{Re}^{-1/2}\sqrt{\Delta p}, \quad \text{i.e.} \ \Delta\delta/\delta \sim \sqrt{\Delta p}. \tag{1.3.8}$$

Thus, the major contribution to the displacement effect of the boundary layer upon the external potential flow is made by the viscous sublayer. The contribution of the main part of the boundary layer where $u = O(1)$, as we have seen, has the order of magnitude $\mathrm{Re}^{-1/2}\Delta p$. Therefore, this flow region plays a passive role and (in the first approximation) is only displaced by the viscous sublayer, so that the slope of the streamlines determined by the development of the latter is transferred to the external potential flow without change.

Using (1.3.5)–(1.3.8), it is not difficult to estimate the order of magnitude of the pressure gradient and the curvature of the free streamline resulting from the displacement effect of the sublayer:

$$\frac{\Delta p}{\Delta x} \sim (\Delta p)^{-1/2}, \quad \kappa \sim \frac{\Delta \delta}{(\Delta x)^2} \sim \mathrm{Re}^{-1/2}(\Delta p)^{-5/2}. \tag{1.3.9}$$

Now if we bear in mind that, on the basis of the original relations (1.3.1) and (1.3.2), both these values must be of the order of $k(\Delta x)^{-1/2}$ within the region considered, then, from (1.3.9) and the earlier relations (1.3.5)–(1.3.8), we may obtain estimates for all the variables in the viscous sublayer of the interaction region:

$$\Delta x = O(\mathrm{Re}^{-3/8}), \quad y = \mathrm{Re}^{-1/2}Y = O(\mathrm{Re}^{-5/8}),$$
$$u = O(\mathrm{Re}^{-1/8}), \quad v = O(\mathrm{Re}^{-3/8}), \quad \Delta p = O(\mathrm{Re}^{-1/4}). \tag{1.3.10}$$

Substituting these estimates into (1.3.4), for instance, we find that

$$k = O(\mathrm{Re}^{-1/16}). \tag{1.3.11}$$

Thus, we see that as the Reynolds number tends to infinity, the pressure gradient and the curvature of the streamlines in the external flow increase indefinitely as $\mathrm{Re}^{1/8}$, and the length of the interaction zone decreases as $\mathrm{Re}^{-3/8}$.

1.4 The Preseparation Flow

Now let us turn to a detailed asymptotic analysis of the flow near the separation point at high Reynolds number, that is, to the description of the system of asymptotic expansions which in combination represent the solution to be found as $\mathrm{Re} \to \infty$.[8]

We start with consideration of the flow ahead of the interaction region, for $x < 0$. The boundary layer along the body surface develops here under the effect of the pressure gradient, which is determined for $x \to -0$ by the expression (1.3.2), where the constant k, as we have found, tends to zero as $\mathrm{Re} \to \infty$ according to the law $\mathrm{Re}^{-1/16}$. Therefore, we take

$$k = \mathrm{Re}^{-1/16}c_1, \quad c_1 > 0 \tag{1.4.1}$$

and represent the solution of the boundary-layer equation for the stream function (1.1.11), together with the corresponding pressure distribution,

---

[8] See Sychev (1972).

in the form of the expansions

$$\Psi = \Psi_0(x, Y) + \text{Re}^{-1/16}\Psi_1(x, Y) + O(\text{Re}^{-1/8}),$$
$$p_e = p_0(x) + \text{Re}^{-1/16}p_1(x) + O(\text{Re}^{-1/8}).$$

(1.4.2)

Here $\Psi_0(x, Y)$ represents the solution of the boundary-value problem (1.1.11), (1.1.12), (1.1.9) for the boundary layer in a flow which satisfies the Brillouin–Villat condition ($k = 0$), that is, in the presence of a given favorable pressure gradient $p_0'(x)$. As $x \to -0$, it may be expressed in the form of the expansions

$$\Psi_0 = \Psi_{00}(Y) + (-x)\Psi_{01}(Y) + O\big[(-x)^{7/6}\big],$$
$$p_0 = p_{00} + O\big[(-x)^{3/2}\big].$$

(1.4.3)

The appearance in these expansions of singular terms, whose orders are indicated (see Messiter and Enlow, 1973), results from the singularity in the pressure distribution (1.3.2).

Let us recall that the position of the separation point for a free streamline in an inviscid flow with a given pressure p_{00} in the separation zone depends upon the value of the parameter k. For $k = O(\text{Re}^{-1/16})$, this point will be displaced downstream from its original position (which corresponds to $k = 0$) through a distance proportional to $\text{Re}^{-1/16}$. In the relations (1.4.2), (1.4.3) and everywhere in the following, we shall assume that the value $x = 0$ corresponds to the new position of the separation point.

Since the shear stress at the body surface is positive when $k = 0$, then for the first term of (1.4.3) we have near the wall:

$$\Psi_{00} = \tfrac{1}{2}a_0 Y^2 + O(Y^{9/2}), \quad Y \to 0.$$

(1.4.4)

The equation for the function $\Psi_1(x, Y)$ obtained through the substitution of (1.4.2) into (1.1.11) will have the form

$$\frac{\partial\Psi_0}{\partial Y}\frac{\partial^2\Psi_1}{\partial x\partial Y} + \frac{\partial^2\Psi_0}{\partial x\partial Y}\frac{\partial\Psi_1}{\partial Y} - \frac{\partial\Psi_0}{\partial x}\frac{\partial^2\Psi_1}{\partial Y^2} - \frac{\partial^2\Psi_0}{\partial Y^2}\frac{\partial\Psi_1}{\partial x} + \frac{dp_1}{dx} = \frac{\partial^3\Psi_1}{\partial Y^3},$$

(1.4.5)

where, according to (1.3.2) and (1.4.1),

$$p_1 = -2c_1(-x)^{1/2} + \cdots, \quad x \to -0.$$

(1.4.6)

Since as $x \to -0$, the viscous term on the right-hand side must remain of a finite order of magnitude, while the pressure gradient increases indefinitely, the solution of equation (1.4.5) as $x \to -0$ will correspond to locally inviscid flow in a region with transverse scale $Y = O(1)$ (which

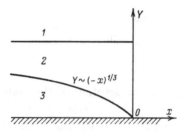

Fig. 1.5 Schematic picture of the flow structure ahead of the interaction region.

in the following will be referred to as the main part of the boundary layer) and will not satisfy the no-slip condition at the solid boundary. To satisfy this condition it is necessary to introduce a viscous sublayer into the analysis, in which the flow should be determined by a balance of the pressure gradient and the inertia and friction forces.[9] The flow in the preseparation region is shown schematically in Figure 1.5.[10] Since according to (1.4.4), the wall shear stress a_0 is different from zero, the convection terms in equation (1.4.5), for example the last one, will be of the same order as the viscous terms on the right-hand side in a region with transverse dimension $Y \sim (-x)^{1/3}$, and will be of the same order as the pressure gradient if the changes in $\Psi_1(x,Y)$ in this region are proportional to $(-x)^{1/2}$. Therefore, the solution for the viscous sublayer as $x \to -0$ is sought in the form

$$\Psi_1 = (-x)^{1/2} f_1(\eta) + O[(-x)^{4/3}], \quad \eta = Y/(-x)^{1/3}. \qquad (1.4.7)$$

Using the matching condition for the solutions in the main part of the boundary layer 2 and in the sublayer 3, on the basis of (1.4.4) and (1.4.7) we can represent the function $\Psi_0(x,Y)$ in the sublayer as

$$\Psi_0 = \tfrac{1}{2} a_0 (-x)^{2/3} \eta^2 + O[(-x)^{3/2}], \quad x \to -0. \qquad (1.4.8)$$

Now substituting the expressions (1.4.7) and (1.4.8) for the leading terms of the expansions of the functions $\Psi_0(x,Y)$ and $\Psi_1(x,Y)$ into (1.4.5)–(1.4.6), we obtain the equation for the function $f_1(\eta)$ in the following form:

$$f_1''' - \frac{a_0}{3}\eta^2 f_1'' + \frac{a_0}{2}\eta f_1' - \frac{a_0}{2}f_1 = c_1. \qquad (1.4.9)$$

The boundary conditions for this equation are the no-slip condition at

[9] Let us note that a similar situation arises every time that a boundary layer is subjected to a large local pressure gradient (Neiland and Sychev, 1966).

[10] Region 1 in this figure represents the external potential flow.

the solid wall

$$f_1(0) = f_1'(0) = 0, \qquad (1.4.10)$$

and also the condition of the absence of exponential growth of the function $f_1(\eta)$ as $\eta \to \infty$, which makes it possible to perform matching of the solution in the sublayer with the solution for the main part of the boundary layer. The above-mentioned conditions determine uniquely the dependent variable, which for large η is represented asymptotically as

$$f_1 = a_1 \eta^{3/2} + b_1 \eta + O(1), \quad \eta \to \infty. \qquad (1.4.11)$$

Integrating (1.4.9) gives the values[11] for the constants in (1.4.11):

$$a_1 = -5.48 \, c_1 a_0^{-1/2}, \quad b_1 = 4.92 \, c_1 a_0^{-2/3}. \qquad (1.4.12)$$

The asymptotic representation of the function $f_1(\eta)$ as $\eta \to \infty$ allows us, on the basis of the matching principle for the solutions in the viscous sublayer and in the main part of the boundary layer, to write, using (1.4.11) and (1.4.7), the expression for the function $\Psi_1(x, Y)$ as $Y \to 0$, which will have the form

$$\Psi_1 = a_1 Y^{3/2} + (-x)^{1/6} b_1 Y + O[(-x)^{1/2}]. \qquad (1.4.13)$$

Using this expression, we can write the following expansion for the function $\Psi_1(x, Y)$ when $Y = O(1)$:

$$\Psi_1 = \Psi_{10}(Y) + (-x)^{1/6} \Psi_{11}(Y) + O[(-x)^{1/2}], \quad x \to -0. \qquad (1.4.14)$$

By substituting this expansion together with (1.4.3) into (1.4.5)–(1.4.6), we obtain an equation for the function $\Psi_{11}(Y)$

$$\Psi_{00}' \Psi_{11}' - \Psi_{00}'' \Psi_{11} = 0,$$

the solution of which has the form

$$\Psi_{11} = A_1 \Psi_{00}'. \qquad (1.4.15)$$

---

[11] The solution of equation (1.4.9) may be expressed through confluent hypergeometric functions, and the precise values of the constants a_1 and b_1 will be

$$a_1 = -\left[\tfrac{4}{3} \Gamma\left(\tfrac{1}{6}\right) / \Gamma\left(\tfrac{2}{3}\right)\right] c_1 a_0^{-1/2},$$
$$b_1 = 2 \left[6^{1/3} \Gamma\left(\tfrac{2}{3}\right)\right] c_1 a_0^{-2/3}.$$

From the condition (1.4.13), and taking into account (1.4.4), the constant $A_1 = b_1/a_0$ is determined. Also it should be noted that

$$\Psi_{10} = a_1 Y^{3/2} + \cdots, \quad Y \to 0. \qquad (1.4.16)$$

The solution obtained for the stream function $\Psi_1(x, Y)$ in the main part of the boundary layer in essence corresponds to a simple displacement or pushing aside of this region by the viscous sublayer. In fact, if we write out the leading term of the corresponding change in streamline slope $(\Delta\vartheta)$ in the region $Y = O(1)$ for $x \to -0$, then using (1.4.3), (1.4.14), and (1.4.15) we find that

$$\mathrm{Re}^{1/2}\Delta\vartheta = -\mathrm{Re}^{-1/16}\frac{\partial\Psi_1}{\partial x}\bigg/\frac{\partial\Psi_0}{\partial Y} + \cdots$$

$$= \mathrm{Re}^{-1/16}\tfrac{1}{6}(-x)^{-5/6}b_1/a_0 + \cdots. \qquad (1.4.17)$$

This expression does not depend upon Y and, consequently, in the first approximation the slope of the streamlines induced by the sublayer is in effect transmitted across the main part of the boundary layer to the external potential flow without any change.

Thus, the appearance of singular behavior in the expansion (1.4.14) for the function $\Psi_1(x, Y)$, resulting from the term of order $(-x)^{1/6}$, is caused by the development of the viscous sublayer next to the wall under the influence of the adverse pressure gradient (1.4.6).

Using (1.4.2), (1.4.7), and (1.4.8), we now write the expansion for the stream function in the sublayer, where $\eta = O(1)$:

$$\Psi = \Psi_0(x, Y) + \mathrm{Re}^{-1/16}\Psi_1(x, Y) + O(\mathrm{Re}^{-1/8});$$
$$\Psi_0 = \tfrac{1}{2}a_0(-x)^{2/3}\eta^2 + \ldots, \quad \Psi_1 = (-x)^{1/2}f_1(\eta) + \ldots, \qquad (1.4.18)$$
$$\eta = Y/(-x)^{1/3}, \quad x \to -0.$$

This expansion is no longer valid for $|x| = O(\mathrm{Re}^{-3/8})$, when the terms written here become of the same order of magnitude. Therefore, it is necessary for us now to analyze separately a neighborhood of the separation point with longitudinal scale of order $\mathrm{Re}^{-3/8}$, which must be the interaction region and the origin of the separated flow, as was established in the preceding section.

However, before we describe the flow in this region, another comment should be made. From the expression (1.3.2) for the pressure gradient and the estimate (1.3.11) obtained for the coefficient k, it turns out that for $|x| = O(\mathrm{Re}^{-1/16})$, the orders of magnitude of the first and third terms of the expansion (1.3.2) are comparable. Therefore, generally

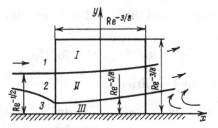

Fig. 1.6 Schematic representation of the flow structure in the interaction region.

speaking, to consider the flow in the region with a longitudinal scale of order $\mathrm{Re}^{-1/16}$, it is necessary to take into account the term $O\left(|x|^{1/2}\right)$ in (1.3.2). However, for the study of separation that is not at all necessary, since for $\mathrm{Re}^{1/16}x \to -0$ the corresponding solution for the leading terms of the expansion, with the accuracy indicated, would coincide with the two-layer asymptotic solution of the original boundary-layer equations obtained as $x \to -0$.

1.5 The Interaction Region

On the basis of the analysis of the flow in the vicinity of the separation point performed in Section 1.3, or using the matching conditions with the solution in the preseparation region (Section 1.4), we now can write the leading terms of the asymptotic expansions for the dependent flow variables in the interaction region with longitudinal scale $|x| = O(\mathrm{Re}^{-3/8})$. The structure of the flow field in this region is shown schematically in Figure 1.6. It is usually called a three-layer or "triple-deck" structure. The main part II of the interaction region is the continuation of zone 2 of the pre-separation region (Figure 1.5) and it still has thickness of order $\mathrm{Re}^{-1/2}$. There is no effect of viscosity in the leading terms here. Next to the body surface there is a viscous sublayer III allowing satisfaction of the no-slip condition and responsible for the displacement effect of the boundary layer upon the external potential flow I. The sublayer in the interaction region is a continuation of the layer next to the wall in the preseparation flow (Figure 1.5, region 3), whose thickness as $x \to -0$ decreases in proportion to $\mathrm{Re}^{-1/2}(-x)^{1/3}$ and for $|x| = O(\mathrm{Re}^{-3/8})$ becomes a quantity of order $\mathrm{Re}^{-5/8}$, which corresponds to the estimate (1.3.10) obtained earlier.

For the following discussion of the solutions in all three subregions, it is necessary to introduce independent variables of order unity for each of them. In the main part of the boundary layer, such variables will be, as we have seen,

$$x^* = \text{Re}^{3/8}x, \quad Y = \text{Re}^{1/2}y.$$ (1.5.1)

The solution here may be written as

$$\psi = \text{Re}^{-1/2}\Psi = \text{Re}^{-1/2}\tilde{\Psi}_0(x^*,Y) + \text{Re}^{-9/16}\tilde{\Psi}_1(x^*,Y) + \cdots,$$
$$p = p_{00} + \text{Re}^{-1/4}\tilde{P}_0(x^*,Y) + O(\text{Re}^{-3/8}).$$ (1.5.2)

This follows from consideration of the inner limit of the outer asymptotic expansions (1.4.2), taking into account (1.4.3), (1.4.14), and (1.4.6). If we substitute (1.5.1) and (1.5.2) into the original Navier–Stokes equations (1.1.2), (1.1.3) and keep the leading terms, we obtain for $\tilde{\Psi}_0(x^*,Y)$ the equation for an inviscid boundary layer without pressure gradient[12]

$$\frac{\partial\tilde{\Psi}_0}{\partial Y}\frac{\partial^2\tilde{\Psi}_0}{\partial x^*\partial Y} - \frac{\partial\tilde{\Psi}_0}{\partial x^*}\frac{\partial^2\tilde{\Psi}_0}{\partial Y^2} = 0.$$ (1.5.3)

From the integral of this equation

$$\frac{\partial\tilde{\Psi}_0}{\partial Y} = H_0(\tilde{\Psi}_0)$$ (1.5.4)

it follows that

$$\frac{\partial}{\partial Y}\left(\frac{\partial\tilde{\Psi}_0}{\partial x^*} \bigg/ \frac{\partial\tilde{\Psi}_0}{\partial Y}\right) = 0.$$ (1.5.5)

Thus, the longitudinal velocity component (1.5.4) in zone II is preserved along each streamline, and the slope of the streamlines (1.5.5) does not change across this zone. In other words, the velocity profile remains unchanged here throughout the entire region of interaction. The subregion II is merely displaced by the viscous sublayer III next to the wall, and as a result the problem is reduced to the analysis of this sublayer and the resulting process of boundary-layer interaction with the external potential flow.

In the sublayer the independent variables of order unity will be

$$x^* = \text{Re}^{3/8}x, \quad Y^* = \text{Re}^{5/8}y = \text{Re}^{1/8}Y.$$ (1.5.6)

---

[12] Strictly speaking, all the asymptotic expansions should be substituted into the Navier–Stokes equations, but it can be easily observed that, in those cases when the longitudinal scale of the region analyzed exceeds its transverse scale (so that in the limit as $\text{Re} \to \infty$, the region becomes arbitrarily thin), for the leading terms of the expansions the Navier–Stokes equations will be reduced to Prandtl's equations.

We shall still consider the stream function ψ as the dependent variable; it will be of order $\mathrm{Re}^{-3/4}$ here. This follows both from the estimate (1.3.10) and from the condition of matching with the solution (1.4.18) for sublayer 3. Therefore, we seek the solution for the region of the viscous sublayer III in the form

$$\psi = \mathrm{Re}^{-3/4}\Psi_0^*(x^*, Y^*) + O(\mathrm{Re}^{-7/8}),$$
$$p = p_{00} + \mathrm{Re}^{-1/4}P_0^*(x^*, Y^*) + O(\mathrm{Re}^{-3/8}). \tag{1.5.7}$$

Substituting (1.5.6) and (1.5.7) into the original equations (1.1.2) and (1.1.3), we obtain for the stream function $\Psi_0^*(x^*, Y^*)$ the boundary-layer equation with a pressure gradient

$$\frac{\partial \Psi_0^*}{\partial Y^*}\frac{\partial^2 \Psi_0^*}{\partial x^* \partial Y^*} - \frac{\partial \Psi_0^*}{\partial x^*}\frac{\partial^2 \Psi_0^*}{\partial Y^{*2}} + \frac{\partial P_0^*}{\partial x^*} = \frac{\partial^3 \Psi_0^*}{\partial Y^{*3}}, \quad \frac{\partial P_0^*}{\partial Y^*} = 0. \tag{1.5.8}$$

The solution of this equation must satisfy the condition of no slip at the solid boundary

$$\Psi_0^* = \frac{\partial \Psi_0^*}{\partial Y^*} = 0 \quad \text{at} \quad Y^* = 0, \tag{1.5.9}$$

and also the conditions of matching with the solutions in the surrounding regions. Matching with the solution (1.4.18) for the sublayer of the preseparated flow, we find that as $x^* \to -\infty$

$$\Psi_0^* = \frac{1}{2}a_0(-x^*)^{2/3}\left[\frac{Y^*}{(-x^*)^{1/3}}\right]^2 + (-x^*)^{1/2}f_1\left[\frac{Y^*}{(-x^*)^{1/3}}\right] + \cdots. \tag{1.5.10}$$

The condition of matching the solution in the sublayer with the solution for the main region II, where the velocity profile according to (1.5.4) and (1.5.5) remains unchanged, will obviously be determined by the solution for the upstream region 2 as $x \to -0$, $Y \to 0$. Then on the basis of (1.4.2)–(1.4.4) and (1.4.13), we obtain the expression

$$\Psi_0^* = \frac{1}{2}a_0 Y^{*2} + a_1 Y^{*3/2} + \cdots, \quad Y^* \to \infty, \tag{1.5.11}$$

which coincides in its leading terms with the asymptotic representation (1.5.10) for the function $\Psi_0^*(x^*, Y^*)$ as $x^* \to -\infty$.

The displacement effect of the sublayer (which is transmitted across zone II without change) is defined by the asymptotic expression

$$\vartheta = \mathrm{Re}^{-1/4}G(x^*) + \cdots, \quad G(x^*) = \lim_{Y^* \to \infty}\left\{-\frac{\partial \Psi_0^*}{\partial x^*}\bigg/\frac{\partial \Psi_0^*}{\partial Y^*}\right\}. \tag{1.5.12}$$

Hence it follows that in the main part of the interaction region II, the transverse velocity component will be of order $\mathrm{Re}^{-1/4}$, and thus for the leading terms of the expansion of the stream function (1.5.2) we obtain

$$\frac{\partial \tilde{\Psi}_0}{\partial x^*} = \frac{\partial \tilde{\Psi}_1}{\partial x^*} = 0. \tag{1.5.13}$$

Then on the basis of the second momentum equation (1.1.2), it is easy to ascertain that

$$\frac{\partial \tilde{P}_0}{\partial Y} = 0 \quad \text{and} \quad \tilde{P}_0 = P_0^*(x^*); \tag{1.5.14}$$

that is, the transverse pressure gradient in the boundary layer will be absent in the interaction region in the first approximation. (It has magnitude of order $\mathrm{Re}^{-3/8}$.)

Thus, the longitudinal pressure gradient in equation (1.5.8) for the viscous sublayer will be determined from the solution of the problem for the outer region of potential flow I subjected to the displacement effect of the boundary layer (1.5.12). To determine the external potential flow, one may apply the method of small perturbations and (on the basis of (1.5.12)) look for a suitable solution in the form

$$u = 1 + \mathrm{Re}^{-1/4} u_1(x^*, y^*) + \cdots, \quad v = \mathrm{Re}^{-1/4} v_1(x^*, y^*) + \cdots,$$
$$p = p_{00} + \mathrm{Re}^{-1/4} p_1(x^*, y^*) + \cdots. \tag{1.5.15}$$

As independent variables of order unity we take

$$x^* = \mathrm{Re}^{3/8} x, \quad y^* = \mathrm{Re}^{3/8} y, \tag{1.5.16}$$

since both the longitudinal and transverse scales of the outer region I must be the same.[13]

If we substitute (1.5.15) and (1.5.16) into the Navier–Stokes equations (1.1.2), we obtain the system of linearized Euler equations:

$$\frac{\partial u_1}{\partial x^*} + \frac{\partial p_1}{\partial x^*} = 0, \quad \frac{\partial v_1}{\partial x^*} + \frac{\partial p_1}{\partial y^*} = 0, \quad \frac{\partial u_1}{\partial x^*} + \frac{\partial v_1}{\partial y^*} = 0.$$

The solution may be obtained using the classical small-disturbance theory. Introducing the analytic function

$$w(z) = p_1 + i v_1$$

of the complex variable $z = x^* + i y^*$, we can represent it in the form

$$w(z) = \Phi(z) + i 2 c_1 z^{1/2}.$$

---

[13] Otherwise we shall obtain "degenerate" equations that are not adequate for the problem considered.

Then the condition of matching this solution as $|z| \to \infty$ with the solution for the main region 1 of potential flow corresponding to the expansions (1.3.1), (1.3.2), together with (1.4.1), will have the form

$$\Phi \to 0, \quad |z| \to \infty. \qquad (1.5.17)$$

On the real axis, taking into account (1.5.12), we have

$$\mathrm{Im}\ \Phi|_{y^*=0} = -2c_1 x^{*1/2} H(x^*) + G(x^*), \qquad (1.5.18)$$

where $H(x^*)$ is the Heaviside function, equal to unity for $x^* > 0$ and vanishing for $x^* < 0$.

The solution of the problem of determining the analytic function $\Phi(z)$ in the upper half-plane under the conditions (1.5.17)–(1.5.18) is well known and has the form (e.g., see Muskhelishvili, 1968)

$$\Phi(z) = \frac{1}{\pi} \int_{-\infty}^{\infty} \frac{G(t) - 2c_1 t^{1/2} H(t)}{t - z} dt.$$

Using the Sokhotski–Plemelj formula for this integral and separating the real and imaginary parts, we obtain an expression for the pressure distribution[14]

$$p_1(x^*, 0) = P_0^*(x^*) = -2c_1 |x^*|^{1/2} H(-x^*)$$
$$+ \frac{1}{\pi} \int_{-\infty}^{\infty} \frac{G(t) - 2c_1 t^{1/2} H(t)}{t - x^*} dt, \qquad (1.5.19)$$

where the function $G(t)$ is determined from (1.5.12). Let us note that the formula for the pressure gradient thus has the form

$$\frac{\partial p_1}{\partial x^*}\bigg|_{y^*=0} = \frac{dP_0^*}{dx^*} = \frac{1}{\pi} \int_{-\infty}^{\infty} \frac{G'(t)}{t - x^*} dt. \qquad (1.5.20)$$

The relations (1.5.19) or (1.5.20), together with (1.5.12), complete the formulation of the problem of boundary-layer interaction with the external potential flow.

1.6 The Flow in the Separation Region

Now let us turn to the analysis of the separated flow for $x > 0$, that is, to the region located downstream from the interaction zone and the origin of separation. This flow is shown schematically in Figure 1.7. Region 2' represents the continuation of the main part of the separated

---

[14] In this expression and everywhere in the following the symbol $\int$ denotes an integral in the sense of the principal value (Cauchy).

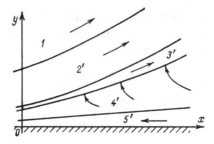

Fig. 1.7 Schematic representation of the flow structure behind the interaction region.

boundary layer, that is, the main part of the mixing layer dividing the external potential flow 1 from the separation zone $4'$. If we introduce an orthogonal curvilinear coordinate system Osn, in which the line $n = 0$ coincides with the zero dividing streamline in the mixing layer, then the leading term of the asymptotic expansion for the stream function ψ as $s \to +0$ can be written as

$$\mathrm{Re}^{1/2}\psi = \widehat{\Psi}_0(N) + \cdots, \quad N = \mathrm{Re}^{1/2}n. \tag{1.6.1}$$

The relations (1.5.13) together with the matching conditions for the solutions (1.6.1) and (1.5.2) show that the velocity profile does not change along the main part of the mixing layer $2'$. But then it will coincide, in the first approximation, with the initial velocity profile in the main part 2 of the boundary layer of the upstream flow (Figure 1.5), and hence

$$\left.\frac{\partial^2 \widehat{\Psi}_0}{\partial N^2}\right|_{N=0} = a_0. \tag{1.6.2}$$

Consequently, the main part of the mixing layer $2'$ will be divided from the separation zone $4'$ by a narrow sublayer $3'$, across which the shear stress must change from a_0 on the boundary with zone $2'$ to zero in the zone of separation. The solution for the sublayer as $s \to +0$ must satisfy a matching condition with the solution for the sublayer in the interaction region. Such a solution may be written as

$$\mathrm{Re}^{1/2}\psi = s^{2/3}g_0(\xi) + \cdots, \quad \xi = N/s^{1/3}. \tag{1.6.3}$$

For the function $g_0(\xi)$, with the help of the substitution of (1.6.3) into the original boundary-layer equation (written in variables s, N), we obtain the equation

$$g_0''' + \tfrac{2}{3}g_0 g_0'' - \tfrac{1}{3}g_0'^2 = 0. \tag{1.6.4}$$

The matching condition for the solution in this sublayer and the solution in the main part of the mixing layer (which has the initial boundary-layer profile) is (as a result of (1.6.2))

$$g_0 = \tfrac{1}{2}a_0\xi^2 + \cdots, \quad \xi \to \infty, \tag{1.6.5}$$

and the condition for matching with the solution for the separation zone $4'$ is determined by the fact that the speed of the recirculating fluid motion in this region must tend to zero as $\mathrm{Re} \to \infty$, as a consequence of which it has the form

$$g_0' \to 0 \text{ as } \xi \to -\infty. \tag{1.6.6}$$

Finally, at the dividing streamline we take, as usual,

$$g_0(0) = 0, \tag{1.6.7}$$

and obtain the well studied problem (1.6.4)–(1.6.7).[15] The solution gives, in particular, the value

$$g_0(-\infty) = -K_0 = -1.2539a_0^{1/3} \tag{1.6.8}$$

(Neiland, 1971a).

Henceforth, we shall assume that the motion of the fluid in the separation zone $4'$ is completely due to the entraining action of the mixing layer. Let us consider this motion in more detail. The equation of the boundary of the separation zone for $x > 0$ has, from (1.3.1) and (1.4.1), the following form:

$$y = y_S = y_{S_0}(x) + \mathrm{Re}^{-1/16}y_{S_1}(x) + O(\mathrm{Re}^{-1/8}),$$
$$y_{S_0} = c_0 x^{5/2} + \cdots, \quad y_{S_1} = \tfrac{4}{3}c_1 x^{3/2} + \cdots, \quad x \to +0. \tag{1.6.9}$$

The terms of this expression become of the same order of smallness when $x = O(\mathrm{Re}^{-1/16})$. Therefore, we introduce into the analysis a region with this longitudinal scale, located immediately behind the interaction zone. Assuming that

$$x = \mathrm{Re}^{-1/16}\tilde{x}, \quad y = \mathrm{Re}^{-5/32}\tilde{y}, \quad \psi = \mathrm{Re}^{-13/24}\tilde{\psi}_0(\tilde{x}, \tilde{y}) + \cdots, \tag{1.6.10}$$

we can, using (1.6.9), (1.6.3), and (1.6.8), represent the equation of the boundary of the separation region, and accordingly the value of the

---

[15] Diesperov (1984) has proved the existence and uniqueness of the solution of this problem.

stream function there, as[16]

$$\tilde{y} = \tilde{y}_S = \left[c_0 \tilde{x}^{5/2} + \tfrac{4}{3} c_1 \tilde{x}^{3/2} \right] + o(1) , \qquad (1.6.11)$$

$$\tilde{\psi}_0(\tilde{x}, \tilde{y}_S) = -K_0 \tilde{x}^{2/3} . \qquad (1.6.12)$$

Substituting (1.6.10) into the Navier–Stokes equations shows that the flow in the separation region appears inviscid and hence is described by a Poisson equation for the stream function which, because the region considered is thin, assumes the form

$$\frac{\partial^2 \tilde{\psi}_0}{\partial \tilde{y}^2} = \omega(\tilde{\psi}_0) . \qquad (1.6.13)$$

The reversed character of the flow in the separation zone, where all streamlines entering into the mixing layer come from the region lying behind the region $\tilde{x} = O(1)$, allows us to determine the function $\omega(\tilde{\psi}_0)$ from the condition as $\tilde{x} \to \infty$. Since according to (1.6.11) and (1.6.12), the derivative $\partial^2 \tilde{\psi}_0 / \partial \tilde{y}^2 |_{\tilde{x} \to \infty}$ decreases in order of magnitude as $\tilde{x}^{-13/3}$, then $\omega(\tilde{\psi}_0) = 0$, and the solution (1.6.13) for the separation zone that satisfies the boundary conditions (1.6.12) and $\tilde{\psi}_0(\tilde{x}, 0) = 0$ has the form

$$\tilde{\psi}_0 = -K_0 \tilde{x}^{2/3} \frac{\tilde{y}}{\tilde{y}_S(\tilde{x})} . \qquad (1.6.14)$$

As $\tilde{x} \to +0$, the largest term in (1.6.11) will be the second, so that

$$\tilde{\psi}_0 = -\tfrac{3}{4} K_0 \tilde{y} / c_1 \tilde{x}^{5/6} + \cdots , \quad \tilde{x} \to +0 . \qquad (1.6.15)$$

As $\tilde{x} \to \infty$, the largest term in (1.6.11) will be the first, and we obtain

$$\tilde{\psi}_0 = -K_0 \tilde{y} / c_0 \tilde{x}^{11/6} + \cdots , \quad \tilde{x} \to \infty . \qquad (1.6.16)$$

The asymptotic expression (1.6.15) allows us to write the condition for matching the solution (1.6.14) with the solution for the flow in the sublayer of the interaction region, and the expression (1.6.16) allows us to determine the flow variables in the separation region on the order of the body scale as $x \to +0$. In fact, writing the expression (1.6.15) with the help of (1.6.10) in the variables (1.5.6), (1.5.7) for the interaction region, we find that

$$\Psi_0^* = -\tfrac{3}{4} K_0 Y^* / c_1 x^{*5/6} + \cdots , \quad x^* \to \infty , \qquad (1.6.17)$$

---

[16] The relation between the independent variables s and x is defined on the basis of (1.6.9): $s = x + O(x^4)$ as $x \to +0$.

and using (1.6.16) we find that the longitudinal velocity component in the separation zone is

$$u = \frac{\partial \psi}{\partial y} = \mathrm{Re}^{-1/2}(-K_0/c_0 x^{11/6}) + \cdots, \quad x \to +0. \qquad (1.6.18)$$

From this it follows, in particular, that the variable part of the pressure coefficient in the separation zone, proportional to the square of this velocity, becomes of the order of magnitude of Re^{-1} at finite distances behind the body.

Since the reverse flow in the separation zone appears inviscid and the solution (1.6.14) does not satisfy the no-slip condition of the fluid at the body surface, it is necessary to introduce another region into the analysis – a viscous boundary layer 5′ in the vicinity of the solid wall. A standard procedure for deriving the boundary-layer equations shows that as $\tilde{x} = \mathrm{Re}^{1/16} x \to +0$ the stream function in this boundary layer may be written as

$$\psi = \mathrm{Re}^{-23/32} x^{1/12} \varphi_0(\zeta) + \cdots, \quad \zeta = \mathrm{Re}^{9/32} y / x^{11/12}. \qquad (1.6.19)$$

This follows from the character of the pressure distribution defined using (1.6.15).

The equations and boundary conditions that are satisfied by the function $\varphi_0(\zeta)$ have the form

$$\varphi_0''' + \frac{1}{12} \varphi_0 \varphi_0'' + \frac{5}{6} \varphi_0'^2 = \frac{15}{32} \frac{K_0^2}{c_1^2},$$

$$\varphi_0(0) = \varphi_0'(0) = 0, \quad \varphi_0'(\infty) = -\frac{3}{4} \frac{K_0}{c_1} \qquad (1.6.20)$$

and hence lead to a well-known problem (Rosenhead, 1963). Matching the solution (1.6.19) with the solution for the sublayer in the interaction region yields the following asymptotic expression:

$$\Psi_0^* = x^{*1/12} \varphi_0 \left[Y^* / x^{*11/12} \right] + \cdots, \quad x^* \to \infty. \qquad (1.6.21)$$

We note that the relations (1.6.17) and (1.6.21) determine the behavior of the reverse flow as $x^* \to \infty$ and play the role of initial conditions for calculation of the flow within the interaction zone.

This completes the analysis and the derivation of all the required relations for the separated flow in the zone located behind the interaction region.

1.7 The Numerical Solution of the Problem

The analysis carried out for laminar fluid flow in the vicinity of a separation point as $\mathrm{Re} \to \infty$ shows that flow separation from a smooth surface always occurs under the action of a large self-induced pressure gradient $\partial p/\partial x \sim \mathrm{Re}^{1/8}$ acting over a small section of the surface $\Delta x \sim \mathrm{Re}^{-3/8}$.

A flow model that assumes flow separation from a smooth surface to occur as the result of an adverse pressure gradient distributed over a finite section of the surface, and which has been the object of numerous studies, does not correspond to reality (Sychev, 1972).

For the analysis of the mechanism of separation, the key is the problem of boundary-layer interaction with the external potential flow, which determines the self-induced character of separation. This problem consists in integrating equation (1.5.8) with boundary conditions (1.5.9) and (1.5.11), and the condition of matching with the solutions (1.5.10) ahead of, and (1.6.3), (1.6.17), and (1.6.21) behind, the interaction region, as well as the condition (1.5.20), which defines the relation between the displacement effect of the boundary layer (1.5.12) and the pressure gradient.

It is easy to establish that substituting the variables

$$x^* = a_0^{-5/4}\bar{X}, \quad Y^* = a_0^{-3/4}\bar{Y}, \quad \Psi_0^* = a_0^{-1/2}\bar{\Psi},$$
$$P_0^* = a_0^{1/2}\bar{P}, \quad K_0 = a_0^{1/3}\bar{K} \tag{1.7.1}$$

reduces all the relations mentioned to a form including only one parameter:

$$\alpha = 2c_1/a_0^{9/8}. \tag{1.7.2}$$

For convenience, the complete system of relations describing the flow within the interaction region in similarity variables is written below:

$$\frac{\partial\bar{\Psi}}{\partial\bar{Y}}\frac{\partial^2\bar{\Psi}}{\partial\bar{X}\partial\bar{Y}} - \frac{\partial\bar{\Psi}}{\partial\bar{X}}\frac{\partial^2\bar{\Psi}}{\partial\bar{Y}^2} + \frac{d\bar{P}}{d\bar{X}} = \frac{\partial^3\bar{\Psi}}{\partial\bar{Y}^3}, \tag{1.7.3}$$

$$\frac{d\bar{P}}{d\bar{X}} = \frac{1}{\pi}\int_{-\infty}^{\infty}\frac{\bar{G}'(t)dt}{t-\bar{X}}, \quad \bar{G}(\bar{X}) = \lim_{\bar{Y}\to\infty}\left\{-\frac{\partial\bar{\Psi}}{\partial\bar{X}}\Big/\frac{\partial\bar{\Psi}}{\partial\bar{Y}}\right\}; \tag{1.7.4}$$

$$\bar{\Psi} = \frac{\partial\bar{\Psi}}{\partial\bar{Y}} = 0, \quad \bar{Y} = 0,$$

$$\bar{\Psi} = \tfrac{1}{2}\bar{Y}^2 + O(\bar{Y}^{3/2}), \quad \bar{Y} \to \infty,$$

$$\bar{\Psi} = \tfrac{1}{2}\bar{Y}^2 + \tfrac{1}{2}\alpha(-\bar{X})^{1/2}\bar{f}_1(\bar{\eta}) + \cdots, \quad \bar{\eta} = \frac{\bar{Y}}{(-\bar{X})^{1/3}}, \quad \bar{X} \to -\infty,$$

$$\left.\begin{array}{ll} \bar{\Psi} = \bar{X}^{2/3}\bar{g}_0(\bar{\xi}) + \cdots, & \bar{\xi} = \frac{\bar{Y} - \frac{2}{3}\alpha\bar{X}^{3/2}}{\bar{X}^{1/3}} = O(1), \\[2mm] \bar{\Psi} = -\frac{3}{2}\left(\frac{K}{\alpha}\right)\bar{Y}/\bar{X}^{5/6} + \cdots, & \bar{Y}/\bar{X}^{3/2} = O(1), \\[2mm] \bar{\Psi} = \bar{X}^{1/12}\bar{\varphi}_0(\bar{\zeta}) + \cdots, & \bar{\zeta} = \bar{Y}/\bar{X}^{11/12} = O(1). \end{array}\right\} \bar{X} \to \infty$$

$$(1.7.5)$$

In these relations

$$\bar{f}_1(\bar{\eta}) = a_0 c_1^{-1} f_1(a_0^{-1/3}\bar{\eta}),$$
$$\bar{g}_0(\bar{\xi}) = a_0^{-1/3} g_0(a_0^{-1/3}\bar{\xi}), \qquad (1.7.6)$$
$$\bar{\varphi}_0(\bar{\zeta}) = a_0^{19/48} \varphi_0(a_0^{19/48}\bar{\zeta}).$$

If the solution of the stated problem (1.7.3)–(1.7.5) exists and is unique, it should correspond to some well-defined numerical value of the parameter α. Then the solution obtained in the variables (1.7.1) will become universal, and these variables will be the local similarity variables for different flows in the vicinity of a separation point.

Numerical analysis of the stated interaction problem was first carried out by Smith (1977). Using an iterative process of alternate construction of solutions of the boundary-layer equations (1.7.3) and determination of the pressure gradient from (1.7.4), he succeeded in establishing both the existence and uniqueness of the problem considered. The numerical value of the parameter α was found to be 0.44. To simplify the solution procedure, the velocity of the reverse flow within the separation region was taken to be zero, which allowed avoiding the difficulties associated with the necessity of satisfying downstream flow conditions (as $\bar{X} \to \infty$).

Calculations carried out later by Korolev (1980a) on the basis of a more precise numerical method (see Chapter 7) with consideration of the reverse flow yielded the close value $\alpha = 0.42$. This was also confirmed by work of van Dommelen and Shen (1984), who obtained $\alpha = 0.415 \pm 0.005$.

As an illustration of the results found by Korolev, Figure 1.8 shows a sketch of the streamlines near the separation point.[17]

Smith's (1977) work also showed the first comparison of the asymptotic solution of the problem with the results of a numerical solution of the complete Navier–Stokes equations, for the case of separated flow past a circular cylinder at Reynolds numbers up to 100 (Dennis and Chang, 1970), and also with experimental measurements of skin friction on the cylinder surface for Reynolds numbers of about 200 (Dimopoulos and Hanratty, 1968). The agreement between them, in spite of the

---

[17] The constant λ in this figure is introduced for the conversion from Korolev's (1980a) variables to the variables (1.7.1).

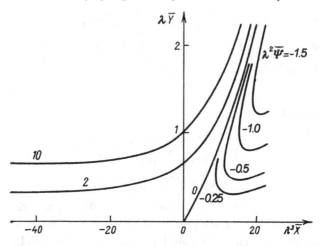

Fig. 1.8 Sketch of the streamlines near the separation point; $\lambda = (4\alpha)^{2/7}$.

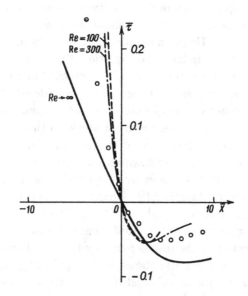

Fig. 1.9 Comparison of the asymptotic skin-friction distribution (Re $\to \infty$) with the results of numerical analyses at Re = 100, 300 (Fornberg, 1980) and experimental measurements (Varty and Currie, 1984) shown by open circles.

low values of the Reynolds number, was found to be fairly good. Later, Fornberg (1980) obtained a numerical solution of the same problem using the complete Navier–Stokes equations for Reynolds numbers up to 300. His results seem to be the most accurate at present. Figure 1.9

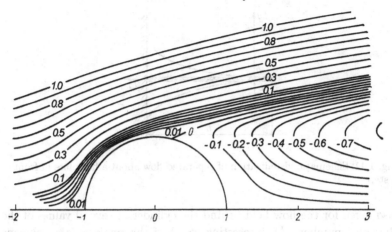

Fig. 1.10 Sketch of the streamlines in the region of separation behind a circular cylinder at Re = 200 (Fornberg, 1980).

shows curves of dimensionless skin-friction distribution along the surface of a circular cylinder $\left(\bar{\tau}(\bar{X}) = \partial^2\bar{\Psi}/\partial\bar{Y}^2|_{\bar{Y}=0}\right)$ near the separation point ($\bar{X} = 0$), obtained by Korolev (1980a) on the basis of the asymptotic theory (Re $\to \infty$), and Fornberg's numerical results at Re = 100 and 300. Also plotted here are results of experimental measurements carried out by Varty and Currie (1984) at Re $= 1.2 \cdot 10^4$. The diameter of the cylinder was chosen in all cases as the characteristic length L in defining the Reynolds number. In order to reduce the results of the numerical analysis and of the experiments to similarity variables, it is necessary to know the numerical value of the constant a_0 which represents the value of the dimensionless skin-friction coefficient τ ahead of the interaction region. This value by convention was taken equal to the maximum value of the dimensionless skin-friction coefficient, since there is a subsequent decrease related to the pressure rise caused by the interaction process. For Fornberg's data the numerical value of a_0 corresponding to the average value $\tau_{\max}$ in the range of Reynolds numbers from 20 to 300 was taken as 1.7. For treating the experimental data of Varty and Currie, the value $a_0 = 2.81$, adopted on the basis of Achenbach's (1968) work, was used.

Taking into account that $\mathrm{Re}^{-1/16}$ is the small parameter of the asymptotic theory, the agreement between its results and those obtained from experimental data and numerical analysis may be considered satisfactory.

In the cited work of Fornberg (1980) the results of the calculation are

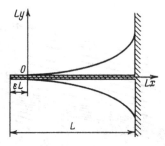

Fig. 1.11 Schematic visualization of separated flow about a flat plate in front of a step.

also given for the flow field behind the cylinder at several values of the Reynolds number. An interesting result of this analysis is an obvious tendency of the streamlines to concentrate in the region of the mixing layer as the Reynolds number increases. This confirms the flow model adopted earlier according to which all streamlines of the reverse flow in the separation region merge into the mixing layer. To illustrate this, Figure 1.10 shows a diagram of the streamlines in the separation region at $\mathrm{Re} = 200$.

1.8 The Effect of Surface Curvature

Until now we have considered the flow over a surface where the radius of curvature in the vicinity of the separation point was of the order of the characteristic dimension of the body; that is, the dimensionless curvature κ_0 in the original equation (1.3.1) was assumed to be a quantity of order unity. Let us now consider situations where this condition is not met.

First of all, we treat the case of separation from a flat surface ($\kappa_0 = 0$). Although such separation represents a special case of separation from a smooth surface, the direct application of the theory to this case needs to be considered separately. To illustrate this, one can speak about boundary-layer separation from a flat plate which takes place as a result of the action of an adverse pressure gradient arising from some external cause, for instance, a step located at a finite distance from the leading edge (Figure 1.11).

The point is that the limiting state of the flow field as $\mathrm{Re} \to \infty$ in this case cannot correspond to the solution of the problem of a flow past a blunt body according to Kirchhoff's scheme with the Brillouin–Villat condition. Therefore, following the work of Sychev (1978a), we draw the following conclusion: the limiting state of the flow field in the vicinity of the separation point corresponds to another well-known solution in the theory of potential flows with free streamlines, a flow with the formation of a stagnation zone ahead of the step, beginning with a cusp as the point of flow reversal. Such flows were first studied by Chaplygin (1899).

If we take the distance from the leading edge of the plate to the obstacle causing separation as the characteristic length L, and the flow speed at the boundary of the stagnation zone (which is not equal to the speed at infinity) as the characteristic speed U_0, then the form of the stagnation zone and the pressure distribution ahead of it will still be determined through the dimensionless relations (1.3.1) and (1.3.2), in which κ_0 should be set equal to zero, but the coefficient k now is to be considered as a quantity of order unity. However, from physical considerations, it is clear that to reduce this problem to the previous one, we should not take the plate length as the characteristic length, but rather the distance from its leading edge to the separation point, since it is this distance which characterizes the state of the boundary layer in the vicinity of the separation point. Let us set this distance equal to εL (Figure 1.11). Then, by introducing into the analysis a Reynolds number

$$\mathrm{Re}_\varepsilon = \frac{U_0 \varepsilon L}{\nu} = \varepsilon \, \mathrm{Re} \qquad (1.8.1)$$

and changing from the dimensionless variables x, y to the variables

$$\bar{x} = x/\varepsilon, \quad \bar{y} = y/\varepsilon, \qquad (1.8.2)$$

we can reduce all the relations of the problem under consideration to the corresponding equations and boundary conditions of the previous problem (Sychev, 1978a). In particular, the relations (1.3.1) and (1.3.2) rewritten in the variables (1.8.2) will then include a constant of order $\mathrm{Re}_\varepsilon^{-1/16}$ as the coefficient of the first term. Hence it follows that $\varepsilon = O(\mathrm{Re}_\varepsilon^{-1/8}) = O(\mathrm{Re}^{-1/9})$. After some transformations with the use of similarity variables, one can obtain the following expression for the distance of the separation point from the leading edge of the plate (Ruban and Sychev, 1979):

$$\varepsilon = \sigma_0 \mathrm{Re}^{-1/9}, \quad \sigma_0 = \left(\frac{\alpha a_0^{9/8}}{2k} \right)^{16/9},$$

where α is the similarity parameter (1.7.2) and a_0 is the dimensionless skin-friction coefficient in the upstream flow (which in this case is determined by the Blasius, 1908, solution), numerically equal to 0.3321.

Thus, the separation point of the boundary layer approaches the leading edge of the plate as the Reynolds number increases, and in the limit as Re → ∞, separation will occur at this edge. This result agrees well with experimental observations (see, for instance, Schlichting, 1979, p. 36), and also with numerical solutions of the full Navier–Stokes equations (Suh and Liu, 1990; McLachlan 1991a).[18]

Now let us describe the opposite case – separation from a highly curved surface. It is easy to determine that as the curvature of the wall (κ_0) increases, the theory of self-induced separation from a smooth surface is no longer valid for values of κ_0 of order $Re^{1/8}$. In fact, as we have seen, it is this order of magnitude that is typical of the curvature of the streamlines in a region of boundary-layer interaction with an external flow (Section 1.5). Therefore, when $\kappa_0 = O(Re^{1/8})$, the body curvature will begin to produce an effect which compensates (in the case of a convex surface) for the displacement effect of the boundary layer upon the external flow. As a result, the positive pressure gradient provoking flow separation will be determined not only by the process of boundary-layer interaction with the inviscid flow but also directly by the form of the body surface. An extreme case of such a situation will be the separation from a corner, which is analyzed in detail in the following chapter.

---

[18] Zubtsov (1985) carried out an analysis of boundary-layer separation from a plane surface caused by steps of small relative height: $h(Re) \to 0$, Re → ∞.

2

Flow Separation from Corners
of a Body Contour

If a solid-body contour (Figure 2.1) has a sharp corner forming a convex angle $\gamma < \pi$, then an incompressible fluid flow around it cannot be unseparated. It is well known that for unseparated flow over a convex corner an unrealistic situation arises, when the speed of fluid elements increases without limit as the corner is approached. According to potential-flow theory, the speed is proportional to $r^{-\alpha}$, where $\alpha = (\pi - \gamma)/(2\pi - \gamma)$, and r is the distance to the corner O. The pressure decreases as the point O is approached and, according to Bernoulli's law, must become negative in some vicinity of the point.

The physical reason for flow separation from a corner is the viscosity of the medium. If the flow around the corner remained unseparated, then the fluid acceleration ahead of the point O would be accompanied by a deceleration downstream of that point. The boundary layer next to the wall immediately behind the corner would then be subjected to an infinitely large adverse pressure gradient, which would lead to flow separation. There can be an exception in the case of slight surface bending, when the adverse pressure gradient proves to be insufficient for boundary-layer separation. This case will be considered in Section 3, devoted to determining the conditions for the onset of separation. As will be shown, the unseparated state of the flow near the corner is maintained up to angles $\pi - \gamma = O(\mathrm{Re}^{-1/4})$. A further increase of the surface bending angle $\pi - \gamma$ leads to boundary-layer separation. Therefore, in analyzing the flow near a convex corner with $\pi - \gamma = O(1)$ in Section 1, we shall assume from the very beginning that the flow is accompanied by separation at the corner.[1]

---

[1] For flow over a concave corner, flow separation also appears for $|\pi - \gamma| = O(\mathrm{Re}^{-1/4})$. If $|\pi - \gamma|$ exceeds $\mathrm{Re}^{-1/4}$ in order of magnitude, then the separation point is displaced upstream from the corner O, and the theory of

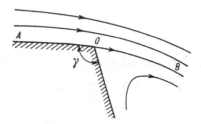

Fig. 2.1 Flow separation from a corner of a body contour.

2.1 Separation from a Corner with the Formation of an Extensive Stagnation Zone

A classical example of a flow with separation from a corner of a body contour is the flow of Kirchhoff (1869), who considered a flat plate set in a stream of an ideal fluid (Figure 2.2) and constructed a corresponding solution of the Euler equations, assuming that the pressure along the free streamlines OB and $O'B'$ is constant and coincides with the pressure in the oncoming flow. We restrict our attention to the vicinity of the corner; however, we shall carry out the flow analysis in this vicinity on a more general basis, including consideration of the case when the contour AO has some curvature, and the angle γ between the tangents to the contour of the body at the point O differs from zero (Figure 2.1). As the basic equations of motion we shall, as in the previous chapter, use the Navier–Stokes equations. The problem consists in finding an asymptotic solution as $\mathrm{Re} \to \infty$.

Let us introduce dimensionless variables: all distances will be referred to the characteristic dimension L of the body, and the velocity components to the limiting value U_0 of the velocity on the free streamline into which the mixing layer between the flow separating from the corner and the stagnation zone degenerates as $\mathrm{Re} \to \infty$. The pressure p as usual will be referred to twice the dynamic pressure ρU_0^2, and the Reynolds number will be defined as $\mathrm{Re} = LU_0/\nu$.

The region of potential flow. For the analysis of the main inviscid-flow region (region 1) it is convenient to use a rectangular Cartesian system of coordinates $Ox'y'$ whose origin coincides with the corner of the body contour, with the axis Ox' directed along the tangent to the surface AO (Figure 2.3). The velocity components in this system of

boundary-layer separation from a smooth section of a body contour described in the preceding chapter becomes valid.

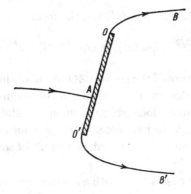

Fig. 2.2 Flow past a flat plate according to the scheme of Kirchhoff.

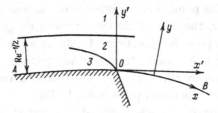

Fig. 2.3 Asymptotic structure of characteristic flow regions ahead of a corner of a body contour.

coordinates will be denoted by u' and v', and we note that the leading terms of the expansions of the dependent flow variables as Re $\to \infty$

$$u' = u_0(x', y') + \cdots, \quad v' = v_0(x', y') + \cdots, \quad p = p_0(x', y') + \cdots,$$
$$(2.1.1)$$

satisfy the Euler equations. Just as in the case of separation from a smooth surface, the solution of these equations must satisfy the condition of no flow through the surface AO and the condition of constant pressure along the free streamline OB. This analogy results in a similar form for the solution around the separation point. The complex velocity

$$w_0(z) = u_0 - iv_0, \quad \text{where } z = x' + iy',$$

is represented as $|z| \to 0$ by the asymptotic expansion (Imai, 1953)

$$w_0 = 1 + 2kiz^{1/2} - \left(\tfrac{10}{3}k^2 - i\kappa_0\right) z + O(|z|^{3/2}).$$
$$(2.1.2)$$

The pressure on the surface AO ahead of the separation point ($x' \to -0$) behaves as

$$p_0 = p_{00} + 2k(-x')^{1/2} - \tfrac{16}{3}k^2(-x') + O[(-x')^{3/2}],$$
$$(2.1.3)$$

and the free streamline $y' = y_S(x')$ has the form

$$y_S = -\tfrac{4}{3}k(x')^{3/2} - \tfrac{1}{2}\kappa_0(x')^2 + O[(x')^{5/2}], \quad x' \to +0. \qquad (2.1.4)$$

Here κ_0 is the curvature of the surface AO ahead of the separation point $(x' = -0)$, and p_{00} denotes the pressure in the stagnation region. The dependence of the above local solution upon the global flow picture is now expressed through the parameter k, whose value for a given position of the separation point may be found as a result of solving the problem for region 1 as a whole.

In the case of flow past a body with a smooth contour, when the separation point is unknown beforehand, solution of the problem for region 1 only allows determination of a relationship between the parameter k and the position of this point. On the other hand, as was described in the preceding chapter, the flow within the interaction region located close to the separation point is self-consistent only for a specific value of the parameter k, which determines the whole flow picture.

A distinctive feature of separation from a corner is the fact that here the position of the separation point is fixed. Therefore, in starting the analysis of the flow in the neighborhood of a corner, we can take the parameter k to be known. Depending upon the particular form of the body surface, it may have different values. Henceforth, we shall pay attention primarily to the study of flows with positive k of order unity[2] (among which in particular belongs Kirchhoff flow). Section 2 is devoted to the discussion of other possibilities.

The boundary layer ahead of the corner. We introduce an orthogonal curvilinear coordinate system Oxy, where x is measured from the corner of the body contour along the zero streamline, and y along the normal to it; for $x < 0$ the zero streamline coincides with the body contour AO, and downstream from the corner it becomes the dividing streamline OB (Figure 2.3). The velocity components in this coordinate system will be designated by u, v. We also introduce a dimensionless stream function ψ, defining it by the relations

$$\frac{\partial \psi}{\partial x} = -(1 + \kappa y)v, \quad \frac{\partial \psi}{\partial y} = u,$$

where κ is the curvature of the zero streamline.

---

[2] One should note that the sign of the parameter k is changed here to the opposite of that in Chapter 1.

The asymptotic analysis of the Navier–Stokes equations in the boundary layer is based upon the limit process

$$x = O(1), \quad Y = \text{Re}^{1/2} y = O(1), \quad \text{Re} \to \infty.$$

For the leading term of the stream-function expansion

$$\psi = \text{Re}^{-1/2} \Psi_0(x, Y) + \cdots \qquad (2.1.5)$$

in the region of flow under consideration, the Prandtl equation holds:

$$\frac{\partial \Psi_0}{\partial Y} \frac{\partial^2 \Psi_0}{\partial x \partial Y} - \frac{\partial \Psi_0}{\partial x} \frac{\partial^2 \Psi_0}{\partial Y^2} = -\frac{dp_e}{dx} + \frac{\partial^3 \Psi_0}{\partial Y^3}. \qquad (2.1.6)$$

The boundary conditions are the condition of no slip at the surface

$$\Psi_0 = \frac{\partial \Psi_0}{\partial Y} = 0 \quad \text{at } Y = 0 \qquad (2.1.7)$$

and the condition of matching the longitudinal velocity component in the boundary layer and the external-flow region (region 1)

$$\frac{\partial \Psi_0}{\partial Y} = U_e(x) \quad \text{at } Y = \infty.$$

In addition, one should impose a condition at some initial cross section of the boundary layer. However, the form of this condition does not affect the subsequent analysis and therefore it is not specified here.

The pressure distribution $p_e(x)$ and tangential velocity component $U_e(x)$ at the outer edge of the boundary layer are found from the solution for region 1. In the vicinity of the corner, they are expressed by formulas (2.1.2), (2.1.3). If we rewrite these formulas in the variables of the orthogonal coordinate system Oxy, then as $x \to -0$ we obtain

$$\begin{aligned} p_e &= p_{00} + 2k(-x)^{1/2} - \tfrac{16}{3} k^2(-x) + O[(-x)^{3/2}], \\ U_e &= 1 - 2k(-x)^{1/2} + \tfrac{10}{3} k^2(-x) + O[(-x)^{3/2}]. \end{aligned} \qquad (2.1.8)$$

A detailed analysis of the properties of the boundary layer ahead of a corner ($x \to -0$) was performed by Ackerberg (1970, 1971, 1973). He found that under the influence of an "extremely favorable" pressure gradient

$$\frac{dp_e}{dx} = -\frac{k}{(-x)^{1/2}} + O(1) \qquad (2.1.9)$$

the boundary layer ahead of the corner splits into two distinct flow regions (Figure 2.3). In one of them, the wall region 3, the effect of friction forces remains significant at arbitrarily small values of $|x|$. The second one, region 2, occupying the major part of the boundary layer, turns out to be locally inviscid as $x \to -0$. If we consider the viscous flow, then the following relations among the orders of the quantities in equation (2.1.6) should be satisfied:

$$\frac{\partial \Psi_0}{\partial Y} \frac{\partial^2 \Psi_0}{\partial x \partial Y} \sim \frac{dp_e}{dx} \sim \frac{\partial^3 \Psi_0}{\partial Y^3}. \qquad (2.1.10)$$

Taking account of the expression (2.1.9) for the pressure gradient, we find that the thickness of the viscous region decreases as $x \to -0$ according to the rule

$$Y = O[(-x)^{3/8}]. \qquad (2.1.11)$$

In the rest of the boundary layer viscosity does not greatly affect the flow.

From (2.1.10) it also follows that $\Psi_0 = O[(-x)^{5/8}]$ in the viscous region. Thus, the leading term of the asymptotic expansion of the function $\Psi_0(x, Y)$ as $x \to -0$ can be written here as

$$\Psi_0 = (8k)^{1/4}(-x)^{5/8} f_0(\eta) + \cdots, \qquad \eta = \left(\frac{k}{2}\right)^{1/4} \frac{Y}{(-x)^{3/8}}. \qquad (2.1.12)$$

The coefficients in front of the function $f_0(\eta)$ and in the expression for the independent variable η have been chosen so that the equation for $f_0(\eta)$ does not include the parameter k. If we substitute (2.1.12) into the boundary-layer equation (2.1.6) and take the limit as $x \to -0$, assuming that the variable η is of order unity, then we obtain

$$f_0''' - \tfrac{5}{4} f_0 f_0'' + \tfrac{1}{2} f_0'^2 + 1 = 0. \qquad (2.1.13)$$

From the no-slip condition (2.1.7) it follows that

$$f_0(0) = f_0'(0) = 0. \qquad (2.1.14)$$

As was pointed out by Ackerberg (1970), to determine uniquely the solution of equation (2.1.13) with the boundary conditions (2.1.14), it is sufficient to require as a third condition that the asymptotic expansion of the function $f_0(\eta)$ as $\eta \to \infty$ should grow no faster than some power of η.[3] Equation (2.1.13) was later studied by McLeod (1972), who proved theorems of existence and uniqueness for it: the solution of equation (2.1.13) with the boundary conditions (2.1.14) and the condition of algebraic growth of the function $f_0(\eta)$ as $\eta \to \infty$ exists and is unique if

---

[3] Since in the main part of the boundary layer (region 2), the fluid speed $\partial \Psi_0 / \partial Y$ is of order unity, the violation of this condition would make it impossible to match the solution in region 3 with that in region 2.

it also satisfies the supplementary condition $f_0' \geq 0$, which implies the absence of reverse flow in region 3.

A numerical solution of equation (2.1.13) satisfying these conditions was constructed by Ackerberg (1970). He noted that as $\eta \to \infty$ the asymptotic expansion of the function $f_0(\eta)$ has the form

$$f_0 = A_0 \eta^{5/3} + B_0 \eta^{2/3} + \frac{9}{5A_0} \eta^{1/3} + O(\eta^{-1/3}).$$ (2.1.15)

The constants A_0 and B_0 may be determined by integrating equation (2.1.13) from some sufficiently large value of the variable η down to $\eta = 0$ and selecting A_0 and B_0 to satisfy the boundary conditions (2.1.14). It was found that

$$A_0 = 1.9507, \quad B_0 = -1.5776, \quad f_0''(0) = 3.0140.$$

With this we conclude the analysis of region 3 and pass on to region 2, which occupies the main part of the boundary layer (Figure 2.3). The asymptotic analysis of equation (2.1.6) within this region is based upon the limit process

$$Y = O(1), \quad x \to -0.$$

The simplest way to determine the form of the asymptotic expansion of the function $\Psi_0(x, Y)$ in region 2 is to do the following. We substitute (2.1.15) into (2.1.12) and use the variable Y instead of η. As a result we obtain

$$\Psi_0 = (2k^2)^{1/3} A_0 Y^{5/3} + (2^7 k^5)^{1/12} B_0 (-x)^{3/8} Y^{2/3}$$
$$+ (4k)^{1/3} \frac{9}{5A_0} (-x)^{1/2} Y^{1/3} + \cdots.$$

According to the matching principle for asymptotic expansions, this will imply that in region 2

$$\Psi_0 = \Psi_{00}(Y) + (-x)^{3/8} \Psi_{01}(Y) + O[(-x)^{1/2}]$$ (2.1.16)

and that as $Y \to 0$ the asymptotic representations of the functions Ψ_{00} and Ψ_{01} have the form

$$\Psi_{00} = (2k^2)^{1/3} A_0 Y^{5/3} + \cdots, \quad \Psi_{01} = (2^7 k^5)^{1/12} B_0 Y^{2/3} + \cdots.$$ (2.1.17)

Substituting the expansion (2.1.16) into the boundary-layer equation (2.1.6) and using the expression (2.1.9) for the pressure gradient, we obtain the equation

$$\Psi_{00}' \Psi_{01}' - \Psi_{00}'' \Psi_{01} = 0.$$ (2.1.18)

The solution that satisfies the conditions (2.1.17) has the form

$$\Psi_{01} = \left(\frac{2}{k}\right)^{1/4} \frac{3B_0}{5A_0} \Psi_{00}'(Y). \qquad (2.1.19)$$

If we now make use of the solutions (2.1.16), (2.1.19) and calculate the slope ϑ of the streamlines, it turns out that its magnitude

$$\mathrm{Re}^{1/2}\vartheta = -\frac{\partial\Psi_0}{\partial x}\bigg/\frac{\partial\Psi_0}{\partial Y} + \cdots = \left(\frac{2}{k}\right)^{1/4}\frac{9B_0}{40A_0}(-x)^{-5/8} + \cdots \quad (2.1.20)$$

remains unchanged across region 2.

The analysis cited above shows that as $x \to -0$ the thickness of the viscous region 3 tends to zero, and consequently the limiting $(x = -0)$ velocity profile in the boundary layer is determined by the solution for region 2:

$$u_0(Y) = \Psi_{00}'(Y).$$

Since the solution $\Psi_0(x, Y)$ of the boundary-layer equation at each point (x, Y) depends upon the boundary conditions everywhere upstream from this point, the local analysis performed does not allow us to determine the function $u_0(Y)$ uniquely. Nevertheless, its behavior near the surface is known. As follows from (2.1.17),

$$u_0 = \tfrac{5}{3}(2k^2)^{1/3} A_0 Y^{2/3} + \cdots \quad \text{as } Y \to 0. \qquad (2.1.21)$$

For any fixed point of the contour AO lying upstream of the point O, the dimensionless skin friction

$$\tau = \frac{\partial u}{\partial Y}\bigg|_{Y=0}$$

has a finite value. Therefore

$$u = O(Y) \quad \text{for } Y \to 0, \quad x < 0.$$

Comparing this relation with the formula (2.1.21) we come to the conclusion that a highly favorable pressure gradient leads to an intense acceleration of fluid elements in the flow region next to the wall. According to (2.1.12) the magnitude of the skin friction increases with the approach to the corner as

$$\tau = (2k^3)^{1/4} f_0''(0)(-x)^{-1/8} + \cdots \quad \text{as } x \to -0.$$

The higher sensitivity of the slower fluid in the wall layer to the singular pressure gradient is manifested by the fact that it is this layer

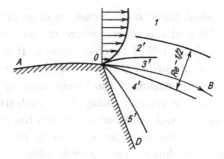

Fig. 2.4 Asymptotic structure of characteristic flow regions behind a corner of a body contour.

(region 3) that determines the displacement effect of the entire boundary layer. From expression (2.1.12) it follows that the inclination angle of the streamlines

$$Re^{1/2}\vartheta = \left(\frac{2}{k}\right)^{1/4} \frac{5f_0 - 3\eta f_0'}{8f_0'}(-x)^{-5/8} + \cdots$$

changes from zero at the wall ($\eta = 0$) to

$$Re^{1/2}\vartheta = \left(\frac{2}{k}\right)^{1/4} \frac{9B_0}{40A_0}(-x)^{-5/8} + \cdots$$

at the outer edge ($\eta \to \infty$) of region 3, and after that maintains its magnitude (2.1.20) up to the outer edge of the boundary layer.

The mixing layer and the flow within the stagnation region. In order to examine the mixing layer that divides the inviscid fluid flow separating from the corner (region 1 in Figure 2.4) from the stagnation zone (region 4′), we shall use the orthogonal coordinate system Oxy introduced earlier. We recall that x is measured along the dividing streamline, and y along the normal to it. Just as in the boundary layer next to the wall, the leading term in the expansion of the stream function in the mixing layer will be

$$\psi = Re^{-1/2}\Psi_0(x, Y) + \cdots \quad \text{as } Re \to \infty. \tag{2.1.22}$$

Substituting (2.1.22) into the original Navier–Stokes equations and assuming, as usual, the pressure to be constant in the first approximation along the mixing layer, we find that the function $\Psi_0(x, Y)$ satisfies equation (2.1.6) with $dp_e/dx = 0$. Before we formulate the boundary conditions for this equation, let us note that the solution within the wall boundary layer ahead of the corner has a singularity. This singularity, as will be shown below, leads to the formation of an interaction region, located in the vicinity of the corner, where the solution obtained for regions 1, 2, and 3 (Figure 2.3) becomes invalid. It should be noted, however, that main part of the boundary layer in the interaction region, which is a continuation of the locally inviscid region 2, must be inviscid. Therefore the velocity change along each streamline during the transition from the wall boundary layer into the mixing layer can be related to the increment of pressure by Bernoulli's integral. Since the pressure drop in the interaction region is infinitesimal,[4] the initial velocity profile for the mixing layer will in the leading approximation coincide with the limiting velocity profile for the boundary layer, which can be written as

$$\Psi_0 = \Psi_{00}(Y) \quad \text{at } x = +0, \quad Y \geq 0.$$

The velocity at the outer edge of the mixing layer is equal to unity:

$$\frac{\partial \Psi_0}{\partial Y} = 1 \quad \text{at } Y = \infty,$$

and the condition of matching with the solution in the stagnation region, where the fluid velocity tends to zero as Re $\to \infty$, has the form

$$\frac{\partial \Psi_0}{\partial Y} = 0 \quad \text{at } Y = -\infty. \tag{2.1.23}$$

Finally, it is necessary to set a kinematic condition

$$\Psi_0 = 0 \quad \text{at } Y = 0, \tag{2.1.24}$$

which is associated with the choice of the coordinate system to be used.

Now let us construct an asymptotic solution of the boundary-value problem formulated above as $x \to +0$. As the boundary layer is leaving the corner, it comes into contact with the fluid at rest in the stagnation zone, and this gives rise to the viscous region 3', which separates the main part of the mixing layer from the stagnation zone (Figure 2.4). Proceeding from the requirement of agreement of the orders of the convective and viscous terms

$$\frac{\partial \Psi_0}{\partial Y} \frac{\partial^2 \Psi_0}{\partial x \partial Y} \sim \frac{\partial^3 \Psi_0}{\partial Y^3}$$

in the boundary-layer equation (2.1.6), and using for the stream function the estimate $\Psi_0 = O(Y^{5/3})$, which follows from the form (2.1.17) of the initial velocity profile, it is easy to find that the width of the viscous

---

[4] According to the expression (2.1.8) it is estimated as $\Delta p = O[(\Delta x)^{1/2}]$.

region 3' decreases as $x \to +0$ according to the rule $Y = O(x^{3/8})$, and the stream function $\Psi_0(x, Y)$ here is of order $x^{5/8}$. Therefore, the solution in region 3' is represented as

$$\Psi_0 = (8k)^{1/4} x^{5/8} g_0(\xi) + \cdots, \quad \xi = \left(\frac{k}{2}\right)^{1/4} \frac{Y}{x^{3/8}}. \qquad (2.1.25)$$

The function $g_0(\xi)$ is a solution of the following boundary-value problem:

$$g_0''' + \tfrac{5}{4} g_0 g_0'' - \tfrac{1}{2} g_0'^2 = 0, \quad g_0'(-\infty) = g_0(0) = 0, \\ g_0 = A_0 \xi^{5/3} + O(\xi^{2/3}) \quad \text{as } \xi \to \infty. \qquad (2.1.26)$$

The equation for $g_0(\xi)$ is obtained by substitution of (2.1.25) into (2.1.6), with subsequent passage to the limit as $x \to +0$. The boundary conditions at $\xi = -\infty$ and $\xi = 0$ result from (2.1.23) and (2.1.24), and the condition as $\xi \to \infty$ represents the result of matching with the solution in the main part of the mixing layer (region 2'), where the leading term of the asymptotic expansion of the function $\Psi_0(x, Y)$ coincides with $\Psi_{00}(Y)$.

As was shown by Diesperov (1984), the solution of the boundary-value problem (2.1.26) that satisfies the auxiliary condition $g_0' > 0$ does exist and is unique. The asymptotic expansion of the function $g_0(\xi)$ as $\xi \to \infty$ has the form

$$g_0 = A_0 \xi^{5/3} + \widehat{B}_0 \xi^{2/3} + O(\xi^{-1/3}).$$

If $\xi \to -\infty$, then

$$g_0 = -K_0 + O\left[\exp\left(\tfrac{5}{4} K_0 \xi\right)\right].$$

Numerical solution of the boundary-value problem (2.1.26) yielded the following values for the constants $\widehat{B}_0$ and K_0:

$$\widehat{B}_0 = 0.428, \quad K_0 = 1.181.$$

Let us turn to region 2', which occupies the main part of the mixing layer (Figure 2.4). The asymptotic analysis of equation (2.1.6) in this region is associated with the limit process

$$x \to +0, \quad Y = O(1).$$

As we have already noted, the leading term of the asymptotic expansion of the stream function $\Psi_0(x, Y)$ here is $\Psi_{00}(Y)$. The second term of the expansion can be found in exactly the same way as was done for region 2. As a result, we obtain

$$\Psi_0 = \Psi_{00}(Y) + \left(\frac{2}{k}\right)^{1/4} \frac{3\widehat{B}_0}{5A_0} x^{3/8} \Psi_{00}'(Y) + \cdots .$$

To satisfy ourselves of the correctness of condition (2.1.23) and the requirement of constant pressure along the mixing layer, let us consider the stagnation zone in more detail (region 4', Figure 2.4). As in the case of separation from a smooth surface, we assume that the fluid motion in region 4' is due to the entraining effect of the mixing layer. Then from matching with the solution for the mixing layer, it follows that $\psi = O(\mathrm{Re}^{-1/2})$ in region 4'. The velocity components have the same order of smallness, and the pressure change is estimated as the square of this value: $p - p_{00} = O(\mathrm{Re}^{-1})$.

It is convenient to seek the solution in region 4' in a polar coordinate system r, φ. Let us place its origin at the point O and measure the angle φ from the tangent, taken through the point O, to the body surface OD exposed to the reverse flow (Figure 2.4). The tangent to the dividing streamline OB is written in this system of coordinates as $\varphi = \beta$, where $\beta = \pi - \gamma$.

The asymptotic analysis of the Navier–Stokes equations in region 4' is related to the limit process

$$r = O(1), \quad \varphi = O(1), \quad \mathrm{Re} \to \infty .$$

We represent the solution in the form

$$\psi = \mathrm{Re}^{-1/2} \tilde{\psi}_0(r, \psi) + \cdots, \quad p - p_{00} = \mathrm{Re}^{-1} \tilde{p}_0(r, \varphi) + \cdots . \quad (2.1.27)$$

The equations for the functions $\tilde{\psi}_0(r, \varphi)$ and $\tilde{p}_0(r, \varphi)$ are the Euler equations. Therefore, in region 4' the law of vorticity conservation along streamlines holds:

$$\Delta\tilde{\psi}_0 = \omega(\tilde{\psi}_0). \quad (2.1.28)$$

If $r \to 0$, the condition of matching with the solution in region 3' can be written as

$$\tilde{\psi}_0(r, \beta) = -(8k)^{1/4} r^{5/8} K_0 + \cdots . \quad (2.1.29)$$

The second boundary condition is that of no flow through the surface OD:

$$\tilde{\psi}_0(r, 0) = 0. \quad (2.1.30)$$

From the form of the condition (2.1.29), we represent the solution of equation (2.1.28) as

$$\tilde{\psi}_0 = r^{5/8}\Phi(\varphi) + \cdots \quad \text{as } r \to 0. \tag{2.1.31}$$

It should be mentioned that the entraining effect of the mixing layer grows as the corner is approached — as $r \to 0$, the fluid speed in region 4' increases, as follows from (2.1.31), in proportion to $r^{-3/8}$. At the same time the vorticity $\omega(\tilde{\psi}_0)$ maintains its value along each streamline. Therefore, as in the case of separation from a smooth surface, equation (2.1.28) at small values of r is equivalent to the Laplace equation. The solution that satisfies conditions (2.1.29), (2.1.30) has the form

$$\tilde{\psi}_0 = -\frac{(8k)^{1/4}K_0}{\sin\left(\frac{5}{8}\beta\right)}r^{5/8}\sin\left(\tfrac{5}{8}\varphi\right) + \cdots \quad \text{as } r \to 0. \tag{2.1.32}$$

Since the fluid flow in region 4' is inviscid, the pressure distribution can be obtained from Bernoulli's equation:

$$\tilde{p}_0 = -\left[\frac{5}{8}\frac{(2k)^{1/4}K_0}{\sin\left(\frac{5}{8}\beta\right)}\right]^2 r^{-3/4} + \cdots \quad \text{as } r \to 0. \tag{2.1.33}$$

The solution (2.1.32) for region 4' does not satisfy the no-slip condition on the surface OD. It is therefore necessary also to analyze region 5', a viscous boundary layer, located close to this surface (Figure 2.4). The fluid flow in region 5', as well as that in the boundary layer at the surface AO, is directed toward the corner of the body contour. In either case the pressure gradient is favorable. However, there is a considerable difference between these two flows. On the surface AO it is only the pressure gradient that tends to infinity. The pressure itself remains bounded as $x \to -0$. Under these conditions, as noted above, the boundary layer splits into two distinct parts (Figure 2.3). The action of friction forces turns out to be important only in region 3 next to the wall. However, the solution for this region does not satisfy the boundary condition at the outer edge of the boundary layer. Therefore, there arises the necessity to introduce an intermediate locally inviscid region 2.

On the surface OD, not only the pressure gradient but the pressure (2.1.33) itself tends to infinity as the corner is approached. In this case the boundary layer (region 5' in Figure 2.4) remains intact and does not split into additional flow regions.

Let us introduce the orthogonal curvilinear coordinate system $O\tilde{x}\tilde{y}$, placing the origin at the corner O and directing the axis $O\tilde{x}$ along the contour OD and the axis $O\tilde{y}$ along the normal to it. As $\tilde{x} \to 0$, the solution for region $5'$ will be written as

$$\psi = -\mathrm{Re}^{-3/4}d^{1/2}\tilde{x}^{5/16}\varphi_0(\zeta) + \cdots .$$

The independent variable ζ and the constant d are calculated according to the formulas

$$\zeta = d^{1/2}\frac{\mathrm{Re}^{1/4}\tilde{y}}{\tilde{x}^{11/16}}, \quad d = \frac{5}{8}\frac{(8k)^{1/4}K_0}{\sin\left(\frac{5}{8}\beta\right)} .$$

The function $\varphi_0(\zeta)$ is determined by the following boundary-value problem:

$$\varphi_0''' - \tfrac{5}{16}\varphi_0\varphi_0'' + \tfrac{3}{8}\left(1 - \varphi_0'^2\right) = 0,$$
$$\varphi_0(0) = \varphi_0'(0) = 0, \quad \varphi_0'(\infty) = 1.$$

The solution of this problem exists and is unique under one auxiliary condition: $\varphi_0' > 0$ in the interval $(0, \infty)$. Proof of the corresponding theorem is given, for instance, in an article by Gabutti (1984).

Concluding the analysis of the outer regions surrounding the interaction region, we determine those perturbations that arise in the outer potential part of the flow (region 1) due to the displacement effect of the boundary layer. To take this effect into account, it is necessary, in addition to the leading terms of the expansion (2.1.1) of the flow variables in region 1, to retain the terms of order $\mathrm{Re}^{-1/2}$:

$$u' = u_0(x', y') + \cdots + \mathrm{Re}^{-1/2}u_2(x', y') + \cdots ,$$
$$v' = v_0(x', y') + \cdots + \mathrm{Re}^{-1/2}v_2(x', y') + \cdots , \qquad (2.1.34)$$
$$p = p_0(x', y') + \cdots + \mathrm{Re}^{-1/2}p_2(x', y') + \cdots .$$

Here, we use the rectangular Cartesian coordinate system $Ox'y'$ introduced previously, with the origin at the corner O and the axis Ox' directed along the tangent to the contour AO (Figure 2.3). Later, we shall show that the expansions (2.1.34) also contain intermediate terms of order $\mathrm{Re}^{-4/9}$.

Substituting (2.1.34) into the Navier–Stokes equations, it is simple to convince oneself that the perturbations of the velocity components u_2, v_2 possess the same property as their leading approximations u_0, v_0: if we set $w_2(z) = u_2 - iv_2$, then the function $w_2(z)$ will be an analytic function of the complex variable $z = x' + iy'$. A second consequence of this substitution is the linearized Bernoulli equation, which relates the pressure perturbation to the velocity components:

$$p_2 = -(u_0u_2 + v_0v_2). \qquad (2.1.35)$$

In solving the problem for region 1 in the leading approximation, we took the no-flow condition at the body surface AO and the condition of constant pressure along the dividing streamline OB to be the boundary conditions for the complex velocity $w_0(z) = u_0 - iv_0$. To obtain the appropriate boundary conditions for the function $w_2(z)$ it is necessary to match the solution in the boundary layer with that for the mixing layer. The angle of inclination of the streamlines in region 2 is determined by the formula (2.1.20). Therefore, the first boundary condition will be

$$v_2(x', 0) = \left(\frac{2}{k}\right)^{1/4} \frac{9B_0}{40A_0}(-x')^{-5/8} + \cdots \quad \text{as } x' \to -0. \quad (2.1.36)$$

Owing to the fact that the pressure variations in region $4'$ are of order Re^{-1}, the pressure distribution along the outer edge of the mixing layer is determined by the difference Δp across this layer. From the second equation of the system of Navier–Stokes equations (see (1.1.2), Chapter 1), it follows that $\Delta p = O(\mathrm{Re}^{-1/2}\kappa)$. Here, κ is the curvature of the dividing streamline

$$\kappa = -\frac{y_S''}{(1 + y_S'^2)^{3/2}}.$$

If one uses the formula (2.1.4), it is easy to show that the pressure perturbations in region 1 satisfy the following condition at the boundary with the mixing layer:

$$p_2(x', 0) = O[(x')^{-1/2}] \quad \text{as } x' \to +0.$$

Using Bernoulli's equation (2.1.35), one may write this condition in a different form:

$$u_2(x', 0) = O[(x')^{-1/2}] \quad \text{as } x' \to +0. \quad (2.1.37)$$

Conditions (2.1.36) and (2.1.37) determine uniquely the leading term of the expansion of the function $w_2(z)$ as $|z| \to 0$:

$$w_2 = -i\left(\frac{2}{k}\right)^{1/4} \frac{9B_0}{40A_0 \cos \frac{5}{8}\pi} z^{-5/8} + O\left(|z|^{-1/2}\right).$$

Taking the real part u_2 of this expression for $\arg z = \pi$ and substituting the result into (2.1.35), we conclude that the pressure perturbation generated on the surface AO due to the displacement effect of the boundary layer increases while approaching the corner as

$$p_2 = \left(\frac{2}{k}\right)^{1/4} \frac{9B_0}{40A_0} \tan \frac{5}{8}\pi(-x')^{-5/8} + \cdots \quad \text{as } x' \to -0. \quad (2.1.38)$$

The interaction region. From formula (2.1.38) it follows that the pressure gradient induced at the surface AO by the displacement effect of the boundary layer increases as $O[\mathrm{Re}^{-1/2}(-x)^{-13/8}]$ when $x \to -0$. The original pressure gradient (2.1.9) imposed on the boundary layer is of order $O[(-x)^{-1/2}]$. It is clear that there exists a neighborhood of the corner in which the induced pressure gradient becomes of the same order as the original pressure gradient. This neighborhood has the extent

$$\Delta x = O(\mathrm{Re}^{-4/9}). \tag{2.1.39}$$

Here, the pressure gradient acting upon the boundary layer is no longer given in advance, but must be determined as the result of a combined solution of the problems for the viscous wall layer and the outer potential part of the flow.

One can also obtain the estimate (2.1.39) for the length of the interaction region with the help of an analysis like that of Section 3, Chapter 1. The estimates of the effects of the physical processes occurring in the interaction region upon which this analysis is based hold also in the present case. It is necessary only to replace the relation (1.3.6) of Chapter 1 by $u \sim Y^{2/3}$, taking into account the change in the form of the initial velocity profile (2.1.21) ahead of the interaction region. Then instead of the relation $\Delta x = O(\mathrm{Re}^{-3/8})$, which determines the length of the interaction region on a smooth surface, we obtain (2.1.39).

As has been shown, the boundary layer ahead of the interaction region has a two-layer structure. It includes a locally inviscid region 2, which occupies the main part of the boundary layer, and a viscous wall region 3, which in the leading approximation determines the displacement effect of the whole boundary layer. Region 3 "gives birth" to the viscous wall layer of the interaction region (region III), and region 2 to the middle layer (region II in Figure 2.5).

Let us first consider region III. Its length is determined by the relation (2.1.39), and its thickness is what the thickness (2.1.11) of region 3 would be at a distance $\Delta x = O(\mathrm{Re}^{-4/9})$ from the corner: $y = O(\mathrm{Re}^{-2/3})$. Thus, the asymptotic analysis of the Navier–Stokes equation in region III is associated with the limit process

$$x^* = \mathrm{Re}^{4/9}x = O(1), \quad Y^* = \mathrm{Re}^{2/3}y = O(1), \quad \mathrm{Re} \to \infty. \tag{2.1.40}$$

Taking $|x| = O(\mathrm{Re}^{-4/9})$ in the solution (2.1.5), (2.1.12) for region 3, we conclude that $\psi = O(\mathrm{Re}^{-7/9})$ in region III . In a similar way, from (2.1.8) we obtain $p - p_{00} = O(\mathrm{Re}^{-2/9})$. Therefore, the solution of the Navier–

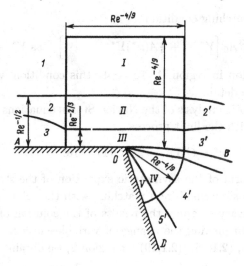

Fig. 2.5 Asymptotic structure of the interaction region.

Stokes equations in region III will be sought in the form

$$\psi = \text{Re}^{-7/9}\Psi_0^*(x^*, Y^*) + \cdots, \quad p = p_{00} + \text{Re}^{-2/9}P_0^*(x^*) + \cdots . \quad (2.1.41)$$

From the second equation of the system of Navier–Stokes equations (see (1.1.2), Chapter 1) it follows that, for the approximation considered, the pressure remains constant across region III. In this connection the function $P_0^*(x^*)$, being a part of the expansion (2.1.41) for the pressure, was from the very beginning written as a function of the single argument x^*. Substituting (2.1.41) into the Navier–Stokes equations, we obtain

$$\frac{\partial \Psi_0^*}{\partial Y^*}\frac{\partial^2 \Psi_0^*}{\partial x^* \partial Y^*} - \frac{\partial \Psi_0^*}{\partial x^*}\frac{\partial^2 \Psi_0^*}{\partial Y^{*2}} = -\frac{dP_0^*}{dx^*} + \frac{\partial^3 \Psi_0^*}{\partial Y^{*3}}. \quad (2.1.42)$$

The boundary conditions are the no-slip condition at the surface AO:

$$\Psi_0^* = \frac{\partial \Psi_0^*}{\partial Y^*} = 0 \quad \text{at } Y^* = 0, \quad x^* < 0, \quad (2.1.43)$$

the condition of matching with the solution (2.1.5), (2.1.12) in region 3

$$\Psi_0^* = (8k)^{1/4}(-x^*)^{5/8} f_0(\eta) + \cdots \quad \text{as } x^* \to -\infty, \quad (2.1.44)$$

where

$$\eta = \left(\frac{k}{2}\right)^{1/4} \frac{Y^*}{(-x^*)^{3/8}},$$

and also the matching condition

$$\Psi_0^* = (2k^2)^{1/3} A_0 \left[Y^{*5/3} + \tfrac{5}{3} A(x^*) Y^{*2/3} + \cdots \right] \quad \text{as } Y^* \to \infty \quad (2.1.45)$$

with the solution in region II. To verify this condition we analyze region II in more detail.

The asymptotic analysis of the Navier–Stokes equations in this region is associated with the limit process

$$x^* = O(1), \quad Y = O(1), \quad \text{Re} \to \infty.$$

Whereas the form of the asymptotic expansion of the stream function in region III was obtained from matching with the solution in region 3, now it is necessary to apply the results of the solution of the problem in region 2. On making the change of variables $x = \text{Re}^{-4/9} x^*$ in the solution (2.1.5), (2.1.16), (2.1.19) for region 2, we obtain

$$\psi = \text{Re}^{-1/2} \Psi_{00}(Y) + \text{Re}^{-2/3} (-x^*)^{3/8} \left(\frac{2}{k} \right)^{1/4} \frac{3B_0}{5A_0} \Psi'_{00}(Y) + \cdots .$$

According to the matching principle for asymptotic expansions, the above expression should coincide with the asymptotic representation of the stream function in region II as $x^* \to -\infty$. Therefore, we seek the solution in region II in the form

$$\psi = \text{Re}^{-1/2} \tilde{\Psi}_0(x^*, Y) + \text{Re}^{-2/3} \tilde{\Psi}_1(x^*, Y) + \cdots ,$$
$$p = p_{00} + \text{Re}^{-2/9} \tilde{P}_0(x^*, Y) + \cdots . \tag{2.1.46}$$

On substituting (2.1.46) into the Navier–Stokes equations, we obtain the following equation for the leading term of the expansion of the stream function:

$$\frac{\partial \tilde{\Psi}_0}{\partial Y} \frac{\partial^2 \tilde{\Psi}_0}{\partial x^* \partial Y} - \frac{\partial \tilde{\Psi}_0}{\partial x^*} \frac{\partial^2 \tilde{\Psi}_0}{\partial Y^2} = 0.$$

Rewriting it in the form

$$\frac{\partial}{\partial Y} \left(\frac{\partial \tilde{\Psi}_0}{\partial x^*} \Big/ \frac{\partial \tilde{\Psi}_0}{\partial Y} \right) = 0,$$

we conclude that the inclination angle of the streamlines

$$\vartheta = -\frac{\partial \psi}{\partial x} \Big/ \frac{\partial \psi}{\partial y} = -\text{Re}^{-1/18} \frac{\partial \tilde{\Psi}_0}{\partial x^*} \Big/ \frac{\partial \tilde{\Psi}_0}{\partial Y} + \cdots \tag{2.1.47}$$

remains unchanged across region II and coincides with the inclination angle of the streamlines at the outer edge of region III. Since $\vartheta =$

$O(\mathrm{Re}^{-2/9})$ in region III, then in the present approximation the inclination angle of the streamlines (2.1.47) is identically equal to zero. Along with it, the derivative $\partial \tilde{\Psi}_0 / \partial x^*$ also vanishes. As for the function $\tilde{\Psi}_0(x^*, Y)$ itself, it is determined from matching with the solution in region 2:

$$\tilde{\Psi}_0 = \Psi_{00}(Y) . \qquad (2.1.48)$$

For the functions $\tilde{P}_0(x^*, Y)$ and $\tilde{\Psi}_1(x^*, Y)$, the equations are

$$\frac{\partial \tilde{P}_0}{\partial Y} = 0, \quad \Psi'_{00} \frac{\partial^2 \tilde{\Psi}_1}{\partial x^* \partial Y} - \Psi''_{00} \frac{\partial \tilde{\Psi}_1}{\partial x^*} = 0 .$$

According to the first of these the pressure does not change across region II and hence it coincides with the pressure in region III:

$$\tilde{P}_0 = P_0^*(x^*) . \qquad (2.1.49)$$

The solution of the second equation that satisfies the condition of matching with the solution in region 2 has the form

$$\tilde{\Psi}_1 = A(x^*) \Psi'_{00}(Y) . \qquad (2.1.50)$$

Here, $A(x^*)$ is an arbitrary function; from matching with the solution in region 2 we know only that

$$A = (-x^*)^{3/8} \left(\frac{2}{k} \right)^{1/4} \frac{3 B_0}{5 A_0} + o(1) \quad \text{as } x^* \to -\infty .$$

It remains to carry out the matching of the expansions (2.1.46), (2.1.48), (2.1.50) of the stream function in region II with the expansion (2.1.41) of the stream function in region III, and formula (2.1.45) will be proved.

The solution (2.1.46), (2.1.48), (2.1.50) for region II may be written as

$$\psi = \mathrm{Re}^{-1/2} \Psi_{00}[Y + \mathrm{Re}^{-1/6} A(x^*)] + \cdots . \qquad (2.1.51)$$

From (2.1.51), it follows that all streamlines in region II repeat each other in shape, the longitudinal velocity component

$$u = \Psi'_{00}[Y + \mathrm{Re}^{-1/6} A(x^*)] + \cdots$$

remaining constant along each streamline. Thus, the profile of the longitudinal velocity component does not change with a change in x^*. The displacement effect of the viscous sublayer (region III) leads only to the fact that at every point x^* the inviscid part of the boundary layer (region II) is displaced as a whole along normals to the surface. The function $A(x^*)$ which characterizes the displacement effect of the viscous sublayer must be determined from solution of equation (2.1.42). However, in contrast to the classical statement of a problem for the boundary-layer equation, in this case the pressure gradient dP_0^*/dx^* is not given. Therefore, to complete the problem (2.1.42)–(2.1.45) it is also necessary to examine the flow in region I (Figure 2.5).

The asymptotic analysis of the Navier–Stokes equations in region I is associated with the limit process

$$x'_* = x'\mathrm{Re}^{4/9} = O(1), \quad y'_* = y'\mathrm{Re}^{4/9} = O(1), \quad \mathrm{Re} \to \infty.$$

Here, as in region 1, we shall use a Cartesian system of coordinates $Ox'y'$. We represent the flow variables in the form of the expansions

$$
\begin{aligned}
u' &= 1 + \mathrm{Re}^{-2/9}u_1^*(x'_*, y'_*) + \cdots, \\
v' &= \mathrm{Re}^{-2/9}v_1^*(x'_*, y'_*) + \cdots, \\
p &= p_{00} + \mathrm{Re}^{-2/9}p_1^*(x'_*, y'_*) + \cdots.
\end{aligned}
\tag{2.1.52}
$$

Substituting (2.1.52) into the Navier–Stokes equations, we conclude that the function

$$w_1^*(z_*) = u_1^* - iv_1^*$$

is an analytic function of the complex variable $z_* = x'_* + iy'_*$, and that p_1^* satisfies the linearized Bernoulli equation

$$p_1^* = -u_1^*. \tag{2.1.53}$$

Let us formulate the boundary conditions for the function $w_1^*(z_*)$. The first of them,

$$w_1^* = 2kiz_*^{1/2} + \cdots \quad \text{as } |z_*| \to \infty \tag{2.1.54}$$

is a consequence of the condition of matching with the solution (2.1.1), (2.1.2) in region 1. It is also necessary to carry out the matching with the solution in region II. According to (2.1.46), (2.1.48), (2.1.50) the inclination angle of the velocity vector in region II is determined as

$$\vartheta = -\frac{\partial\psi}{\partial x} \bigg/ \frac{\partial\psi}{\partial y} = -\mathrm{Re}^{-2/9}A'(x^*) + \cdots. \tag{2.1.55}$$

On the other hand, in region I

$$\vartheta' = \frac{v'}{u'} = \mathrm{Re}^{-2/9}v_1^*(x'_*, y'_*) + \cdots. \tag{2.1.56}$$

Comparing formulas (2.1.55) and (2.1.56), it is necessary to keep in mind that the first is written in a curvilinear system of coordinates Oxy and determines the inclination angle of the velocity vector with respect to the axis Ox. The second is written in a Cartesian system of coordinates $Ox'y'$ and determines the inclination angle of the velocity vector with respect to the axis Ox'. For $x'_* < 0$ the difference between these values, which is equal to the inclination angle of the body surface to the axis Ox', is estimated as $O(\mathrm{Re}^{-4/9})$ and does not affect the result of matching the expansions (2.1.55) and (2.1.56):

$$v_1^*(x'_*, 0) = -A'(x'_*) \quad \text{for } x'_* < 0. \tag{2.1.57}$$

The situation is different downstream from the separation point. As follows from the solution (2.1.4) for region 1, the angle between the tangent to the axis Ox and the axis Ox' decreases as $x' \to 0$ according to the rule $O[(x')^{1/2}]$. Within the interaction region this angle assumes a value of the same order $O(\mathrm{Re}^{-2/9})$ as the inclination angle of the velocity vector.

We write the equation of the dividing streamline in the interaction region as

$$y' = \mathrm{Re}^{-2/3} F(x'_*) + \cdots.$$

Then, in addition to condition (2.1.57), we obtain

$$v_1^*(x'_*, 0) = F'(x'_*) - A'(x'_*) \quad \text{for } x'_* > 0. \tag{2.1.58}$$

A peculiarity of the interaction region considered is manifested in the fact that its viscous layer (region III), separating from the body surface at point O, immediately comes into contact with region IV, lying in the stagnation zone (Figure 2.5). The motion of the fluid in region IV follows the same laws as that in region 4'. Taking this similarity into account we will not dwell upon the construction of the solution of the Navier–Stokes equations for region IV. We shall only point out that in this region, of extent $r = O(\mathrm{Re}^{-4/9})$, the pressure changes are, according to the condition of matching with the solution (2.1.27), (2.1.33) for region 4', estimated as[5]

$$p - p_{00} = O(\mathrm{Re}^{-2/3}).$$

Comparing this estimate with the expansion (2.1.41) for the pressure in region III, we conclude that

$$P_0^*(x^*) = 0 \quad \text{for } x^* > 0. \tag{2.1.59}$$

---

[5] A more detailed analysis of region IV, and also of the viscous wall region V (Figure 2.5), is given in the article by Ruban (1974).

If we now take advantage of the condition of matching the expansion (2.1.52) for the pressure in region I with the expansion (2.1.46), (2.1.49) for the pressure in region II

$$p_1^*(x_*', 0) = P_0^*(x_*')$$

and Bernoulli's equation (2.1.53), then we obtain the condition

$$u_1^*(x_*', 0) = 0 \quad \text{for } x_*' > 0, \qquad (2.1.60)$$

which allows formulation of the problem in region I as a mixed boundary-value problem in the theory of functions of a complex variable: to find in the upper half-plane of the variable z_* an analytic function $w_1^*(z_*)$ with its imaginary part given by (2.1.57) on the negative real axis and its real part given by (2.1.60) on the positive real axis, if one knows that as $|z_*| \to \infty$ the function $w_1^*(z_*)$ satisfies the condition (2.1.54).

Using the Keldysh–Sedov (1937) method, one can show that the solution of the given problem satisfying the auxiliary condition $w_1^* = o(z_*^{-1/2})$ in the vicinity of the point $z_* = 0$ exists and is unique. It is written (Ruban, 1976a) in terms of an integral of Cauchy type:

$$w_1^*(z_*) = iz_*^{1/2} \left[2k + \frac{1}{\pi} \int_{-\infty}^0 \frac{G(s)}{\sqrt{-s}(s - z_*)} ds \right] . \qquad (2.1.61)$$

Here, $G(s)$ is the inclination angle of the velocity vector at the outer edge of region III: $G(s) = -A'(s)$.

In order to find the limiting value of the function $w_1^*(z_*)$ on the path of integration (i.e., as $z_* \to x_*'$, where $x_*' < 0$), one has to use the Sokhotski–Plemelj formula. Then for the pressure in region III, we obtain the expression

$$P_0^*(x^*) = \sqrt{-x^*} \left[2k + \frac{1}{\pi} \fint_{-\infty}^0 \frac{G(s)}{\sqrt{-s}(s - x^*)} ds \right] , \qquad (2.1.62)$$

which closes the interaction problem.

On writing the pressure with the help of formula (2.1.62), we have the possibility of considering the flow ahead of the corner independently of the flow for $x^* > 0$. The interaction problem in this case will include equation (2.1.42) for the stream function in region III with boundary conditions (2.1.43)–(2.1.45) and the relation (2.1.62), which connects the inclination angle of the velocity vector at the outer edge of region III with the pressure in this region.

We assume

$$\Psi_0^* = k^{-4/9}\bar{\Psi}, \quad P_0^* = k^{4/9}\bar{P}, \quad A = k^{-2/3}\bar{A},$$
$$x^* = k^{-10/9}\bar{X}, \quad Y^* = k^{-2/3}\bar{Y}, \tag{2.1.63}$$

and then the interaction problem is written in the form

$$\frac{\partial\bar{\Psi}}{\partial\bar{Y}}\frac{\partial^2\bar{\Psi}}{\partial\bar{X}\partial\bar{Y}} - \frac{\partial\bar{\Psi}}{\partial\bar{X}}\frac{\partial^2\bar{\Psi}}{\partial\bar{Y}^2} = -\frac{d\bar{P}}{d\bar{X}} + \frac{\partial^3\bar{\Psi}}{\partial\bar{Y}^3},$$

$$\bar{P} = \sqrt{-\bar{X}}\left[2 - \frac{1}{\pi}\int_{-\infty}^{0}\frac{\bar{A}'(s)}{\sqrt{-s}(s-\bar{X})}ds\right],$$

$$\bar{\Psi} = \frac{\partial\bar{\Psi}}{\partial\bar{Y}} = 0 \quad \text{at } \bar{Y} = 0, \tag{2.1.64}$$

$$\bar{\Psi} = 2^{1/3}A_0\left[\bar{Y}^{5/3} + \tfrac{5}{3}\bar{A}(\bar{X})\bar{Y}^{2/3} + \cdots\right] \quad \text{as } \bar{Y}\to\infty,$$

$$\bar{\Psi} = 2^{3/4}(-\bar{X})^{5/8}f_0(\eta) + \cdots \quad \text{as } \bar{X}\to-\infty,$$

where

$$\eta = 2^{-1/4}\frac{\bar{Y}}{(-\bar{X})^{3/8}}.$$

Thus, the transformations (2.1.63) permit elimination of the dependence of the flow variables upon the parameter k. This means that all flows of the type considered are similar to each other in the interaction region.

The above analysis of the fluid flow in the interaction region was carried out by Ruban (1974). A numerical solution of the interaction problem (2.1.64) was obtained in his later work (Ruban, 1976a). It was found that the skin friction

$$\tau = \mathrm{Re}^{1/18}k^{8/9}\bar{\tau}(\bar{X}), \quad \text{where } \bar{\tau}(\bar{X}) = \left.\frac{\partial^2\bar{\Psi}}{\partial\bar{Y}^2}\right|_{\bar{Y}=0},$$

remains positive along the entire surface of the body up to the corner O. Moreover, $\bar{\tau}(\bar{X})$ increases monotonically as this point is approached (Figure 2.6a). At the same time the pressure decreases (Figure 2.6b), so that the pressure gradient $d\bar{P}/d\bar{X}$ is ignorable for all $\bar{X} < 0$. Ahead of the corner it has finite magnitude, at the point itself it undergoes a discontinuity, and according to (2.1.59) it becomes equal to zero for $\bar{X} > 0$.

Solving the interaction problem (2.1.64) allows determination of the function $\Psi_0^*(x^*, Y^*)$, not in the whole region III but only upstream from the section $x^* = 0$. Also determined is the function

$$\Psi_{00}^*(Y^*) = \lim_{x^*\to-0}\Psi_0^*(x^*, Y^*).$$

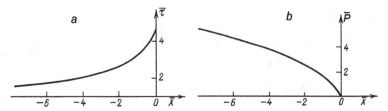

Fig. 2.6 Graphs showing skin friction (a) and pressure (b) vs. distance to the corner.

Knowing this, it is possible to continue the solution of equation (2.1.42) downstream from the corner. The initial condition here will be written as

$$\Psi_0^* = \Psi_{00}^*(Y^*) \quad \text{at } x^* = +0, \quad Y^* \geq 0;$$

the pressure gradient in equation (2.1.42) should be taken as zero, and the no-slip condition

$$\frac{\partial \Psi_0^*}{\partial Y^*} = 0 \quad \text{at } Y^* = 0$$

should be replaced by the condition of matching with the solution in the stagnation region IV:

$$\frac{\partial \Psi_0^*}{\partial Y^*} = 0 \quad \text{at } Y^* = -\infty.$$

The second boundary condition (2.1.43) at $Y^* = 0$ and the condition (2.1.45) at the outer edge of region III remain unchanged. We note that, for a unique determination of the solution of the boundary-value problem under consideration, it is sufficient to keep only the first term in the expansion of $\Psi_0^*(x^*, Y^*)$ as $Y^* \to \infty$ in the condition (2.1.45). The second term is written in order to show the rule by which one calculates the function $A(x^*)$.

Now let us come back to the expression (2.1.61) for the function $w_1^*(z_*)$. Taking $z_* = x_*' > 0$ in this expression and using the condition (2.1.58), we obtain the formula

$$F'(x_*') = A'(x_*') - \sqrt{x_*'} \left[2k + \frac{1}{\pi} \int_{-\infty}^{0} \frac{G(s)}{\sqrt{-s}(s - x_*')} ds \right].$$

This allows us to determine the spatial position of the dividing streamline.

Let us discuss in more detail the flow immediately behind the corner of the body contour. For $x^* > 0$, region III turns into an ordinary mixing

layer and obeys the same rules as the mixing layer (regions 2' and 3')
outside the interaction region. If $x^* \rightarrow +0$, the main part of region III
becomes locally inviscid; the action of the friction forces significantly
affects only the flow within the thin mixing layer, which is generated at
the boundary with the stagnation region IV. Comparing the convective
term of equation (2.1.42) with its viscous term

$$\frac{\partial \Psi_0^*}{\partial Y^*} \frac{\partial^2 \Psi_0^*}{\partial x^* \partial Y^*} \sim \frac{\partial^3 \Psi_0^*}{\partial Y^{*3}}$$

and taking into consideration the form of the initial velocity profile

$$\frac{d\Psi_{00}^*}{dY^*} = k^{8/9} \bar{\tau}(-0) Y^* + \cdots \quad \text{as } Y^* \rightarrow 0,$$

we conclude that in the viscous layer $Y^* = O(x^{*1/3})$, and the function
$\Psi_0^*(x^*, Y^*)$ is represented as

$$\Psi_0^* = x^{*2/3} g_0^*(\xi^*) + \cdots \quad \text{as } x^* \rightarrow +0, \tag{2.1.65}$$

where

$$\xi^* = \frac{Y^*}{x^{*1/3}}.$$

It is of interest to point out that the stream function is represented in
the same way in the mixing layer (region 3') which is generated downstream from the boundary-layer separation point on a smooth surface
(see (1.6.3), Chapter 1).

Completing the analysis of the flow within the interaction region, let
us note that its introduction does not allow us to eliminate entirely the
singularity in the solution of the hydrodynamic equations. Although
the pressure gradient remains bounded within the interaction region, as
we have already mentioned, it has a discontinuity at the corner of the
body contour. The solution (2.1.65) is also singular downstream of the
corner. Consequently, some supplementary flow regions should lie in
its vicinity. The first of them seems[6] to be a region with characteristic
dimensions $|x| = O(\text{Re}^{-1/2})$, $|y| = O(\text{Re}^{-1/2})$. Within this region the
pressure drop across the boundary layer becomes significant and causes
the elimination of the discontinuity in pressure gradient. At the same
time the skin friction ahead of the corner remains (in the first approximation) unchanged. The solution (2.1.65) also holds downstream from

---

[6] So far there is no detailed study of this problem, which is why the statement
made should be treated as a hypothesis. It is based on Veldman's work
(1976), which is devoted to the analysis of the inner flow regions around the
trailing edge of a flat plate (Chapter 3).

the corner. The singularity in this solution (and also in the solutions for regions IV and V) is smoothed out in a region with characteristic dimensions

$$|x'| = O(\mathrm{Re}^{-7/9}), \quad |y'| = O(\mathrm{Re}^{-7/9}). \qquad (2.1.66)$$

Here the fluid flow is described by the full system of Navier–Stokes equations with the local Reynolds number equal to unity.

The existence of such a flow region may be predicted in advance with the help of the following argument. In a small neighborhood of the corner the shape of the body is represented by two rays going out from point O. Consequently, the only characteristic dimension of the fluid flow in this vicinity is the viscous length ν/U_0. This determines the extent (2.1.66) of the region considered. The formal basis for introducing this region is the loss of uniform validity of the solution (2.1.65) for the mixing layer. According to (2.1.65) and (2.1.40), at a distance x' from the corner the thickness of the mixing layer is estimated as

$$y' = O\left[\mathrm{Re}^{-14/27}(x')^{1/3}\right].$$

When $|x'| = O(\mathrm{Re}^{-7/9})$, the longitudinal and transverse dimensions of the mixing layer are of the same order. Therefore, equation (2.1.42) describing the fluid flow in region III loses validity and is replaced by the complete Navier–Stokes equations.

Although the solution of the corresponding boundary-value problem for the Navier–Stokes equations has not been constructed at present, there are reasons to assume that within a region with the dimensions (2.1.66) flow separation from the body surface does not occur at the corner but rather behind it. This conclusion results from the work of Bogolepov (1985a), for instance, who analyzed the flow in the base region of a rectangular body. Having constructed a numerical solution of the Navier–Stokes equations for this flow, he showed that the elements of viscous fluid approaching the corner O of the contour of a body at first bend around this point and only afterward leave the body surface, as shown in Figure 2.7.

Another consequence of the above theory is worth mentioning. According to (2.1.52), the asymptotic expansion of the complex velocity in region I has the form

$$u' - iv' = 1 + \mathrm{Re}^{-2/9} w_1^*(z_*) + \cdots \quad \text{as } \mathrm{Re} \to \infty. \qquad (2.1.67)$$

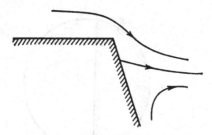

Fig. 2.7 Schematic representation of the fluid flow in the region of the full Navier–Stokes equations.

As follows from (2.1.61), for $|z_*| \to \infty$

$$w_1^* = 2kiz_*^{1/2} - \frac{iz_*^{-1/2}}{\pi} \int_{-\infty}^0 \frac{G(s)}{\sqrt{-s}} ds + O(|z_*|^{-5/8}). \qquad (2.1.68)$$

Substituting (2.1.68) into (2.1.67) and carrying out the change of variable $z_* = \mathrm{Re}^{4/9} z$, we obtain the expression

$$u' - iv' = 1 + 2kiz^{1/2} - \mathrm{Re}^{-4/9}\frac{iz^{-1/2}}{\pi} \int_{-\infty}^0 \frac{G(s)}{\sqrt{-s}} ds + O(\mathrm{Re}^{-1/2}),$$

from which it follows that in region 1, where $z = O(1)$, the asymptotic expansion of the complex velocity has the form

$$u' - iv' = w_0(z) + \mathrm{Re}^{-4/9}w_1(z) + \mathrm{Re}^{-1/2}w_2(z) + \cdots . \qquad (2.1.69)$$

The leading term $w_0(z)$ of this expansion is found, as was mentioned at the beginning of this chapter, as a result of solving a classical problem in the theory of ideal fluid flows with free streamlines. The third term $\mathrm{Re}^{-1/2}w_2(z)$ characterizes the displacement effect of the boundary layer. However, this effect does not determine the main perturbation of the "inviscid" solution $w_0(z)$ of the problem for region 1; more significant is the influence of the interaction region, which is expressed through the intermediate term $\mathrm{Re}^{-4/9}w_1(z)$ in the expansion (2.1.69).

2.2 Flows with Small Values of the Parameter k

When a boundary layer separates from the smooth surface of a solid body, the parameter k that appears in the expansion (2.1.2) of the complex velocity near the separation point is uniquely determined by the solution of a local problem for the interaction region. It follows from the analysis carried out in Chapter 1 that in this case the parameter k

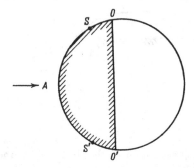

Fig. 2.8 An example of flow with a small value of the parameter k.

is negative[7] and has a value of order $\mathrm{Re}^{-1/16}$. In the case of separation from a corner point of a body contour, the parameter k is found as a result of the analysis of the flow in region 1. Here, depending upon the body configuration, it may change within a rather wide range. The theory described above was based upon the assumption that the parameter k has a positive value of order unity. It is of interest to find out what changes this theory undergoes as the parameter k decreases and how, as a result of these changes, the theory of separation from a corner is transformed into the theory of boundary-layer separation from a smooth surface.[8]

A flow with any arbitrarily small value of the parameter k can be obtained, for example, by placing into an incompressible fluid flow a cylindrical body having as its cross section a segment of a circle. In order to show this, let us first consider the flow around a circular cylinder (Figure 2.8). Let a point S lying on the surface of the cylinder, and the symmetrically located point S', be chosen according to the Brillouin–Villat condition. We recall that according to this condition the parameter k in the expression (2.1.2) must be equal to zero. The Brillouin–Villat condition defines the position of the point of flow separation in the limit when $\mathrm{Re} = \infty$. If the Reynolds number, although large, does not become infinite, the point of flow separation is displaced downstream from the point S at a distance of order $\mathrm{Re}^{-1/16}$.

Now let us place the segment AOO' in an incompressible fluid flow. The character of the flow around it depends on the relative locations

---

[7] Let us recall that, for the notation adopted in the present chapter, the sign of the parameter k has been changed to the opposite of that in Chapter 1.

[8] At the same time as the Russian publication of this book, a study of this theme was published by Bogdanova and Ryzhov (1986, 1987).

of the point O and the Brillouin–Villat point S. If the point O lies downstream from the point S and the length of the arc SO exceeds $\text{Re}^{-1/16}$ in order of magnitude, then flow separation will occur on the smooth part of the segment surface between points S and O. In this case the flow in the vicinity of the separation point is described by the theory which was set forth in Chapter 1. On the other hand, if the point O is located upstream from the point S and the distance between these points is finite, then the theory described in Section 1 of this chapter is valid. We shall start our analysis of the flow with just this case, and decreasing the distance between points O and S, we shall follow the change of flow regimes in the vicinity of the corner of the body contour.

Thus, let the point O be placed ahead of the point S. Then it is evident that separation will take place at the corner points O and O'. Taking into account the symmetry of the flow, we shall further examine one of these points, the point O. In its vicinity, the expansion (2.1.2) for the complex velocity is still correct. The parameter k, which appears in (2.1.2) and determines the magnitude of the pressure gradient (2.1.9) ahead of the separation point, is positive in this case and depends upon the distance between points O and S. The smaller this distance, the smaller is the value of the parameter k.

For the study of such flow regimes it is necessary to take

$$k = \epsilon(\text{Re})c_1\,, \qquad (2.2.1)$$

where c_1 is a positive constant and

$$\epsilon(\text{Re}) \to 0 \quad \text{as } \text{Re} \to \infty\,.$$

Then the expansion of the stream function in the boundary layer ahead of the separation point will be written (by analogy with the expansion (1.4.2) in Chapter 1) as

$$\psi = \text{Re}^{-1/2}\Psi_0(x,\,Y) + \text{Re}^{-1/2}\epsilon\Psi_1(x,\,Y) + \cdots\,. \qquad (2.2.2)$$

The leading term of this expansion determines the state of the boundary layer in a flow satisfying the Brillouin–Villat condition. Therefore, the function $\Psi_0(x,Y)$ exactly coincides with the leading term of the expansion (1.4.2) of Chapter 1. The second term of the expansion (2.2.2) reflects the response of the boundary layer to the effect of the singular pressure gradient. For the function $\Psi_1(x,Y)$ equation (1.4.5) of Chapter 1 remains valid. However, it is important to note that now the

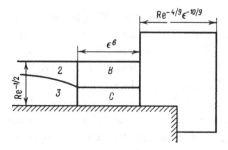

Fig. 2.9 Asymptotic structure of the characteristic flow regions under the condition that $\epsilon(\mathrm{Re})$ exceeds $\mathrm{Re}^{-1/16}$ in order of magnitude.

pressure gradient[9]

$$\frac{dp_1}{dx} = -c_1(-x)^{-1/2} + \cdots \quad \text{as } x \to -0$$

becomes favorable.

As in the case of separation from a smooth surface, the boundary layer considered includes two distinct flow regions: a locally inviscid region 2 and a viscous wall region 3 (Figure 2.9). If $x \to -0$, then within region 3 (see equations (1.4.7), (1.4.8) in Chapter 1)

$$\Psi_0 = \tfrac{1}{2}a_0(-x)^{2/3}\eta^2 + \cdots, \quad \Psi_1 = -(-x)^{1/2}f_1(\eta) + \cdots \qquad (2.2.3)$$

Here, the independent variable η is considered to have a value of order unity:

$$\eta = \frac{Y}{(-x)^{1/3}} = O(1). \qquad (2.2.4)$$

The function $f_1(\eta)$ satisfies the equation (1.4.9) of Chapter 1. If $\eta \to \infty$, then $f_1(\eta)$ has the expansion

$$f_1 = a_1\eta^{3/2} + b_1\eta + O(1). \qquad (2.2.5)$$

Region 3, comprising the decelerated flow near the wall, is the most sensitive to the effect of the singular pressure gradient and provides the principal contribution to the displacement effect of the boundary layer. Therefore, by calculating the inclination angle of the velocity vector at the outer boundary of region 3, we determine at the same time its value at the outer edge of the main part of the boundary layer. Let us designate by $\Delta\vartheta$ the increment of the inclination angle of the velocity

[9] Cf. formula (1.4.6) of Chapter 1.

vector relative to its magnitude in a flow satisfying the Brillouin–Villat condition. Then it follows from (2.2.2), (2.2.3), and (2.2.5) that

$$\Delta\vartheta = -\mathrm{Re}^{-1/2}\epsilon(-x)^{-5/6}\frac{b_1}{6a_0} + \cdots \quad \text{as } x \to -0.$$

This quantity is of the same order as the pressure perturbation

$$p - p_{00} = \mathrm{Re}^{-1/2}\epsilon(-x)^{-5/6}\frac{b_1}{6\sqrt{3}a_0} + \cdots \quad \text{as } x \to -0,$$

which is induced as a result of the displacement effect of the boundary layer. The induced pressure gradient then has the following estimate:

$$\frac{dp}{dx} = O[\mathrm{Re}^{-1/2}\epsilon(-x)^{-11/6}]. \tag{2.2.6}$$

As the corner of the body contour is approached, the displacement effect of the boundary layer becomes stronger. This leads to a distortion of the initial pressure gradient, and finally to the formation of the interaction region. The extent of this region

$$\Delta x = O(\mathrm{Re}^{-3/8}) \tag{2.2.7}$$

is usually determined as a result of comparing the initial pressure gradient (2.1.9), (2.2.1) with the induced gradient (2.2.6). It is of course assumed that everywhere outside the interaction region both the expression (2.1.9) for the initial pressure gradient and the estimate (2.2.6) for the induced gradient remain valid. At the same time, if $\epsilon(\mathrm{Re})$ exceeds $\mathrm{Re}^{-1/16}$ in order of magnitude, then the asymptotic expansion (2.2.2) of the stream function, which was used for the derivation of the formula (2.2.6), loses validity in a wider region. In fact, within the viscous wall layer (region 3) the terms in the expansions (2.2.2), (2.2.3) become quantities of the same order as soon as

$$|x| = O(\epsilon^6). \tag{2.2.8}$$

This implies that one should consider separately the region C (Figure 2.9), whose length is determined by the relation (2.2.8), and whose thickness is estimated in accordance with (2.2.4) as $Y = O(\epsilon^2)$.

The asymptotic analysis of the Navier–Stokes equations in region C is associated with the limit process

$$x_C = \epsilon^{-6}x = O(1), \quad Y_C = \epsilon^{-2}Y = O(1), \quad \mathrm{Re} \to \infty.$$

From the condition of matching with the solution (2.2.2), (2.2.3) for region 3, it follows that the leading term of the asymptotic expansion of the stream function in region C is

$$\psi = \mathrm{Re}^{-1/2}\epsilon^4\Psi_{0C}(x_C, Y_C) + \cdots \quad \text{as } \mathrm{Re} \to \infty. \tag{2.2.9}$$

As long as the displacement effect of the boundary layer is weak, the formula (2.1.9) for the pressure gradient ahead of the separation point remains valid. In region C it takes the form

$$\frac{dp_e}{dx} = -\epsilon^{-2}\frac{c_1}{(-x_C)^{1/2}} + \cdots. \tag{2.2.10}$$

Substituting (2.2.9), (2.2.10) into the Navier–Stokes equations, we obtain

$$\frac{\partial\Psi_{0C}}{\partial Y_C}\frac{\partial^2\Psi_{0C}}{\partial x_C\partial Y_C} - \frac{\partial\Psi_{0C}}{\partial x_C}\frac{\partial^2\Psi_{0C}}{\partial Y_C^2} = \frac{c_1}{(-x_C)^{1/2}} + \frac{\partial^3\Psi_{0C}}{\partial Y_C^3}.$$

The boundary conditions are the no-slip condition

$$\Psi_{0C} = \frac{\partial\Psi_{0C}}{\partial Y_C} = 0 \quad \text{for } Y_C = 0, \quad x_C < 0,$$

the condition of matching with the solution (2.2.2), (2.2.3) in region 3

$$\Psi_{0C} = \tfrac{1}{2}a_0 Y_C^2 + \cdots \quad \text{as } x_C \to -\infty \tag{2.2.11}$$

and the condition of matching with the solution in region B (Figure 2.9), which occupies the main part of the boundary layer

$$\Psi_{0C} = \tfrac{1}{2}a_0 Y_C^2 + \cdots \quad \text{as } Y_C \to \infty. \tag{2.2.12}$$

It is easy to see that in region C we deal with the problem of a boundary layer in a strong favorable pressure gradient, which has been already analyzed in Section 1. As has been demonstrated, the asymptotic representation of the solution to this problem as $x_C \to -0$ does not depend upon the form of the initial condition (2.2.11) and the condition (2.2.12) at the outer boundary of region C. Therefore, despite the changes in these conditions, we may use the results of the previous analysis, and then for the inclination angle of the velocity vector at the outer edge of the boundary layer we obtain

$$\vartheta = \mathrm{Re}^{-1/2}\epsilon^{-4}\left(\frac{2}{c_1}\right)^{1/4}\frac{9B_0}{40A_0}(-x_C)^{-5/8} + \cdots \quad \text{as } x_C \to -0.$$

The induced pressure gradient is now expressed in the form

$$\frac{dp}{dx} = \mathrm{Re}^{-1/2}\epsilon^{-10}\left(\frac{2}{c_1}\right)^{1/4}\frac{9B_0}{64A_0}\tan\frac{5}{8}\pi(-x_C)^{-13/8} + \cdots.$$

Comparing this with the initial pressure gradient (2.2.10), we conclude that the length of the interaction region is estimated as

$$\Delta x = O(\mathrm{Re}^{-4/9}\epsilon^{-10/9}).\qquad(2.2.13)$$

It is shortest (2.1.39) for $k = O(1)$, and now for $k = O(\mathrm{Re}^{-1/16})$ its length reaches its maximum value (2.2.7). At this moment the intervening region C ceases to exist as a separate flow region, since its length (2.2.8) becomes the same as the length of the interaction region.

Hence, to consider flow separation from a corner for small values of the parameter k, it is necessary first of all to single out the class of flows for which $\epsilon(\mathrm{Re})$ exceeds $\mathrm{Re}^{-1/16}$ in order of magnitude. Throughout all of this range for the parameter k, the boundary layer ahead of the interaction region follows the same rules as for a flow with $k = O(1)$. The flow within the interaction region possesses the same property; more precisely speaking, one can verify the following assertion: the theory of flow separation from a corner of a body contour, described in Section 1, has a wider range of application than was originally presumed for its derivation – the theory describes not only flows with $k = O(1)$ but also flows for which the parameter k exceeds $\mathrm{Re}^{-1/16}$ in order of magnitude. The boundary-value problem (2.1.64) for the interaction region remains valid, and the stretching of the scales of the variables that is observed as the parameter k decreases, and is expressed, in particular, by the relation (2.2.13), is realized according to the transformations (2.1.63).

The second distinct flow regime occurs for $\epsilon(\mathrm{Re}) = \mathrm{Re}^{-1/16}$. In this case the equations and boundary conditions that determine the stream function in the viscous wall layer of the interaction region are written in the same way as for the problem of flow separation from a smooth surface (see Chapter 1 (1.5.8)–(1.5.11)). As for the pressure, the formula (2.1.62)[10] is to be used. The point is that downstream from the corner the viscous layer as before is in contact with the stagnation zone, where the pressure in the approximation considered remains constant. Therefore, when $k = O(\mathrm{Re}^{-1/16})$ for the interaction region, we obtain the following boundary-value problem:

$$\frac{\partial\bar{\Psi}}{\partial\bar{Y}}\frac{\partial^2\bar{\Psi}}{\partial\bar{X}\,\partial\bar{Y}} - \frac{\partial\bar{\Psi}}{\partial\bar{X}}\frac{\partial^2\bar{\Psi}}{\partial\bar{Y}^2} = -\frac{d\bar{P}}{d\bar{X}} + \frac{\partial^3\bar{\Psi}}{\partial\bar{Y}^3},$$

$$\bar{P} = \sqrt{-\bar{X}}\left[\alpha + \frac{1}{\pi}\int_{-\infty}^{0}\frac{\bar{G}(s)}{\sqrt{-s}(s - \bar{X})}ds\right],$$

---

[10] In the present case the parameter k in this formula must be replaced by c_1.

$$\bar{G} = -\lim_{\bar{Y} \to \infty} \left(\frac{\partial \bar{\Psi}}{\partial \bar{X}} \middle/ \frac{\partial \bar{\Psi}}{\partial \bar{Y}} \right), \qquad (2.2.14)$$

$$\bar{\Psi} = \frac{\partial \bar{\Psi}}{\partial \bar{Y}} = 0 \quad \text{at } \bar{Y} = 0, \quad \bar{X} < 0,$$

$$\bar{\Psi} = \tfrac{1}{2}\bar{Y}^2 + \cdots \quad \text{as } \bar{X} \to -\infty \text{ and } \bar{Y} \to +\infty.$$

Here

$$\bar{X} = \mathrm{Re}^{3/8} a_0^{5/4} x, \quad \bar{Y} = \mathrm{Re}^{5/8} a_0^{3/4} y,$$

and we designate by $\bar{\Psi}(\bar{X}, \bar{Y})$, $\bar{P}(\bar{X}, \bar{Y})$ the coefficients in the leading terms of the expansions of the stream function and the pressure for the viscous wall layer of the interaction region:

$$\psi = \mathrm{Re}^{-3/4} a_0^{-1/2} \bar{\Psi}(\bar{X}, \bar{Y}) + \cdots,$$

$$p = p_{00} + \mathrm{Re}^{-1/4} a_0^{1/2} \bar{P}(\bar{X}, \bar{Y}) + \cdots.$$

The parameter α is calculated from the formula

$$\alpha = 2c_1 / a_0^{9/8}.$$

When the point O is located sufficiently close to the Brillouin–Villat point and the parameter α is small, the solution of the problem (2.2.14) may be represented as an expansion in powers of α:

$$\bar{\Psi} = \frac{1}{2}\bar{Y}^2 + \alpha \bar{\Psi}_1(\bar{X}, \bar{Y}) + \cdots,$$

$$\bar{P} = \alpha \bar{P}_1(\bar{X}) + \cdots, \quad \bar{G} = \alpha \bar{G}_1(\bar{X}) + \cdots. \qquad (2.2.15)$$

Substituting these expansions into (2.2.14), we arrive at the linear problem

$$\bar{Y} \frac{\partial^2 \bar{\Psi}_1}{\partial \bar{X} \partial \bar{Y}} - \frac{\partial \bar{\Psi}_1}{\partial \bar{X}} = -\frac{d\bar{P}_1}{d\bar{X}} + \frac{\partial^3 \bar{\Psi}_1}{\partial \bar{Y}^3},$$

$$\bar{P}_1 = \sqrt{-\bar{X}} \left[1 + \frac{1}{\pi} \int_{-\infty}^{0} \frac{\bar{G}_1(s)}{\sqrt{-s}(s - \bar{X})} ds \right], \quad \bar{G}_1 = -\lim_{\bar{Y} \to \infty} \left(\frac{1}{\bar{Y}} \frac{\partial \bar{\Psi}_1}{\partial \bar{X}} \right),$$

$$\bar{\Psi}_1 = \frac{\partial \bar{\Psi}_1}{\partial \bar{Y}} = 0 \quad \text{at } \bar{Y} = 0, \quad \bar{X} < 0,$$

$$\bar{\Psi}_1 = o(\bar{Y}^2) \quad \text{as } \bar{X} \to -\infty \text{ and } \bar{Y} \to +\infty.$$

The solution was obtained with the help of the Wiener–Hopf method.[11] In particular, the skin-friction perturbation on the surface of the body ahead of the corner ($\bar{X} < 0$) is expressed as

$$\bar{\tau}_1 = \frac{\partial^2 \bar{\Psi}_1}{\partial \bar{Y}^2} \bigg|_{\bar{Y}=0} = \frac{\pi^{1/2} \gamma^{1/8}}{3^{5/6}} \int_0^\infty \frac{\exp[\gamma^{3/4} \bar{X} t + I(t)]}{t^{5/6}(1 + t^{8/3})} dt,$$

[11] See the work by Brown and Stewartson (1970).

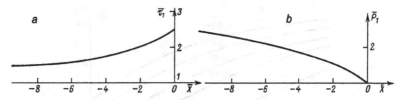

Fig. 2.10 The results of the solution of the linear problem: (a) a graph of the function $\bar{\tau}_1(\bar{X})$; (b) a graph of the function $\bar{P}_1(\bar{X})$.

and the pressure gradient is

$$\frac{d\bar{P}_1}{d\bar{X}} = -\frac{\gamma^{3/8}}{2\pi^{1/2}} \int_0^\infty \frac{\exp[\gamma^{3/4}\bar{X}t + I(t)]}{t^{1/2}(1 + t^{8/3})} dt.$$

Here $\gamma = 3^{2/3}/\Gamma(1/3)$ is a positive constant, and $I(t)$ is defined by the integral

$$I(t) = \frac{2}{3\pi} \int_0^\infty \frac{s^{1/3}\ln(s + t)}{1 - \sqrt{3}s^{4/3} + s^{8/3}} ds.$$

Graphs of the functions $\bar{\tau}_1(\bar{X})$ and $\bar{P}_1(\bar{X})$ are shown in Figure 2.10.

As follows from the expansion (2.2.15) for $\bar{\Psi}$, in the viscous wall layer the profile of the longitudinal velocity component remains unchanged in the leading approximation up to the corner of the body contour. The effect of the stagnation region upon the flow ahead of the corner is manifested only through subsequent terms in the expansion of the stream function in the interaction region. This property of flows with small values of the parameter k was first discovered by Daniels (1977), who studied the limiting case of flow separation from the corner of a body contour with the parameter k identically equal to zero, and found that substantial changes in the flow variables begin in a region requiring the full Navier–Stokes equations, which occupies a neighborhood of the corner with characteristic dimensions

$$|x'| = O(\text{Re}^{-3/4}), \quad |y'| = O(\text{Re}^{-3/4}).$$

A numerical solution of the problem (2.2.14) for finite values of the parameter α was obtained by Kravtsova (1993). She used a rapidly converging iterative procedure based upon the "quasisimultaneous" method of Veldman (1981). Chapter 7 describes it in detail. The results of the calculations performed are shown in Figures 2.11, 2.12, and 2.13. If the point O lies ahead of the point S (Figure 2.8) and the parameter α is positive, the skin friction $\bar{\tau}(\bar{X}) = \frac{\partial^2 \bar{\Psi}}{\partial \bar{Y}^2}\Big|_{\bar{Y}=0}$ increases monotonically as the

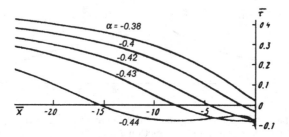

Fig. 2.11 Skin-friction distribution in the interaction region.

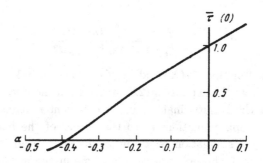

Fig. 2.12 Dependence of the minimum value of skin friction on α.

corner is approached, in the same way as in the case of finite values of k considered in the preceding Section 1. If the point O is displaced downstream of the Brillouin–Villat point ($\alpha < 0$), the skin friction decreases with increasing $\bar{X}$ (see Figure 2.11). As the distance between the points O and S grows, the minimum value of the skin friction $\bar{\tau}(0)$ drops, so that there exists a critical value of the parameter $\alpha = \alpha_* = -0.3877$, for which the skin friction at the corner reaches zero (Figure 2.12). Further reduction of the parameter α leads to a displacement of the separation point upstream relative to the corner. In particular, at $\alpha = 0.42$ the streamlines assume the configuration shown in Figure 2.13.

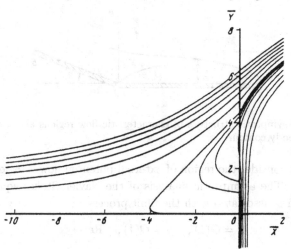

Fig. 2.13 Sketch of streamlines in the interaction region for $\alpha = 0.42$.

2.3 The Onset of Separation near a Corner of a Body Contour with Small Turning Angle

For simplicity we consider the flow over a solid body formed by two flat plates AO and OB (Figure 2.14). Let the first one be aligned parallel to the velocity of the oncoming flow and the second be deflected through a small angle θ. As before, we shall use dimensionless variables: the velocity components are referred to the speed U_∞ of the oncoming flow, the pressure increase relative to its value at infinity is referred to twice the dynamic pressure ρU_∞^2, and we take the length of the plate AO as the characteristic geometric length L. We introduce a rectangular Cartesian coordinate system Oxy, placing its origin at the corner O, and directing the axis Ox parallel to the plate AO and the axis Oy normal to it. The velocity components in this system of coordinates are denoted by u, v.

We take

$$\theta = \epsilon(\mathrm{Re})\theta_0,\tag{2.3.1}$$

where

$$\epsilon(\mathrm{Re}) \to 0 \quad \text{as } \mathrm{Re} = U_\infty L/\nu \to \infty,$$

and θ_0 is a constant that can assume either positive or negative values. The problem is to find the asymptotic solution of the Navier–Stokes equations for this flow as $\mathrm{Re} \to \infty$ and to define the conditions under which flow separation first occurs.

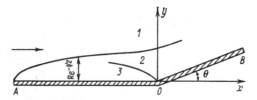

Fig. 2.14 Asymptotic structure of characteristic flow regions ahead of a sharp bend in a body contour.

We first consider the region of external potential flow – region 1 (Figure 2.14). The asymptotic analysis of the Navier–Stokes equations in this region is associated with the limit process

$$x = O(1), \quad y = O(1), \quad \mathrm{Re} \to \infty.$$

We represent the expansions of the dependent flow variables in the form

$$u = 1 + \epsilon u_1(x, y) + \cdots, \quad v = \epsilon v_1(x, y) + \cdots,$$
$$p = \epsilon p_1(x, y) + \cdots. \tag{2.3.2}$$

Then for u_1, v_1, p_1 we obtain the linearized Euler equations, from which it follows that the complex velocity

$$w_1(z) = u_1 - iv_1$$

is an analytic function of the variable $z = x + iy$ and the pressure perturbation satisfies the linearized Bernoulli equation

$$p_1 = -u_1. \tag{2.3.3}$$

Since we are considering flow regimes such that separation either is completely absent or is localized within a small neighborhood of the corner, the boundary condition at the surface of a solid body is the condition of no normal flow

$$v_1(x, 0) = 0 \quad \text{for } -1 < x < 0,$$
$$v_1(x, 0) = \theta_0 \quad \text{for } 0 < x < l,$$

where l is the dimensionless length of the plate OB. This condition determines uniquely the leading term of the expansion of the function $w_1(z)$ in the vicinity of the corner:

$$w_1 = \frac{\theta_0}{\pi} \ln z + d - i\theta_0 + o(1) \quad \text{as } |z| \to 0. \tag{2.3.4}$$

As for the second term $d - i\theta_0$ of this expansion, its real part – the constant d – can be determined only as a result of solving the flow problem for the entire region 1, taking into account all the boundary conditions.

Taking the real part of the expression (2.3.4) and using Bernoulli's equation (2.3.3), we conclude that the pressure at the outer edge of the boundary layer has a logarithmic singularity

$$p_1 = -\frac{\theta_0}{\pi}\ln|x| - d + o(1) \quad \text{as } |x| \to 0,$$

and the pressure gradient behaves as

$$\frac{dp_1}{dx} = -\frac{\theta_0}{\pi}\frac{1}{x} + O(1) \quad \text{as } |x| \to 0. \tag{2.3.5}$$

The boundary layer ahead of the corner. The asymptotic analysis of the Navier–Stokes equations for the boundary layer that is generated at the surface of the plate AO is based on the limit process

$$x = O(1), \quad Y = y\text{Re}^{1/2} = O(1), \quad \text{Re} \to \infty.$$

Then the expansion of the stream function has the following form:

$$\psi = \text{Re}^{-1/2}\Psi_0(x, Y) + \epsilon\text{Re}^{-1/2}\Psi_1(x, Y) + \cdots . \tag{2.3.6}$$

The leading term of this expansion corresponds to the boundary layer on a flat plate with a zero value for the corner angle of the surface, when the classical solution of Blasius (1908) holds:

$$\Psi_0 = (1 + x)^{1/2}F(\tilde{\eta}), \quad \tilde{\eta} = \frac{Y}{(1 + x)^{1/2}}. \tag{2.3.7}$$

Here, $F(\tilde{\eta})$ is a function satisfying the equation

$$F''' + \frac{1}{2}FF'' = 0,$$

which must be integrated with the boundary conditions

$$F(0) = F'(0) = 0, \quad F'(\infty) = 1.$$

It is known in particular that as $\tilde{\eta} \to 0$

$$F = \frac{1}{2}a_0\tilde{\eta}^2 + O(\tilde{\eta}^5), \quad a_0 = 0.3321. \tag{2.3.8}$$

For further analysis it is important to note that the Blasius solution does not have a singularity at the point $x = 0$ and can be represented near it by an expansion in powers of x:

$$\Psi_0 = \Psi_{00}(Y) + x\Psi_{01}(Y) + \cdots , \tag{2.3.9}$$

where

$$\Psi_{00} = F(Y), \quad \Psi_{01} = \frac{1}{2}F(Y) - \frac{1}{2}YF'(Y).$$

If $Y \to 0$, then

$$\Psi_{00} = \frac{1}{2}a_0Y^2 + O(Y^5), \quad \Psi_{01} = -\frac{1}{4}a_0Y^2 + O(Y^5). \qquad (2.3.10)$$

Now let us consider the second term of the expansion (2.3.6). Substituting (2.3.6) into the Navier–Stokes equations, we obtain

$$\frac{\partial\Psi_0}{\partial Y}\frac{\partial^2\Psi_1}{\partial x \partial Y} + \frac{\partial^2\Psi_0}{\partial x \partial Y}\frac{\partial\Psi_1}{\partial Y} - \frac{\partial\Psi_0}{\partial x}\frac{\partial^2\Psi_1}{\partial Y^2} - \frac{\partial^2\Psi_0}{\partial Y^2}\frac{\partial\Psi_1}{\partial x} = -\frac{dp_1}{dx} + \frac{\partial^3\Psi_1}{\partial Y^3}.$$
$$(2.3.11)$$

In Chapter 1 the asymptotic solution of equation (2.3.11) as $x \to -0$ was constructed for the case when the pressure gradient (see the expression (1.4.6), Chapter 1) increases as $c_1(-x)^{-1/2}$ as the point $x = 0$ is approached. But now dp_1/dx is given by the formula (2.3.5). Despite this change the boundary layer as $x \to -0$ splits as before into two distinct flow regions – a locally inviscid region 2, occupying the main part of the boundary layer, and a viscous wall region 3 (Figure 2.14).

In region 3, the first (and the last) convective term of equation (2.3.11) must agree in order of magnitude with the pressure gradient and also with the viscous term:

$$\frac{\partial\Psi_0}{\partial Y}\frac{\partial^2\Psi_1}{\partial x \partial Y} \sim \frac{1}{x} \sim \frac{\partial^3\Psi_1}{\partial Y^3}.$$

According to (2.3.9) and (2.3.10)

$$\frac{\partial\Psi_0}{\partial Y} = a_0Y + \cdots \quad \text{as } x \to -0, \quad Y \to 0.$$

Therefore, the thickness of region 3 is estimated as $Y = O[(-x)^{1/3}]$, and Ψ_1 here has a value of order unity. This implies that in region 3, where the asymptotic analysis of equation (2.3.11) is associated with the limit process

$$\eta = \frac{Y}{(-x)^{1/3}} = O(1), \quad x \to -0,$$

the leading term of the expansion of the function $\Psi_1(x, Y)$ should be represented in the form

$$\Psi_1 = f_1(\eta) + \cdots \qquad (2.3.12)$$

Making the replacement $Y = (-x)^{1/3}\eta$ in the Blasius solution (2.3.7), and using the asymptotic representation (2.3.8) of the function $F(\tilde{\eta})$ for small values of the argument, we find that in region 3

$$\Psi_0 = \frac{1}{2} a_0 (-x)^{2/3} \eta^2 + \cdots . \qquad (2.3.13)$$

It remains to substitute (2.3.12), (2.3.13), and (2.3.5) into (2.3.11); the result is the following equation for the function $f_1(\eta)$:

$$f_1''' - \frac{1}{3} a_0 \eta^2 f_1'' = \frac{\theta_0}{\pi} . \qquad (2.3.14)$$

For the solution of this equation satisfying the no-slip condition at the solid wall

$$f_1(0) = f_1'(0) = 0$$

to be unique, it is sufficient to introduce as an additional condition the requirement that exponentially growing terms be absent in the asymptotic expansion of the function $f_1(\eta)$ as $\eta \to \infty$. If this condition is met, then

$$f_1 = A_1 \eta + \frac{3\theta_0}{\pi a_0} \ln \eta + O(1) \quad \text{as } \eta \to \infty . \qquad (2.3.15)$$

The constant A_1 is determined in the process of integrating the equation (2.3.14); its value can be expressed through the Euler gamma function

$$A_1 = -\frac{\theta_0}{\pi (3a_0)^{2/3}} \left[\Gamma \left(\tfrac{1}{3} \right) \right]^2 .$$

Now we must repeat the same arguments as in Chapter 1, Section 1.4 (see also Section 1 of this chapter). We substitute (2.3.15) into (2.3.12) and make the replacement $\eta = Y(-x)^{-1/3}$. As a result we obtain the expression

$$\Psi_1 = A_1 (-x)^{-1/3} Y - \frac{\theta_0}{\pi a_0} \ln (-x) + O(1) ,$$

from which it follows that in region 2 (Figure 2.14), where the asymptotic analysis of equation (2.3.11) is associated with the limit process

$$Y = O(1), \quad x \to -0 ,$$

the expansion of function $\Psi_1(x, Y)$ should be represented in the form

$$\Psi_1 = (-x)^{-1/3} \Psi_{10}(Y) + O[\ln(-x)] . \qquad (2.3.16)$$

Substituting (2.3.16) together with the expansion (2.3.9) of the function $\Psi_0(x, Y)$ into (2.3.11), we obtain for $\Psi_{10}(Y)$ the already known equation (2.1.18) from Section 1:

$$\Psi'_{00}\Psi'_{10} - \Psi''_{00}\Psi_{10} = 0.$$

The solution satisfying the condition of matching with the solution in region 3,

$$\Psi_{10} = A_1 Y + \cdots \quad \text{as } Y \to 0,$$

has the following form:

$$\Psi_{10} = \frac{A_1}{a_0}\Psi'_{00}(Y). \qquad (2.3.17)$$

As shown by the above analysis, the boundary layer ahead of the corner possesses the usual properties characteristic of a boundary layer with a singular pressure gradient. The region most sensitive to the effect of such a pressure gradient is the wall region 3, "filled" with decelerated fluid elements. This region introduces the principal contribution to the displacement effect of the entire boundary layer. According to (2.3.6), (2.3.12), and (2.3.13), the increment in the inclination angle of the velocity vector caused by the singular pressure gradient (2.3.5) is calculated in region 3 from the formula

$$\Delta\vartheta = -\epsilon \mathrm{Re}^{-1/2}\frac{\partial\Psi_1}{\partial x}\Big/\frac{\partial\Psi_0}{\partial Y} + \cdots = -\epsilon \mathrm{Re}^{-1/2}\frac{f'_1(\eta)}{3a_0}(-x)^{-4/3} + \cdots.$$

At the surface of the body $\Delta\vartheta = 0$, and at the outer boundary of region 3, where the function $f_1(\eta)$ has the asymptotic representation (2.3.15), the increment in the inclination angle of the velocity vector approaches the form

$$\Delta\vartheta = -\epsilon \mathrm{Re}^{-1/2}\frac{A_1}{3a_0}(-x)^{-4/3} + \cdots \quad \text{as } x \to -0 \qquad (2.3.18)$$

and after that, as follows from (2.3.6), (2.3.9), (2.3.16), and (2.3.17), it remains unchanged across the entire region 2 up to its outer edge.

The interaction region. The necessity of considering the interaction region as a separate flow region is associated with the loss of uniform validity of the solution for the boundary layer ahead of the corner. For the construction of this solution it was assumed that the pressure gradient acting on the boundary layer is expressed by the formula (2.3.5). However, as the corner is approached, the displacement effect of the boundary layer leads to a strong distortion of the original pressure gradient. As follows from (2.3.18), the pressure perturbation is given by the estimate

$$p = O[\epsilon \operatorname{Re}^{-1/2}(-x)^{-4/3}].$$

Comparing the original pressure gradient to that induced by the displacement effect of the boundary layer

$$\frac{\epsilon}{x} \sim \frac{\epsilon \operatorname{Re}^{-1/2}}{(-x)^{7/3}},$$

we conclude that they become of the same order when

$$|x| = O(\operatorname{Re}^{-3/8}).$$

Thus, the length of the interaction region in this case is the same as for the problem of boundary-layer separation from a smooth surface (Chapter 1).

To construct a theory that would describe the transition from an unseparated flow around a corner to a flow with separation, it is necessary to choose the function $\epsilon(\operatorname{Re})$ in a special way. This function appears in the expression (2.3.1) for the angle θ and thus determines the level of the perturbations introduced into the original Blasius boundary layer. Upstream from the interaction region the solution for the boundary layer is written with the help of the expansion (2.3.6). At any finite distance from the corner, the leading term of the expansion (2.3.6) exceeds the second term of this expansion in order of magnitude, so that the perturbations introduced into the boundary layer are found to be insufficient to cause flow reversal. However, as $|x|$ is reduced, the relationship between the orders of magnitude in the expansion (2.3.6) changes. The closer the location of the point considered is to the corner of the body contour, the greater is the distortion of the original Blasius boundary layer. An especially rapid growth of the perturbations is observed in the viscous wall region 3, where the leading term in the expansion (2.3.6) is determined by the formula (2.3.13) and the second term is given by the relation (2.3.12). This process continues up to the interaction region, within which further growth of the perturbations becomes impossible owing to the smoothing of the singularity in the pressure gradient (2.3.5).

Therefore, reverse flow first appears in the viscous wall layer of the interaction region. To determine the function $\epsilon(\operatorname{Re})$, it is necessary to compare the first and the second terms of the expansion (2.3.6),

$$\operatorname{Re}^{-1/2}(-x)^{2/3} \sim \epsilon \operatorname{Re}^{-1/2},$$

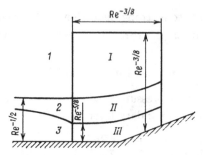

Fig. 2.15 Asymptotic structure of the interaction region.

taking $|x| = O(\mathrm{Re}^{-3/8})$ here. As a result we obtain

$$\epsilon = \mathrm{Re}^{-1/4}.$$

We start the analysis of the interaction region with the main part of the boundary layer (region II, Figure 2.15), for which

$$x^* = \mathrm{Re}^{3/8} x = O(1), \quad Y = O(1).$$

As usual, after making the change of variable $x = \mathrm{Re}^{-3/8} x^*$ in the solution for region 2, we conclude that the expansions of the stream function and the pressure in region II should be represented in the form

$$\psi = \mathrm{Re}^{-1/2} \Psi_{00}(Y) + \mathrm{Re}^{-5/8} \tilde{\Psi}_1(x^*, Y) + \cdots,$$
$$p = \mathrm{Re}^{-1/4} \ln \mathrm{Re} \, \frac{3\theta_0}{8\pi} + \mathrm{Re}^{-1/4} \tilde{P}_0(x^*, Y) + \cdots. \qquad (2.3.19)$$

The results of substituting these expansions into the Navier–Stokes equations are the equations

$$\frac{\partial \tilde{P}_0}{\partial Y} = 0, \quad \Psi'_{00} \frac{\partial^2 \tilde{\Psi}_1}{\partial x^* \partial Y} - \Psi''_{00} \frac{\partial \tilde{\Psi}_1}{\partial x^*} = 0.$$

The first of these shows that the pressure is constant across region II; according to this equation the function $\tilde{P}_0(x^*, Y)$ can be written as a function of the single argument x^*:

$$\tilde{P}_0 = P_0^*(x^*).$$

The solution of the second equation is expressed in the form

$$\tilde{\Psi}_1 = A(x^*) \Psi'_{00}(Y), \qquad (2.3.20)$$

the function $A(x^*)$ remaining arbitrary. Before turning to consideration of the potential-flow region (region I, Figure 2.15), it is necessary to determine the inclination angle of the velocity vector in region II. It follows from (2.3.19), (2.3.20) that

$$\vartheta = -\frac{\partial \psi}{\partial x} \Big/ \frac{\partial \psi}{\partial y} = -\text{Re}^{-1/4} A'(x^*) + \cdots . \qquad (2.3.21)$$

In region I the asymptotic analysis of the Navier–Stokes equations is associated with the limit process

$$x^* = O(1), \quad y^* = \text{Re}^{3/8} y = O(1), \quad \text{Re} \to \infty .$$

We represent the velocity components and the pressure by the expansions

$$u = 1 - \text{Re}^{-1/4} \ln \text{Re}\, \frac{3\theta_0}{8\pi} + \text{Re}^{-1/4} u_1^*(x^*, y^*) + \cdots ,$$
$$v = \text{Re}^{-1/4} v_1^*(x^*, y^*) + \cdots , \qquad (2.3.22)$$
$$p = \text{Re}^{-1/4} \ln \text{Re}\, \frac{3\theta_0}{8\pi} + \text{Re}^{-1/4} p_1^*(x^*, y^*) + \cdots$$

Then for the complex velocity

$$w_1^*(z^*) = u_1^* - i v_1^*$$

we have the following boundary-value problem: to find an analytic function $w_1^*(z^*)$ in the upper half-plane of the variable $z^* = x^* + i y^*$ when its imaginary part is given along the real axis as

$$\text{Im}\, w_1^* = A'(x^*) \quad \text{at } y^* = 0 ,$$

if it is known that

$$w_1^* = \frac{\theta_0}{\pi} \ln z^* + d - i\theta_0 + o(1) \quad \text{as } |z^*| \to \infty .$$

The first of these conditions is obtained as a result of matching the inclination angle (2.3.21) of the velocity vector in region II with the inclination angle

$$\vartheta = \frac{v}{u} = \text{Re}^{-1/4} v_1^*(x^*, y^*) + \cdots$$

in region I. The second condition is a consequence of matching the expansions (2.3.22) with the solutions (2.3.2), (2.3.4) in region 1.

Noting that $w_1^*(z^*)$ increases without bound as $|z^*| \to \infty$, we consider instead the function

$$\Phi(z^*) = -\frac{dw_1^*}{dz^*} = -\frac{\partial u_1^*}{\partial x^*} + i \frac{\partial v_1^*}{\partial x^*} .$$

In accordance with Bernoulli's equation

$$p_1^* = -u_1^*,$$

it may also be written in the form

$$\Phi(z^*) = \frac{\partial p_1^*}{\partial x^*} + i\frac{\partial v_1^*}{\partial x^*}.$$

On the real axis

$$\text{Im } \Phi = -A''(x^*) \quad \text{at } y^* = 0.$$

Moreover, it is known that

$$\Phi = O(|z^*|^{-1}) \quad \text{as } |z^*| \to \infty.$$

Therefore, at any point z^* in the upper half-plane

$$\Phi(z^*) = -\frac{1}{\pi}\int_{-\infty}^{\infty}\frac{A''(s)}{s-z^*}ds. \tag{2.3.23}$$

Applying the Sokhotski-Plemelj formula to the integral on the right-hand side of the relation (2.3.23), we obtain

$$\left.\frac{\partial p_1^*}{\partial x^*}\right|_{y^*=0} = -\frac{1}{\pi}\int_{-\infty}^{\infty}\frac{A''(s)}{s-x^*}ds. \tag{2.3.24}$$

It remains to consider the viscous wall layer of the interaction region (region III, Figure 2.15). The length of this region coincides with the lengths of regions I and II, and the thickness $y = O(\text{Re}^{-5/8})$ is found from the condition of matching with the solution in region 3, where $y = O[\text{Re}^{-1/2}(-x)^{1/3}]$. According to (2.3.6) and (2.3.13), the stream function in region 3 is estimated as $\psi = O[\text{Re}^{-1/2}(-x)^{2/3}]$; for $|x| = O(\text{Re}^{-3/8})$ it becomes of order $\text{Re}^{-3/4}$. Therefore, the asymptotic analysis of the Navier–Stokes equations in region III should be based upon the limit process

$$x^* = O(1), \quad Y^* = \text{Re}^{5/8}y = O(1), \quad \text{Re} \to \infty, \tag{2.3.25}$$

and so the solution in this region is sought in the form

$$\psi = \text{Re}^{-3/4}\Psi_0^*(x^*,Y^*) + \cdots,$$
$$p = \text{Re}^{-1/4}\ln \text{Re}\frac{3\theta_0}{8\pi} + \text{Re}^{-1/4}P_0^*(x^*) + \cdots. \tag{2.3.26}$$

Substituting (2.3.26) into the Navier–Stokes equations, we obtain

$$\frac{\partial \Psi_0^*}{\partial Y^*}\frac{\partial^2 \Psi_0^*}{\partial x^*\partial Y^*} - \frac{\partial \Psi_0^*}{\partial x^*}\frac{\partial^2 \Psi_0^*}{\partial Y^{*2}} = -\frac{dP_0^*}{dx^*} + \frac{\partial^3 \Psi_0^*}{\partial Y^{*3}}. \tag{2.3.27}$$

The boundary conditions for this equation are the condition of matching with the solution (2.3.6), (2.3.13), (2.3.12) in region 3

$$\Psi_0^* = \tfrac{1}{2}a_0(-x^*)^{2/3}\eta^2 + f_1(\eta) + \cdots \quad \text{as } x^* \to -\infty, \qquad (2.3.28)$$

where

$$\eta = \frac{Y^*}{(-x^*)^{1/3}},$$

the condition of matching with the solution (2.3.19), (2.3.20), (2.3.10) in region II

$$\Psi_0^* = \tfrac{1}{2}a_0 Y^{*2} + a_0 A(x^*)Y^* + \cdots \quad \text{as } Y^* \to \infty, \qquad (2.3.29)$$

and also the no-slip condition at the solid surface

$$\Psi_0^* = \frac{\partial \Psi_0^*}{\partial Y^*} = 0 \quad \text{at } Y^* = H(x^*). \qquad (2.3.30)$$

The equation of this surface written in the variables (2.3.25) for region III has the following form:

$$H(x^*) = \begin{cases} 0 & \text{for } x^* < 0, \\ \theta_0 x^* & \text{for } x^* > 0. \end{cases}$$

The interaction problem (2.3.27)–(2.3.30) is closed by the relation (2.3.24), whose left-hand side, as a consequence of the constant pressure across regions II and III, can be replaced by dP_0^*/dx^*.

Substitution of the variables

$$\Psi_0^* = a_0^{-1/2}\bar\Psi, \quad P_0^* = a_0^{1/2}\bar P + \frac{5\theta_0}{4\pi}\ln a_0 - d, \quad x^* = a_0^{-5/4}\bar X,$$

$$Y^* = a_0^{-3/4}\bar Y + H, \quad A = a_0^{-3/4}\bar A - H, \quad H = a_0^{-3/4}\bar H$$

allows us to represent the interaction problem in the following way[12]:

$$\frac{\partial\bar\Psi}{\partial\bar Y}\frac{\partial^2\bar\Psi}{\partial\bar X\partial\bar Y} - \frac{\partial\bar\Psi}{\partial\bar X}\frac{\partial^2\bar\Psi}{\partial\bar Y^2} = -\frac{d\bar P}{d\bar X} + \frac{\partial^3\bar\Psi}{\partial\bar Y^3},$$

$$\frac{d\bar P}{d\bar X} = \frac{1}{\pi}\int_{-\infty}^{\infty}\frac{\bar H''(s) - \bar A''(s)}{s - \bar X}ds,$$

$$\bar\Psi = \frac{\partial\bar\Psi}{\partial\bar Y} = 0 \quad \text{at } \bar Y = 0, \qquad (2.3.31)$$

$$\bar\Psi = \tfrac{1}{2}\bar Y^2 + \bar A(\bar X)\bar Y + \cdots \quad \text{as } \bar Y \to \infty,$$

$$\bar\Psi = \tfrac{1}{2}\bar Y^2 + \cdots \quad \text{as } \bar X \to -\infty.$$

---

[12] In the boundary condition (2.3.28) only the leading term of the expansion of Ψ_0^* as $x^* \to -\infty$ remains, since the second term is determined from the solution of the problem (2.3.31) itself.

Here,

$$\bar{H}(\bar{X}) = \begin{cases} 0 & \text{for } \bar{X} < 0, \\ \alpha\bar{X} & \text{for } \bar{X} > 0. \end{cases}$$

The constant $\alpha = \theta_0 a_0^{-1/2}$ represents a similarity parameter for the flows of the class considered.

Stewartson (1970a) was the first to study the flow around a corner having small turning angle with the help of an asymptotic solution of the Navier–Stokes equations as Re $\to \infty$. He formulated the problem (2.3.31) and analyzed in detail the case of small values of the parameter α, where the functions $\bar{\Psi}(\bar{X}, \bar{Y})$, $\bar{P}(\bar{X})$, and $\bar{A}(\bar{X})$ can be represented as expansions in powers of α:

$$\bar{\Psi} = \frac{1}{2}\bar{Y}^2 + \alpha\bar{\Psi}_1(\bar{X}, \bar{Y}) + \cdots,$$
$$\bar{P} = \alpha\bar{P}_1(\bar{X}) + \cdots, \quad \bar{A} = \alpha\bar{A}_1(\bar{X}) + \cdots.$$

The equation for $\bar{\Psi}(\bar{X}, \bar{Y})$ accordingly can be replaced by the linear equation

$$\bar{Y}\frac{\partial^2 \bar{\Psi}_1}{\partial\bar{X}\partial\bar{Y}} - \frac{\partial\bar{\Psi}_1}{\partial\bar{X}} = -\frac{d\bar{P}_1}{d\bar{X}} + \frac{\partial^3 \bar{\Psi}_1}{\partial\bar{Y}^3}.$$

The solution of the interaction problem linearized in this way was constructed by Stewartson in analytical form by the use of a Fourier transformation (also see Stewartson's note, 1971).

The linear theory is, however, insufficient when one considers the appearance of flow separation, which is associated with a substantial distortion of the original flow:

$$\bar{\Psi} = \frac{1}{2}\bar{Y}^2, \quad \bar{P} = \bar{A} = 0.$$

A numerical solution of the nonlinear interaction problem was first obtained by Ruban (1976b). The numerical method used is described in Chapter 7, and the results of the calculations are shown in Figures 2.16–2.19.

In the case of a concave-corner flow ($\alpha > 0$), deceleration occurs ahead of the corner $\bar{X} = 0$ and is accompanied by a reduction in the skin friction

$$\bar{\tau}(\bar{X}) = \left.\frac{\partial^2 \bar{\Psi}}{\partial\bar{Y}^2}\right|_{\bar{Y}=0},$$

which decreases more rapidly as the value of the parameter α increases (Figure 2.16a). The minimum skin friction is reached at the point $\bar{X} = 0$,

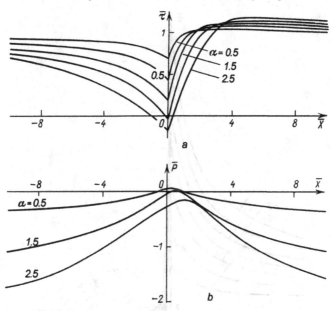

Fig. 2.16 The distribution of skin friction (a) and pressure (b) in the interaction region for a concave corner.

and downstream of this point $\bar{\tau}$ first increases up to values exceeding unity, and later returns to its original value $\bar{\tau} = 1$. An increase in the parameter α leads to a reduction in the minimum value of the skin friction and a displacement of the pressure maximum (Figure 2.16b) downstream from the point $\bar{X} = 0$, so that in the region of recirculating flow, where $\bar{\tau} < 0$, the pressure gradient is always positive. As was shown by Smith and Merkin (1982), who extended Ruban's calculations up to $\alpha = 3.5$, when the parameter α increases, the region of recirculating flow expands; however, the fluid motion within this region remains slow and is characterized by small negative values of the skin friction and by a moderate adverse pressure gradient.

The opposite trends for changes in the flow variables appear in the flow near a convex corner. In this case, instead of flow deceleration ahead of the corner, an acceleration is observed, which is accompanied by an increase in skin friction (Figure 2.17a) and a reduction in pressure (Figure 2.17b). Beyond the corner the skin friction does not change monotonically: at first it drops, and after that, despite a continuing pressure increase, it starts to grow. For all $\alpha \neq 0$ the minimum value of the skin friction, $\bar{\tau}_{\min}$, is found to be less than unity, an increase

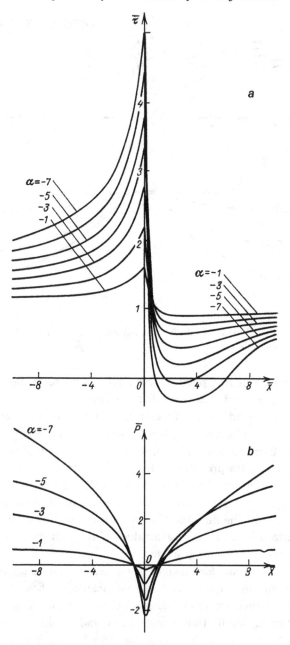

Fig. 2.17 The distribution of skin friction (a) and pressure (b) in the interaction region for a convex corner.

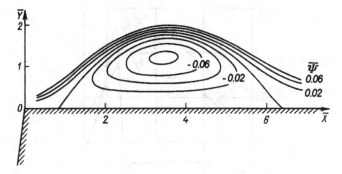

Fig. 2.18 Sketch of streamlines in the vicinity of a corner for $\alpha = -7.0$.

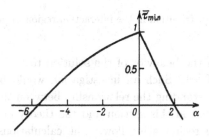

Fig. 2.19 Dependence of the minimum value of skin friction on α.

in the corner angle of the surface causing a decrease in the value $\bar{\tau}_{\min}$. Thus, the fluid deceleration beyond the corner is more intense than its acceleration ahead of this point. Beginning with a certain value of the parameter α, this deceleration results in flow separation. It is of interest to note that in the present case the entire region of recirculating flow is located behind the corner. A sketch of the streamlines in this region is shown in Figure 2.18 for $\alpha = -7.0$.

In plotting the minimum value of the skin friction $\bar{\tau}_{\min}$ versus the parameter α (Figure 2.19), Ruban (1976b) concluded that in the case of a concave corner, flow separation first appears when $\alpha = 2.0$, and in the case of a convex corner when $\alpha = -5.7$. According to the calculation of Smith and Merkin (1982), however, the interval within which the unseparated regime of flow around the corner is preserved is found to be shifted into a larger range of values of the parameter α:

$$-5.21 < \alpha < 2.51 .$$

Unfortunately, neither Ruban (1976b) nor Smith and Merkin (1982)

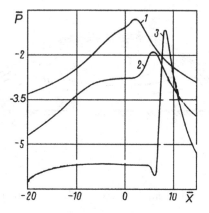

Fig. 2.20 Pressure distribution in the interaction region for a concave corner.

was able to study the behavior of the solution to the problem (2.3.31) at large values of $|\alpha|$. Such an investigation would be of crucial importance for understanding the relationship between the theory of local separation described in this section and the theory of separation with extended zones of recirculating flow. The calculations carried out by Ruban (1976b) and Smith and Merkin (1982) were based on the use of iterations which inevitably became divergent as soon as the region of recirculating flow was sufficiently large. Here, considerable progress was made by Korolev (1991a, 1992) who employed a basically new numerical method, which he called a direct method (Korolev, 1987), for the solution of the problem (2.3.31). A detailed description will be given in Section 4 of Chapter 7, while here we shall discuss the results of the application of this method to the problem (2.3.31).

The distributions of pressure $\bar{P}(\bar{X})$ and skin friction $\bar{\tau}(\bar{X})$ in the flow along a surface with a concave corner are shown in Figures 2.20–2.21. Here, the numerals 1, 2, and 3 designate the curves corresponding to $\alpha = 3.5, 5, 7$, respectively. It is clear that, starting at a certain value of α, a secondary separation appears within the recirculation zone, with direction opposite to the recirculating flow. This corresponds to a region of positive values of the skin friction (Figure 2.21). A sketch of the streamlines with secondary separation is shown in Figure 2.22 for $\alpha = 7$. Here, in the external flow the streamlines are plotted for an interval $\Delta\bar{\Psi} = 15$, in the primary separation region $\Delta\bar{\Psi} = -1.5$, and in the secondary separation region $\Delta\bar{\Psi} = 0.15$.

A further increase in α, as shown by the results of the calculations,

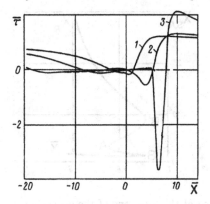

Fig. 2.21 Skin-friction distribution in the interaction region for a concave corner.

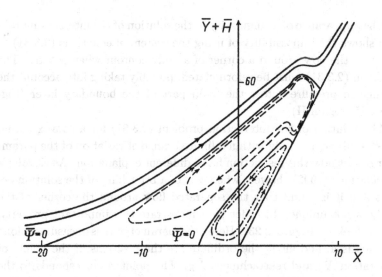

Fig. 2.22 Sketch of streamlines in the interaction region for flow over a concave corner with $\alpha = 7$.

leads to the formation of a singularity in the solution of the interaction problem immediately ahead of the reattachment point. The possible appearance of such a singularity was predicted earlier by Smith (1988a). It develops at a finite value $\alpha = \alpha_*$ when at a certain cross section $\bar{X} = \bar{X}_0$ within the region of reverse flow the pressure becomes discontinuous and the minimum value of the skin friction $\bar{\tau}(\bar{X}_0) \to -\infty$ as $\alpha \to \alpha_* - 0$. From Figures 2.20–2.21, one can see that $\alpha = 7$ is already rather close to the critical value α_*.

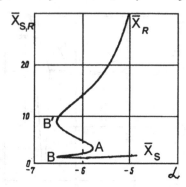

Fig. 2.23 Dependence of the locations of the separation point ($\bar{X}_S$) and reattachment point ($\bar{X}_R$) on α.

The appearance of a singularity in the solution of the interaction problem shows the impossibility of using the system of equations (2.3.31) to describe the flow around a corner of a body contour when $\alpha > \alpha_*$. The problem (2.3.31) must be reformulated, possibly taking into account the change in pressure across the main part of the boundary layer (Figure 2.15, region II).

The solution of the interaction problem (2.3.31) for a convex corner ($\alpha < 0$) also exists only within a limited range of variation of the parameter α, but now this restriction has a different explanation. As Korolev's calculations (1992) showed, when $\alpha < 0$, a branching of the solution occurs and it is found that the separated flow around the corner of the body is not unique. In order to demonstrate this important property, let us look at Figure 2.23. Here, the parameter α is measured along the abscissa, and along the ordinate are the positions of the points of separation $\bar{X}_S$ and reattachment $\bar{X}_R$. The point A corresponds to the flow condition with incipient separation, when $\bar{X}_S$ coincides with $\bar{X}_R$. The part of the curve located below the point A gives the position of the separation point, and the part of the curve above A shows the position of the reattachment point. It is easily seen that the solution exists only up to $\alpha_* = -6.5$, when the separation point reaches B and the reattachment point is at B'. It can be continued further if one returns to larger values of α. This means a nonuniqueness of the solution. It is of interest that the solution corresponding to the limiting value $\alpha = \alpha_*$ has no singularity.

Figure 2.24 shows two possible distributions of the skin friction (solid lines) and pressure (dotted lines) for $\alpha = -6$. The curves 1 correspond

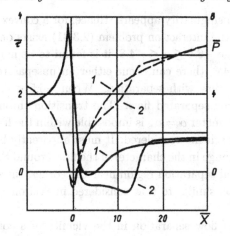

Fig. 2.24 Skin-friction and pressure distributions in the interaction region for flow over a convex corner, for the first and second branches of the solution, with $\alpha = -6$.

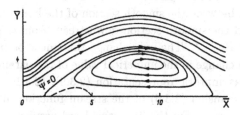

Fig. 2.25 Sketch of streamlines in the interaction region for flow over a convex corner with $\alpha = -6$.

to the first branch of the solution and the curves 2 to the second one. The difference between them is shown, in particular, by the length of the separation zone. The first branch arises through continuous changes from the unseparated regime of flow around the corner, realized for small $|\alpha|$, and corresponds to a shorter separation zone. Figure 2.25 shows a sketch of the streamlines for $\alpha = -6$ corresponding to the second branch of the solution. The streamlines here are plotted for an interval $\Delta\bar{\Psi} = 2$ in the external-flow region and for an interval $\Delta\bar{\Psi} = -0.2$ in the recirculation region. The dashed line is drawn for comparison with the zero streamline at the same surface corner angle but for the first branch of the solution.

It is possible to continue the second branch of the solution only up to $\alpha = -4.8$, when the region of reverse flow becomes long enough that

Smith's (1988a) singularity appears. Hence, for a convex corner ($\alpha < 0$) the solution of the interaction problem (2.3.31) exists only if $\alpha > -6.5$. Within the range $-6.5 \leq \alpha \leq -4.8$, it is found to be nonunique. When $-5.8 < \alpha < -4.8$, there can occur either an unseparated flow around the corner or a flow with separation. When $-6.5 < \alpha < -5.8$, both solutions describe separated flow. The transition through the critical value of the parameter $\alpha = \alpha_*$ is impossible within the framework of the three-layer flow scheme considered. It must apparently be accompanied by an abrupt change in the character of the flow around the corner, when instead of a local separation region near the corner, there will appear a global separation similar to that considered in Section 1 of the present chapter.

The theory of flow separation in the vicinity of a corner in a body contour was, for simplicity, presented in this section for the example of flow around a body made up of two flat plates. However, this theory also remains correct in a more general case, when the contour of the body is represented by a piecewise smooth curve.[13] It is only necessary that the angle between the smooth section of the body surface ahead of the corner and the smooth section downstream from this point have a value of order $\mathrm{Re}^{-1/4}$. The only change that must be introduced into the theory is a change in the numerical value of the parameter a_0, which enters in the asymptotic representation (2.3.9), (2.3.10) of the leading term in the expansion (2.3.6) of the stream function in the boundary layer near the corner. For the flow considered previously this constant, characterizing the value of the skin friction ahead of the interaction region, was determined from the Blasius solution (2.3.7), (2.3.8). Then in a more general case its calculation requires integrating the Prandtl equations for the boundary layer evolving along the surface of the body considered.

2.4 Other Aspects of Multistructured Boundary-Layer Theory[14]

The flow over a surface irregularity. The formation of local separation zones can also be observed in the flow over other irregularities on the surface of a solid body – humps, cavities, steps, etc. If the

---

[13] A study of the flow in the vicinity of a point of discontinuity in surface curvature was carried out by Messiter and Hu (1975).

[14] Section 2.4 was written in cooperation with Vic. V. Sychev.

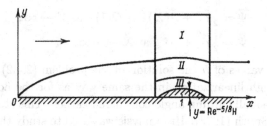

Fig. 2.26 Asymptotic flow structure near an irregularity on the surface of a flat plate.

longitudinal scale of an irregularity is of order $\mathrm{Re}^{-3/8}$ and the transverse scale is of order $\mathrm{Re}^{-5/8}$, in other words, if the surface irregularity can be represented as (Figure 2.26)

$$y = \mathrm{Re}^{-5/8} H\left(\frac{x-1}{\mathrm{Re}^{-3/8}}\right),$$

then in the vicinity of the point $x = 1$ there appears an interaction region quite analogous to that which was considered in the problem of a fluid flow around a corner. This analogy is manifested not only in the fact that the interaction region retains a triple-deck structure, but also in that the previous formulation (2.3.31) of the interaction problem remains valid. For further analysis it will be convenient to represent the expression for the shape of the irregularity in the form

$$\bar{H} = h\bar{F}(\bar{X}). \tag{2.4.1}$$

Here, h is a parameter which characterizes the height of the irregularity, and we shall assume that the function $\bar{F}(\bar{X})$ decreases rapidly as $|\bar{X}| \to \infty$ or that its value becomes zero outside a certain finite segment.

After making a Prandtl transposition transformation (1938)

$$\bar{\bar{X}} = \bar{X}, \quad \bar{\bar{Y}} = \bar{Y} - \bar{H}, \quad \bar{\bar{\Psi}} = \bar{\Psi}, \quad \bar{\bar{P}} = \bar{P},$$
$$\bar{A} = \bar{\bar{A}} + \bar{H}, \quad \bar{\bar{H}} = \bar{H}, \quad \bar{\bar{F}} = \bar{F},$$

we shall write the problem (2.3.31), (2.4.1) in the form

$$\frac{\partial \bar{\bar{\Psi}}}{\partial \bar{\bar{Y}}} \frac{\partial^2 \bar{\bar{\Psi}}}{\partial \bar{\bar{X}} \partial \bar{\bar{Y}}} - \frac{\partial \bar{\bar{\Psi}}}{\partial \bar{\bar{X}}} \frac{\partial^2 \bar{\bar{\Psi}}}{\partial \bar{\bar{Y}}^2} + \frac{d\bar{\bar{P}}}{d\bar{\bar{X}}} = \frac{\partial^3 \bar{\bar{\Psi}}}{\partial \bar{\bar{Y}}^3},$$

$$\frac{d\bar{\bar{P}}}{d\bar{\bar{X}}} = -\frac{1}{\pi}\int_{-\infty}^{\infty} \frac{\bar{\bar{A}}''(s)}{s - \bar{\bar{X}}}\,ds, \tag{2.4.2}$$

$$\bar{\bar{\Psi}} = \frac{\partial \bar{\bar{\Psi}}}{\partial \bar{\bar{Y}}} = 0 \quad \text{at } \bar{\bar{Y}} = \bar{\bar{H}}(\bar{\bar{X}}) = h\bar{\bar{F}}(\bar{\bar{X}}),$$

$$\bar{\bar{\Psi}} = \tfrac{1}{2}\bar{\bar{Y}}^2 + \bar{\bar{A}}(\bar{\bar{X}})\bar{\bar{Y}} + O(1) \quad \text{as } \bar{\bar{Y}} \to \infty,$$

$$\bar{\bar{\Psi}} = \tfrac{1}{2}\bar{\bar{Y}}^2 + \cdots \quad \text{as } \bar{\bar{X}} \to -\infty.$$

For small values of h the solution of the problem (2.4.2) may be obtained through linearization in the same way as for the flow around a corner with small turning angle considered earlier. This study was carried out by Smith (1973). His analysis was used to study the manner in which perturbations decay far from an irregularity, or, in other words, the asymptotic solution of the problem (2.4.2) was found as $|\bar{\bar{X}}| \to \infty$.[15]

For finite values of h the solution of this problem was obtained numerically by Napolitano, Werle, and Davis (1979), Burggraf and Duck (1982), Smith and Merkin (1982), Korolev (1987), and other authors (see the survey by Smith, 1982a, and also Chapter 7). A characteristic feature of the flow around a hump discovered by Korolev is the appearance (starting at a certain value of h) of regions of reverse flow both behind and ahead of it.

The problem (2.4.2) is the most general for flows with local separation regions. At the same time, a decrease or increase (by an order of magnitude) of the longitudinal and transverse scales of an irregularity requires special consideration since it leads to a change in the formulation of this problem. Let us focus briefly on consideration of these questions, following the work of Smith et al. (1981).

First let us analyze the case when the longitudinal and transverse scales of the irregularity are small compared with those considered earlier. The velocity profile approaching such an irregularity has finite skin friction. The flow in the wall sublayer, within which the irregularity is located, is subject to the effect of pressure gradient, and viscous stresses are of importance here. Proceeding from these considerations we replace the variables in the problem (2.4.2) by

$$\bar{\bar{X}} = h^3\widehat{X}, \quad \bar{\bar{Y}} = h\widehat{Y}, \quad \bar{\bar{\Psi}} = h^2\widehat{\Psi}, \quad \bar{\bar{P}} = h^2\widehat{P}, \tag{2.4.3}$$

$$\bar{\bar{H}} = h\widehat{F}(\widehat{X}), \quad \bar{\bar{A}} = A_{00} + h^5\widehat{A}(\widehat{X}), \quad A_{00} = \text{const} = o(h)$$

and take the limit $h \to 0$. As a result we obtain the following problem

$$\frac{\partial\widehat{\Psi}}{\partial\widehat{Y}}\frac{\partial^2\widehat{\Psi}}{\partial\widehat{X}\partial\widehat{Y}} - \frac{\partial\widehat{\Psi}}{\partial\widehat{X}}\frac{\partial^2\widehat{\Psi}}{\partial\widehat{Y}^2} + \frac{d\widehat{P}}{d\widehat{X}} = \frac{\partial^3\widehat{\Psi}}{\partial\widehat{Y}^3},$$

$$\widehat{\Psi} = \frac{\partial\widehat{\Psi}}{\partial\widehat{Y}} = 0 \quad \text{at } \widehat{Y} = \widehat{F}(\widehat{X}) = \widehat{h}_0\widehat{F}_0(\widehat{X}), \tag{2.4.4}$$

---

[15] See also the work of Gittler (1992).

$$\widehat{\Psi} = \tfrac{1}{2}\widehat{Y}^2 + \widehat{P}(\widehat{X}) + o(1) \quad \text{as } \widehat{Y} \to \infty,$$

$$\widehat{\Psi} = \tfrac{1}{2}\widehat{Y}^2 + \cdots \quad \text{as } \widehat{X} \to -\infty.$$

Thus, in this case the problem for the wall region differs from the original problem since instead of the interaction condition in (2.4.2), we have the value $\bar{\bar{A}}(\bar{\bar{X}}) = 0$. The distribution $\widehat{P}(\widehat{X})$ here as previously remains unknown in advance and must be found in the process of solution.

The first such flow regime, called compensatory, was considered by Bogolepov and Neiland (1971). Its characteristic feature is the absence of perturbations extending upstream of the irregularity. Thus if, say, the shape of the irregularity is such that $\widehat{F}(\widehat{X}) = 0$ when $\widehat{X} < 0$, then for all values $\widehat{X} < 0$ the solution of the problem (2.4.4) is $\widehat{\Psi} = \tfrac{1}{2}\widehat{Y}^2$, $\widehat{P} = 0$.

Numerical solutions of the problem (2.4.4) were obtained by Bogolepov (1974) and by Smith (1976), Smith's solution describing a motion with reverse flow. It is of interest that the problem (2.4.4) allows the existence of self-similar solutions for certain special forms of irregularity (Bogolepov and Lipatov, 1982).

In its linearized form ($\widehat{h}_0 \to 0$) the problem was solved earlier by Hunt (1971). In the other limiting case, when $\widehat{h}_0 \to \infty$, a singularity appears at the point of zero skin friction in the solution of the problem (2.4.4). However, unlike Goldstein's (1948) singularity,[16] which appears when the pressure gradient is prescribed, in this case the singularity is removable; that is, the solution may be continued through the point of zero skin friction. This interesting feature of the problem (2.4.4) was discovered and studied in detail by Smith and Daniels (1981). In their work it was found that this flow is realized with an extended separation zone.

In terms of the original physical variables, the longitudinal and transverse scales of the irregularity in the case considered are, according to (2.3.25) and (2.4.3), of order $h^3 \mathrm{Re}^{-3/8}$ and $h\mathrm{Re}^{-5/8}$, respectively. It is easy to see that for $h = \mathrm{Re}^{-1/8}$ these scales are of the same order, namely $\mathrm{Re}^{-3/4}$. Within a region with such scales the flow near the irregularity is now described by the full system of Navier–Stokes equations with a local Reynolds number equal to one. Numerical solutions of the corresponding boundary-value problem were obtained by Bogolepov (1975) and by Kiya and Arie (1975).

For an irregularity with longitudinal scale larger than $O(\mathrm{Re}^{-3/8})$, there occurs a flow with a given pressure gradient. Indeed, proceed-

---

[16] A detailed analysis is given in Chapter 4.

ing from the same considerations as for the formulation of the problem (2.4.4), we represent the solution of the problem (2.4.2) in the form

$$\bar{\bar{X}} = h^{3/5}\tilde{X}, \quad \bar{\bar{Y}} = h\tilde{F}(\tilde{X}) + h^{1/5}\tilde{Y}, \quad \bar{\bar{\Psi}} = h^{2/5}\tilde{\Psi}, \qquad (2.4.5)$$
$$\bar{\bar{P}} = h^{2/5}\tilde{P}, \quad \bar{\bar{H}} = h\tilde{F}(\tilde{X}), \quad \bar{\bar{A}} = -h\tilde{F}(\tilde{X}) + h^{1/5}\tilde{A}(\tilde{X}).$$

After substituting these expressions into (2.4.2) and taking the limit as $h \to \infty$ for the functions with a tilde, we arrive at a problem of the form (2.4.2), with the only difference that instead of the interaction condition we obtain the following expression for the pressure gradient:

$$\frac{d\tilde{P}}{d\tilde{X}} = \frac{1}{\pi} \int_{-\infty}^{\infty} \frac{\tilde{F}''(s)}{s - \tilde{X}} ds, \quad \tilde{F} = \tilde{h}_0 \tilde{F}_0(\tilde{X}) \qquad (2.4.6)$$

and the no-slip condition must be satisfied at $\tilde{Y} = 0$.

Thus, for the flow in the wall sublayer, we arrive at a problem with a prescribed pressure distribution. The solution for small values of the parameter $\tilde{h}_0$ in (2.4.6) was obtained by Smith et al. (1981). In the solution of a nonlinear problem for any irregularity in the form $\tilde{F}_0(\tilde{X})$, there will be a critical value of $\tilde{h}_0$ for which an isolated point of zero skin friction appears in the solution. This point is singular. A detailed description of such flows is given in Chapter 4.

Unsteady flow within the interaction region. First Schneider (1974), then Brown and Daniels (1975), and somewhat later Ryzhov and Terent'ev (1977) broadened the range of application of the interaction theory by including consideration of unsteady flows. They noted that the viscous wall layer in the interaction region, where the fluid motion occurs most slowly, is the most sensitive to unsteady effects. If we denote the dimensionless time by t, referred to L/U_∞, and compare the unsteady term with the convective term

$$\frac{\partial u}{\partial t} \sim u \frac{\partial u}{\partial x},$$

in the first of the system of Navier–Stokes equations (see Chapter 1, equation (1.1.2)), bearing in mind that in the viscous wall layer of the interaction region

$$u = O(\mathrm{Re}^{-1/8}), \quad \Delta x = O(\mathrm{Re}^{-3/8}),$$

as a result we obtain the estimate

$$t = O(\mathrm{Re}^{-1/4})$$

as the characteristic time scale for changes in the flow variables when the motion in region III (Figure 2.26) becomes substantially unsteady. At the same time, in the main part of the boundary layer (region II) and in the external potential flow (region I), where $u = O(1)$, the unsteady term of the first Navier–Stokes equation remains negligible compared with the convective term. Thus, the flow in these regions is quasistationary and possesses the same properties as the corresponding steady flow. Consequently, in order to reformulate the interaction problem (2.3.31) for the case considered, it is only necessary to change the equation for the function $\bar{\Psi}$, writing it as

$$\frac{\partial^2 \bar{\Psi}}{\partial \bar{t} \partial \bar{Y}} + \frac{\partial \bar{\Psi}}{\partial \bar{Y}} \frac{\partial^2 \bar{\Psi}}{\partial \bar{X} \partial \bar{Y}} - \frac{\partial \bar{\Psi}}{\partial \bar{X}} \frac{\partial^2 \bar{\Psi}}{\partial \bar{Y}^2} = -\frac{\partial \bar{P}}{\partial \bar{X}} + \frac{\partial^3 \bar{\Psi}}{\partial \bar{Y}^3},$$

where

$$\bar{t} = \mathrm{Re}^{1/4} a_0^{-3/2} t.$$

The boundary conditions and the expression for the pressure gradient remain unchanged.

Among the studies devoted to the analysis of unsteady flows in the interaction region, first of all one should mention the investigations carried out within the framework of hydrodynamic stability theory. Of course, the asymptotic theory based upon passage to the limit as $\mathrm{Re} \to \infty$ does not allow us to find the critical Reynolds number at which the boundary layer loses its stability. For solving this problem and, in general, for determining the form of the stability curve of the boundary layer near the critical value of the Reynolds number, one might resort to numerical integration of the Orr–Sommerfeld equation. However, the problem of laminar-turbulent transition in the boundary layer is not at all reduced to an analysis of its stability. The contemporary point of view asserts that transition represents an extremely complicated phenomenon. It is a whole sequence of events, the first of which is the transformation of external perturbations into internal boundary-layer oscillations which have the form of Tollmien–Schlichting waves. For low levels of turbulence in the oncoming flow the initial amplitude of the Tollmien–Schlichting wave is small and does not lead to immediate transition. The wave first must be amplified inside the boundary layer, and only after its amplitude reaches a certain level do the nonlinear effects characteristic of the turbulent flow regime begin to play an essential role.

The length of the region where perturbations are amplified, and therefore of the whole transition region, depends essentially upon how rapidly the external perturbations are transformed into Tollmien–Schlichting waves. In connection with this, one of the important problems of transition theory is the problem of boundary-layer receptivity to external perturbations (Morkovin, 1969). From a mathematical point of view the receptivity problem is found to be considerably more complicated than the stability problem. For this reason asymptotic methods have found wide application here.

The first example of the application of interaction theory to a problem involving generation of Tollmien–Schlichting waves was given by Terent'ev (1981). He considered the Blasius boundary layer in an incompressible fluid flow, choosing as the source of the perturbations the vibrations of the surface itself, or, more precisely, of a short segment with length $\Delta x = O(\text{Re}^{-3/8})$. If, furthermore, the frequency of the vibrations is $\Omega = O(\text{Re}^{1/4})$, then the flow in the vicinity of the vibrator is described correctly by the theory of boundary-layer interaction with the external flow. This choice is not just accidental. As shown by Smith (1979b), Zhuk and Ryzhov (1980), and also by Mikhailov (1981), the interaction theory describes the asymptotic structure of Tollmien–Schlichting waves at the lower branch of the boundary-layer stability curve.

Terent'ev (1981) considered the case when the vibrating part of the surface oscillates harmonically in a direction normal to the surface:

$$\bar{H}(\bar{t}, \bar{X}) = \sigma h(\bar{X}) e^{i\omega \bar{t}}.$$

Taking the amplitude σ to be small, he represented the unknown flow variables in the form

$$\bar{\Psi} = \tfrac{1}{2}\bar{Y}^2 + \sigma \bar{\Psi}_1(\bar{X}, \bar{Y}) e^{i\omega \bar{t}} + \cdots,$$
$$\bar{P} = \sigma \bar{P}_1(\bar{X}) e^{i\omega \bar{t}} + \cdots, \quad \bar{A} = \sigma \bar{A}_1(\bar{X}) e^{i\omega \bar{t}} + \cdots. \tag{2.4.7}$$

Here, it has been assumed that from the moment when the motion of the vibrator was initiated, enough time has passed for harmonic oscillations of frequency ω to be established in the flow.

After substituting (2.4.7) into the equations of the interaction theory, Terent'ev (1981) formulated a linear problem for the functions $\bar{\Psi}_1(\bar{X}, \bar{Y})$, $\bar{P}_1(\bar{X})$, and $\bar{A}_1(\bar{X})$. The solution was obtained with the help of a Fourier transformation with respect to the variable $\bar{X}$. Special attention was given to the behavior of the solution downstream of the vibrator. As $\bar{X} \to \infty$, it was found that out of the entire spectrum of perturbations introduced into the flow by the vibrator, the most pronounced is the mode corresponding to internal boundary-layer oscillations – the

Tollmien–Schlichting wave. If the frequency ω is smaller than its critical value $\omega_* = 2.298$, then the wave amplitude is damped in the flow downstream. However, as $\omega \to \omega_*$, the damping rate tends to zero and the wave becomes neutral.

A formal application of the Fourier transformation to the solution of the problem for $\omega > \omega_*$ leads to a physically unrealistic result in which the Tollmien–Schlichting wave is present in the flow ahead of the vibrator, but is completely suppressed by the vibrator and is absent downstream of it. One can rectify the situation in the following way. If we give up the condition that oscillations are damped as $\bar{X} \to \infty$ and add to the solution an eigenfunction having the form of a growing Tollmien–Schlichting wave at supercritical frequency $\omega > \omega_*$, the coefficient of this eigenfunction can be chosen so that the oscillations ahead of the vibrator are canceled and then they will be amplified in the flow downstream of it. This approach was formulated by Bogdanova and Ryzhov (1982) in the form of a postulate of continuous dependence of the solution on the vibrator frequency. Later Terent'ev (1984, 1985) showed this to be correct by solving an initial-value problem in the interaction theory, which he called the problem of starting a vibrator. Terent'ev assumed that up to some initial instant of time the vibrator is at rest and the flow is undisturbed. After the vibrator is started, perturbations are produced in the boundary layer, but at every finite value of time they are damped far from their source, so that for the solution of the problem one can use a Fourier transformation with respect to the longitudinal coordinate and a Laplace transformation in time. The result is that in the limit as time tends to infinity an oscillatory state is established at every fixed point in space in the form predicted by the continuity postulate. At the same time further downstream there is a region within the flow, moving away from the vibrator with almost constant velocity and expanding linearly with time, where the oscillations take more complicated form. This region, giving smooth transitions from single Tollmien–Schlichting wave oscillations just downstream of the vibrator to unperturbed flow further downstream, is referred to as a wave packet. It was studied in detail by Terent'ev (1987) and Ryzhov and Terent'ev (1986, 1987).[17] As a result, it was shown that the amplitude of oscillation in the wave packet increases according to an exponential law, the growth rate being found to coincide with the maximum rate of amplification of Tollmien–

---

[17] Also see surveys by Kozlov and Ryzhov (1990) and by Goldstein and Hultgren (1989).

Schlichting waves in the interval between the upper and lower branches of the boundary-layer stability curve. It was also noticed that the form of the wave packet is determined mainly by the boundary-layer stability characteristics and it depends essentially neither on the initial data nor even on the particular physical process leading to the penetration of perturbations into the boundary layer.

Another well-known way of exciting Tollmien–Schlichting waves in the boundary layer was considered by Ruban (1984, 1985) and by Goldstein (1985). It consists in influencing the boundary layer by sound. An effective transformation of external perturbations into Tollmien–Schlichting waves is possible only when the external perturbations are resonant with internal oscillations in the boundary layer, and this resonance, as observed by Kachanov, Kozlov, and Levchenko (1982), presupposes the coincidence not only of the frequency of the oscillations but also of the wave number. In the problem that was considered by Terent'ev (1981), this requirement was not an issue since the frequency and length of a vibrating section of the surface may be chosen independently of each other. However, in the case when a sound wave plays the role of the external perturbation, we can select its frequency in a suitable way, but then the wave length will be uniquely determined by the speed and Mach number of the external flow. In a subsonic flow it considerably exceeds the Tollmien–Schlichting wave length. To provide conditions for resonance and to introduce the necessary length scale into the problem, Ruban (1984, 1985) and Goldstein (1985) supposed that the surface within the flow is not absolutely flat but has an irregularity whose longitudinal scale coincides with the Tollmien–Schlichting wavelength in order of magnitude: $\Delta x = O(\mathrm{Re}^{-3/8})$. Then, representing the perturbation introduced into the boundary layer by the irregularity as a Fourier integral with respect to the longitudinal coordinate, one may be assured that some harmonic in the spectrum can be found that will satisfy the condition of resonance with a Tollmien–Schlichting wave.

To study such a possibility, Ruban (1984, 1985) and Goldstein (1985) used the method of matched asymptotic expansions and, taking the Reynolds number to be sufficiently large, derived from the Navier–Stokes equations an asymptotic system of equations describing the process of sound-wave interaction with a local surface irregularity within the flow. The solution of this system was constructed in the linear approximation. As a result, it was possible to show that a Tollmien–Schlichting wave propagates downstream of the irregularity. Its amplitude is found to be proportional to the Fourier component of the surface shape, which has a wave number corresponding to free oscillations in the boundary layer with the frequency of the incident sound wave.

There are two more approaches to the study of unsteady flows in the interaction region. Both are associated with analysis of the evolution of essentially nonlinear perturbations. If the amplitude of the perturbations exceeds in order of magnitude the level that was assumed in deriving the equations for the interaction theory, and the length of the perturbed region proves to be smaller than $O(\mathrm{Re}^{-3/8})$, then in order to describe the flow one should use the theory proposed by Zhuk and Ryzhov (1982), Smith and Burggraf (1985), and Ryzhov (1990). They found that under these conditions the flow in the interaction region is inviscid in the leading approximation and may be described by the Benjamin–Ono integro-differential equation

$$\frac{\partial \bar{A}}{\partial \bar{t}} + \bar{A}\frac{\partial \bar{A}}{\partial \bar{X}} = \frac{1}{\pi}\int_{-\infty}^{\infty} \frac{\partial^2 \bar{A}/\partial s^2}{s - \bar{X}}ds$$

for the displacement thickness $\bar{A}(\bar{t}, \bar{X})$. This equation admits solutions periodic in $\bar{t}$ and $\bar{X}$ as well as solutions of soliton type, which can be used to study the essentially nonlinear stages of laminar-turbulent transition in the boundary layer.

The second approach is associated with a direct numerical solution of the Cauchy problem for the nonlinear equations of the interaction theory. Examples of such solutions are given in the work of Duck (1985, 1988), Smith (1986), and Peridier, Smith, and Walker (1991b). All of them give evidence of the inevitable appearance of a singularity in the boundary layer at a finite value of time $\bar{t} = \bar{t}_S$. Its structure was studied in detail by Smith (1988b, 1991), and it was found to resemble closely the singularity that arises in the solution of the steady-flow equations of the interaction theory at a finite value of the problem parameter, say the parameter α in the problem of flow past a corner of a body contour (see Section 3).

3

Flow in the Vicinity of the
Trailing Edge of a Thin Airfoil

In this chapter we consider flows over thin bodies at large values of the Reynolds number. The emphasis will be given mainly to a study of the onset of separation, which is accompanied by the appearance of a local zone of reverse flow in the vicinity of the trailing edge. Such flows, like those considered in the preceding sections, are small perturbations of the flow over a flat plate placed at zero angle of attack. Therefore, we start with consideration of the problem of flow over a plate and, in particular, with an analysis of the flow in the vicinity of its trailing edge.

3.1 The Boundary Layer in the Vicinity of the Trailing Edge of a Flat Plate

We consider the steady fluid flow over a flat plate placed parallel to a uniform oncoming flow. Let $Ox^\circ y^\circ$ be a rectangular Cartesian coordinate system whose origin coincides with the trailing edge, so that the plate of length L occupies the segment $-L \leq x^\circ \leq 0$ of the straight line $y^\circ = 0$. We designate by u°, v° the velocity components along the axes Ox° and Oy°, and by p° and ψ° the pressure and stream function. We also introduce dimensionless variables:

$$x = \frac{x^\circ}{L}, \quad y = \frac{y^\circ}{L}, \quad u = \frac{u^\circ}{U_\infty},$$

$$v = \frac{v^\circ}{U_\infty}, \quad p = \frac{p^\circ - p_\infty}{\rho U_\infty^2}, \quad \psi = \frac{\psi^\circ}{L U_\infty},$$

where as the characteristic velocity and pressure we take the values U_∞ and p_∞ in the uniform oncoming flow; we denote the density and the kinematic viscosity coefficient by ρ and ν.

100

In the following we shall, as before, often use the boundary-layer variables Y and Ψ, which are related to the original variables by

$$\Psi = \mathrm{Re}^{1/2}\psi, \quad Y = \mathrm{Re}^{1/2}y, \qquad (3.1.1)$$

where the Reynolds number is $\mathrm{Re} = U_\infty L/\nu$; then, in the region occupied by the boundary layer and the wake behind the plate, the asymptotic representation is

$$\Psi = \Psi_0(x, Y) + o(1), \quad \mathrm{Re} \to \infty. \qquad (3.1.2)$$

Since the appearance of the work of Prandtl (1904) and Blasius (1908), it has been known that the solution of the problem for a boundary layer developing along the surface of a semi-infinite plate is self-similar. This solution remains valid for the boundary layer along a plate whose length is finite, since the trailing edge does not influence the leading term of the asymptotic expansion (3.1.2). This is associated with two circumstances. First, both in the case of a semi-infinite plate and in the case of a plate of finite length, the external potential flow in the first approximation is uniform, and second, the boundary-layer equations are equations of parabolic type.

In the notation introduced above, the Blasius solution has the form

$$\Psi_0 = (x + 1)^{1/2}F[Y/(x + 1)^{1/2}], \quad -1 < x < 0, \qquad (3.1.3)$$

where the function F is found by solving the following problem[1]:

$$2F''' + FF'' = 0, \quad F(0) = F'(0) = 0, \quad F'(\infty) = 1. \qquad (3.1.4)$$

Near the edges of the plate this solution becomes invalid. Thus, as the leading edge is approached, the skin friction and the vertical component of the velocity vector at the outer edge of the boundary layer increase without limit:

$$\tau = \left.\frac{\partial^2\Psi_0}{\partial Y^2}\right|_{Y=0} = (x+1)^{-1/2}F''(0), \quad F''(0) = a_0 = 0.3321\,;$$

$$\left.\mathrm{Re}^{1/2}v\right|_{Y\to\infty} = -\left.\frac{\partial\Psi_0}{\partial x}\right|_{Y\to\infty} = (x+1)^{-1/2}\beta, \quad \beta = 0.8604\,. \qquad (3.1.5)$$

This singularity is elucidated through introduction of a region for which $|x + 1| = O(\mathrm{Re}^{-1})$, $|y| = O(\mathrm{Re}^{-1})$ (Figure 3.1), and the flow here is described by the full Navier–Stokes equations, with a local Reynolds number equal to unity (Carrier and Lin, 1948). A numerical solution

---

[1] See Weyl (1942).

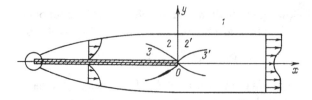

Fig. 3.1 Asymptotic flow structure near a plate of finite length.

of the corresponding boundary-value problem was obtained by van de Vooren and Dijkstra (1970) and also by Yoshizawa (1970).

Now let us consider the flow in the vicinity of the trailing edge of the plate, which was analyzed independently by Stewartson (1969) and by Messiter (1970). As the trailing edge is approached ($x \to -0$), the Blasius solution remains regular, but becomes invalid for $x > 0$ because of the impossibility of satisfying the boundary conditions at the wake plane of symmetry. (Because of the symmetry, here and below we consider only the flow in the upper half plane.)

For $x \to -0$, the solution (3.1.3)–(3.1.5) can be represented by a Taylor series expansion:

$$\Psi_0 = \Psi_{00}(Y) + (-x)\Psi_{01}(Y) + O[(-x)^2],$$
$$\Psi_{00} = F(Y), \quad \Psi_{01} = \tfrac{1}{2}(YF' - F).$$
$$(3.1.6)$$

In the main part of the boundary layer (region 2, Figure 3.1), the solution should remain continuous in passing through the cross section $x = 0$, and consequently

$$\Psi_0(+0, Y) = \Psi_{00}(Y), \quad Y = O(1), \quad Y > 0. \qquad (3.1.7)$$

On the wake axis the no-slip condition must be replaced by the symmetry conditions

$$\Psi = \frac{\partial^2 \Psi}{\partial Y^2} = 0 \text{ at } Y = 0, \quad x > 0. \qquad (3.1.8)$$

However, since the leading term in the expansion of the Blasius function as $Y \to 0$ has the form

$$\Psi_{00} = \frac{a_0}{2}Y^2 + O(Y^5) \qquad (3.1.9)$$

and the value of the dimensionless shear stress $\Psi_{00}''(Y)$ thus has a nonzero value at $Y = 0$, the initial velocity profile (3.1.7) for the wake region does not satisfy the boundary conditions (3.1.8) on the axis. This implies (Goldstein, 1930) that as $x \to +0$ there is a sublayer near the wake axis (region 3′) in which the solution allows the conditions (3.1.8) to be satisfied, while at its outer edge, according to (3.1.9), the shear stress has a nonzero value.

It follows from these considerations, and also from the balance between the inertia and the viscous terms in the boundary-layer equation with zero pressure gradient, that the sublayer solution can be represented in the following form (Goldstein, 1930):

$$\Psi_0 = x^{2/3} g_0(\xi) + O(x^{5/3}), \quad \xi = Y/x^{1/3}. \tag{3.1.10}$$

For the function $g_0(\xi)$ we obtain the equation

$$g_0''' + \tfrac{2}{3} g_0 g_0'' - \tfrac{1}{3} g_0'^2 = 0, \tag{3.1.11}$$

and the boundary conditions (3.1.8) give

$$g_0(0) = g_0''(0) = 0. \tag{3.1.12}$$

For matching with the solution in the main part of the wake it is necessary that $g_0''(\infty) = a_0$, and therefore the asymptotic expansion as $\xi \to \infty$ has the form

$$g_0 = \frac{a_0}{2}(\xi + A_0)^2 + o(1). \tag{3.1.13}$$

The solution of the problem $(3.1.11)$–$(3.1.13)$[2] is found numerically, with the results

$$A_0 = 0.8920 a_0^{-1/3}, \quad g_0'(0) = 1.6109 a_0^{2/3}.$$

Rewriting the expansion (3.1.13) in the variables x, Y and bearing in mind that the leading term of the asymptotic expansion for $x \to +0$, $Y = O(1)$ (region 2′) is the Blasius function, we arrive at the following expansion:

$$\Psi_0 = \Psi_{00}(Y) + x^{1/3} \widehat{\Psi}_{01}(Y) + O(x^{2/3}). \tag{3.1.14}$$

Substituting this into the boundary-layer equation, we find that

$$\widehat{\Psi}_{01} = k_1 \Psi_{00}', \tag{3.1.15}$$

and from matching with the solution in region 3′ we obtain

$$k_1 = A_0. \tag{3.1.16}$$

From the expressions (3.1.14)–(3.1.16) it follows that, when the trailing edge of the plate is approached from the direction of the wake, the

---

[2] The mathematical analysis of this problem was the subject of studies by Weyl (1942), von Mises and Friedrichs (1971), and Diesperov (1986).

magnitude of the vertical component of the velocity vector (or the slope of the streamlines) increases indefinitely:

$$V_0 = \text{Re}^{1/2}v = -\frac{\partial \Psi_0}{\partial x} = -x^{-2/3}\frac{A_0}{3}\Psi_{00}' + O(x^{-1/3}). \qquad (3.1.17)$$

As can be seen from the expressions (3.1.13)–(3.1.17), the singularity in the solution is the result of changes in streamline slope at the outer edge of the viscous sublayer, which in turn are caused by the strong acceleration which fluid elements experience at the "bottom" of the boundary layer while they are passing the section $x = 0$. The latter is obviously related to the fact that the no-slip condition is replaced here by a symmetry condition.

Let us note that the construction of asymptotic expansions of the coordinate type considered was first carried out in the cited work of Goldstein. The application of this method, as we can see in the preceding chapters, has made it possible to study the boundary-layer equations for still other flows with singularities in the boundary conditions.

In the region of the external potential flow the discontinuity in the boundary conditions occurring at the trailing edge of the plate is, first of all, manifested by the change in the pressure distribution, which is caused by changes in the slopes of the streamlines at the outer edge of the boundary layer. As shown by Goldstein's solution, these changes in the vicinity of the trailing edge have a singular character. Using the expressions (3.1.1)–(3.1.6), (3.1.14)–(3.1.17), we represent the complex velocity and the pressure in the region of the external potential flow (region 1) in the following form:

$$\begin{aligned} u - iv &= 1 + \text{Re}^{-1/2}w_1(z) + \cdots, \\ p &= \text{Re}^{-1/2}P_1(x, y) + \cdots, \quad z = x + iy. \end{aligned} \qquad (3.1.18)$$

In the vicinity of the trailing edge ($|z| \to 0$) for $\arg z \to \pi$ and $\arg z \to 0$, these expansions must match, respectively, with the solutions (3.1.6) and (3.1.14)–(3.1.17) for the main part of the boundary layer. Consequently, as $|x| \to 0$

$$\text{Im } w_1 = \begin{cases} O(1) & \text{for } x < 0, \quad y = 0, \\ \frac{A_0}{3}x^{-2/3} + o(x^{-1/3}) & \text{for } x > 0, \quad y = 0. \end{cases}$$

These conditions determine uniquely the leading term in the expansion of the analytic function $w_1(z)$ as $|z| \to 0$:

$$w_1 = \frac{A_0}{3}\left(-\frac{1}{\sqrt{3}} + i\right)z^{-2/3} + o(z^{-1/3}). \qquad (3.1.19)$$

Now taking the real part of this function in (3.1.18) and using Bernoulli's integral, we obtain the following expression for the pressure distribution at the outer edge of the boundary layer as $|x| \to 0$:

$$P_1(x,0) = p_1(x) = \begin{cases} -\dfrac{2A_0}{3\sqrt{3}}(-x)^{-2/3} + o\left[(-x)^{-1/3}\right], & x < 0; \\ \dfrac{A_0}{3\sqrt{3}}x^{-2/3} + o(x^{-1/3}), & x > 0. \end{cases}$$

(3.1.20)

Then the asymptotic representation (3.1.2) in the boundary-layer region at the surface of the plate assumes the form

$$\Psi = \Psi_0(x, Y) + \mathrm{Re}^{-1/2}\Psi_1(x, Y) + \cdots \tag{3.1.21}$$

Substituting this together with (3.1.18) and (3.1.20) into the Navier-Stokes equations, we obtain equation (2.3.11) of Chapter 2 for the function $\Psi_1(x, Y)$.

Thus, according to (3.1.17)–(3.1.20), the changes in the slopes of the streamlines in the wake are transmitted through the region of potential flow to the outer edge of the boundary layer on the plate, leading to the appearance of a singular pressure gradient in the vicinity of the trailing edge.

In Section 3 of the preceding chapter we have already considered the problem of a singular perturbation of the Blasius solution, and we saw that to satisfy the boundary conditions on the body surface as $x \to -0$, it is necessary to introduce a viscous sublayer (region 3) in which $Y = O\left[(-x)^{1/3}\right]$. A similar situation arises in the flow considered here. Therefore, the solution for this region can be represented in the form

$$\begin{aligned} \Psi_0 &= (-x)^{2/3}\frac{a_0}{2}\eta^2 + O\left[(-x)^{5/3}\right], \\ \Psi_1 &= (-x)^{-2/3}f_1(\eta) + o\left[(-x)^{-1/3}\right], \quad \eta = Y/(-x)^{1/3}. \end{aligned}$$

(3.1.22)

Here, as before, the expansion for the function $\Psi_0(x, Y)$ is the Blasius solution taken as $Y \to 0$ and written in the variable η, and the leading term of the expansion for $\Psi_1(x, Y)$ depends on the pressure gradient (3.1.20). On substituting (3.1.22), (3.1.20) into the equation for the function $\Psi_1(x, Y)$, we obtain

$$f_1''' - \frac{a_0}{3}\eta^2 f_1'' - \frac{2}{3}a_0\eta f_1' + \frac{2}{3}a_0 f_1 = -\frac{4A_0}{9\sqrt{3}}. \tag{3.1.23}$$

In addition to the no-slip condition, which has the form

$$f_1(0) = f_1'(0) = 0, \tag{3.1.24}$$

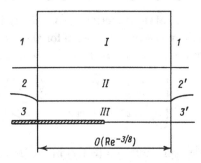

Fig. 3.2 Asymptotic structure of the interaction region.

for the solution to this equation it is necessary to impose a condition requiring the absence of exponential growth as $\eta \to \infty$, from which it follows that

$$f_1''(\infty) = 0. \tag{3.1.25}$$

The latter provides the possibility of matching with the solution in the main part of the boundary layer.

The solution of the problem (3.1.23)–(3.1.25) can be written in explicit form in terms of a confluent hypergeometric function, and then we find that as $\eta \to \infty$

$$f_1 = \frac{2A_0}{3\sqrt{3}a_0} \left[\frac{\Gamma(1/3)}{\Gamma(2/3)} \left(\frac{a_0}{9} \right)^{1/3} \eta - 1 + O(\eta^{-2}) \right].$$

Now comparing the term containing the function $f_1(\eta)$ with the first term of the expansion (3.1.21), (3.1.22) we see that for $|x| = O(\mathrm{Re}^{-3/8})$ they have the same order of magnitude. This implies that the solutions of Blasius and Goldstein become invalid here because of the fact that the changes in the streamline slopes and in the pressure distribution start to influence the character of the flow in the viscous sublayer. Consequently, the changes in the flow variables at the outer edge of the boundary layer and in the viscous sublayer become mutually dependent here; that is, when $|x| = O(\mathrm{Re}^{-3/8})$, an interaction takes place between the potential flow and the flow in the boundary layer.

In the preceding chapters we have already considered flows in which interaction occurs. It is to be noted, however, that Stewartson (1969) and Messiter (1970) were the first to discover and investigate this important effect for incompressible fluid flows, for the study of precisely this flow in the vicinity of the trailing edge of a plate.

Let us proceed to the analysis of the solution in the interaction region.

First of all we note that here the boundary layer consists of two distinct regions: the main part and the viscous sublayer (regions II and III, respectively, in Figure 3.2). In the main part, where $Y = O(1)$ as before (see Chapters 1 and 2), according to (3.1.6) and (3.1.14), the Blasius velocity profile will be preserved in the first approximation. Introducing the inner longitudinal variable

$$x^* = \text{Re}^{3/8} x, \qquad (3.1.26)$$

which has finite values in the interaction region, and transforming to it in the expansions (3.1.14) and (3.1.20), we can represent the solution for this region as follows:

$$\Psi = \Psi_{00}(Y) + \text{Re}^{-1/8} \widetilde{\Psi}_{10}(x^*, Y) + O(\text{Re}^{-1/4}),$$
$$p = \text{Re}^{-1/4} \widetilde{P}_0(x^*, Y) + O(\text{Re}^{-3/8}). \qquad (3.1.27)$$

On substituting these expansions, along with (3.1.1) and (3.1.26), into the original Navier–Stokes equations, we find after integration

$$\widetilde{\Psi}_{10} = A(x^*)\Psi'_{00}(Y), \quad \widetilde{P}_0 = P^*(x^*), \qquad (3.1.28)$$

where the functions $A(x^*)$ and $P^*(x^*)$ for the present remain arbitrary.

Matching with the solutions in regions 2 and 2′ yields

$$(-x^*)^{2/3} P^* \to -\frac{2A_0}{3\sqrt{3}}, \quad x^* \to -\infty;$$
$$x^{*2/3} P^* \to \frac{A_0}{3\sqrt{3}}, \quad x^* \to \infty.$$

Because of the large pressure gradient ($\partial p/\partial x = O(\text{Re}^{1/8})$) the flow in region II is locally inviscid.

Near the body surface and the symmetry plane there is a viscous sublayer III which as usual plays a crucial part in the interaction process. This region is a continuation of regions 3 and 3′; therefore, considering (3.1.10), (3.1.20)–(3.1.22) as outer expansions with respect to the interaction region, and changing to the variable x^*, we represent the solution in region III in terms of the following asymptotic expansions:

$$Y = \text{Re}^{-1/8} Y^*,$$
$$\Psi = \text{Re}^{-1/4} \Psi_0^*(x^*, Y^*) + O(\text{Re}^{-3/8}), \qquad (3.1.29)$$
$$p = \text{Re}^{-1/4} P_0^*(x^*, Y^*) + O(\text{Re}^{-3/8}).$$

Substitution of these expansions, along with (3.1.1) and (3.1.26), into the original equations results in the Prandtl boundary-layer equations:

$$\frac{\partial \Psi_0^*}{\partial Y^*} \frac{\partial^2 \Psi_0^*}{\partial x^* \partial Y^*} - \frac{\partial \Psi_0^*}{\partial x^*} \frac{\partial^2 \Psi_0^*}{\partial Y^{*2}} + \frac{\partial P_0^*}{\partial x^*} = \frac{\partial^3 \Psi_0^*}{\partial Y^{*3}}, \quad \frac{\partial P_0^*}{\partial Y^*} = 0. \quad (3.1.30)$$

Recalling now the expression (3.1.9), we can match the solutions (3.1.27), (3.1.28) for the main part of the boundary layer with the solution in the viscous sublayer. As a result we obtain

$$\Psi_0^* = \frac{a_0}{2} Y^{*2} + a_0 A(x^*) Y^* + O(1) \text{ as } Y^* \to \infty,$$

$$P_0^* = P^*(x^*). \quad (3.1.31)$$

The conditions at the plane of symmetry remain unchanged:

$$\Psi_0^* = 0 \quad \text{at } Y^* = 0,$$

$$\partial \Psi_0^* / \partial Y^* = 0 \quad \text{for } x^* < 0, \quad Y^* = 0, \quad (3.1.32)$$

$$\partial^2 \Psi_0^* / \partial Y^{*2} = 0 \quad \text{for } x^* > 0, \quad Y^* = 0.$$

Moreover, the solution in the sublayer must satisfy the conditions of matching with the Blasius and Goldstein solutions as $|x^*| \to \infty$. According to (3.1.21), (3.1.22) and (3.1.10), this implies that

$$\Psi_0^* = \frac{a_0}{2} Y^{*2} + (-x^*)^{-2/3} f_1(\eta) + \cdots, \quad \eta = Y^*/(-x^*)^{1/3} \text{ as } x^* \to -\infty,$$

$$\Psi_0^* = x^{*2/3} g_0(\xi) + \cdots, \quad \xi = Y^*/x^{*1/3} \text{ as } x^* \to \infty, \quad (3.1.33)$$

where $g_0(\xi)$ and $f_1(\eta)$ are the solutions of the problems (3.1.11)–(3.1.13) and (3.1.23)–(3.1.25).

The derivative of the function $A(x^*)$ in the expansion (3.1.31) determines the slopes of the streamlines at the outer edge of the viscous sublayer:

$$-A'(x^*) = G(x^*) = \lim_{Y^* \to \infty} \left(-\frac{\partial \Psi_0^*}{\partial x^*} \Big/ \frac{\partial \Psi_0^*}{\partial Y^*} \right), \quad (3.1.34)$$

whose values, according to (3.1.28), will be transferred without change across the main part of the boundary layer to the external flow. From the expressions (3.1.27), (3.1.28) we find that

$$V_0^*|_{Y \to \infty} = G(x^*), \quad V_0^* = \mathrm{Re}^{1/4} v. \quad (3.1.35)$$

To close the problem in region III, it is necessary to find the relationship between the changes in pressure and streamline slope at the outer edge of the boundary layer. For this purpose, as we did several times in the preceding chapters, we have to consider a region of the external potential flow whose transverse scale is of the order of the longitudinal scale of the interaction region (region I) in which the expansion (3.1.18) becomes invalid.

Introducing the inner variables

$$x^* = \mathrm{Re}^{3/8}x, \quad y^* = \mathrm{Re}^{3/8}y \qquad (3.1.36)$$

and assuming that the changes in the pressure and the vertical velocity component are, according to (3.1.27) and (3.1.35), of order $\mathrm{Re}^{-1/4}$, we represent the solution in the form

$$
\begin{aligned}
u &= 1 + \mathrm{Re}^{-1/4}u_1^*(x^*, y^*) + O(\mathrm{Re}^{-3/8}), \\
v &= \mathrm{Re}^{-1/4}v_1^*(x^*, y^*) + O(\mathrm{Re}^{-3/8}), \\
p &= \mathrm{Re}^{-1/4}p_1^*(x^*, y^*) + O(\mathrm{Re}^{-3/8}).
\end{aligned}
\qquad (3.1.37)
$$

Substituting these expansions together with (3.1.36) into the Navier–Stokes equations for the unknown functions results in the equations of small-perturbation theory, and therefore

$$w_1^*(z^*) = p_1^* + iv_1^* = -u_1^* + iv_1^* \qquad (3.1.38)$$

is an analytic function of the complex variable $z^* = x^* + iy^*$.

Matching with the solution (3.1.18)–(3.1.20) for the outer potential-flow region, we obtain

$$z^{*2/3}w_1^* \to -\frac{A_0}{3}\left(-\frac{1}{\sqrt{3}} + i\right) \qquad (3.1.39)$$

as $|z^*| \to \infty$; using the expressions (3.1.35) we find that

$$\mathrm{Im}\, w_1^*\big|_{y^*=0} = G(x^*). \qquad (3.1.40)$$

From the solution of the problem (3.1.39), (3.1.40) for the analytic function (3.1.38) (see Chapter 1, Section 5), it follows that

$$p_1^*(x^*, 0) = P^*(x^*) = \frac{1}{\pi}\int_{-\infty}^{\infty}\frac{G(t)\,dt}{t - x^*}. \qquad (3.1.41)$$

In this way we obtain the missing link between the pressure changes and the streamline slopes; moreover, this means that within the interaction region the pressure gradient is self-induced.

Thus the system of relations for the interaction region becomes closed. After the transformations

$$
\begin{aligned}
x^* &= a_0^{-5/4}\bar{X}, \quad Y^* = a_0^{-3/4}\bar{Y}, \quad \Psi_0^* = a_0^{-1/2}\bar{\Psi}, \\
P^* &= a_0^{1/2}\bar{P}, \quad G = a_0^{1/2}\bar{G}, \quad g_0 = a_0^{1/3}\bar{g}, \quad \xi = a_0^{-1/3}\bar{\xi},
\end{aligned}
\qquad (3.1.42)
$$

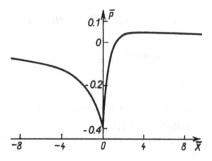

Fig. 3.3 Pressure distribution in the interaction region.

the boundary-value problem (3.1.30)–(3.1.34), (3.1.41), whose solution describes the flow in the viscous sublayer, assumes the form

$$\frac{\partial \bar{\Psi}}{\partial \bar{Y}} \frac{\partial^2 \bar{\Psi}}{\partial \bar{X} \partial \bar{Y}} - \frac{\partial \bar{\Psi}}{\partial \bar{X}} \frac{\partial^2 \bar{\Psi}}{\partial \bar{Y}^2} + \frac{d\bar{P}}{d\bar{X}} = \frac{\partial^3 \bar{\Psi}}{\partial \bar{Y}^3} ,$$

$$\bar{\Psi} = \frac{\partial \bar{\Psi}}{\partial \bar{Y}} = 0 \quad \text{for } \bar{X} < 0, \quad \bar{Y} = 0,$$

$$\bar{\Psi} = \frac{\partial^2 \bar{\Psi}}{\partial \bar{Y}^2} = 0 \quad \text{for } \bar{X} > 0, \quad \bar{Y} = 0, \quad\quad (3.1.43)$$

$$\bar{\Psi} = \frac{\bar{Y}^2}{2} + \cdots \quad \text{as } \bar{X} \to -\infty \quad \text{and } \bar{Y} \to \infty,$$

$$\bar{\Psi} = \bar{X}^{2/3} \bar{g}(\bar{\xi}) + \cdots \quad \text{as } \bar{X} \to \infty,$$

$$\bar{P}(\bar{X}) = \frac{1}{\pi} \int_{-\infty}^{\infty} \frac{\bar{G}(t) dt}{t - \bar{X}}, \quad \bar{G}(\bar{X}) = \lim_{\bar{Y} \to \infty} \left[-\frac{\partial \bar{\Psi}}{\partial \bar{X}} \bigg/ \frac{\partial \bar{\Psi}}{\partial \bar{Y}} \right].$$

A numerical solution of the problem (3.1.43) formulated by Stewartson (1969) and Messiter (1970) was first given by Jobe and Burggraf (1974). Shown in Figures 3.3 and 3.4 are their results for the pressure distribution $\bar{P}(\bar{X})$ as well as the shear-stress distribution $\bar{\tau}(\bar{X}) = \bar{\Psi}_{\bar{Y}\bar{Y}}|_{\bar{Y}=0}$ on the surface of the plate and the velocity $\bar{U}(\bar{X}) = \bar{\Psi}_{\bar{Y}}|_{\bar{Y}=0}$ on the wake plane of symmetry. One can see that under the action of a favorable pressure gradient a certain increase in the skin friction takes place, so that $\bar{\tau}(-0) = a_1 > 1$, $a_1 = 1.343$. This leads to an increase in the drag force Q in comparison with the value given by the Blasius solution. Therefore, the analysis of the flow in the interaction region has allowed us to determine correctly the second term in the expansion of the drag coefficient c_x. Following Stewartson and Messiter, we have the following expansion (for one side of the plate):

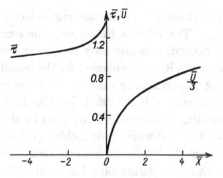

Fig. 3.4 Shear-stress and velocity distributions in the flow plane of symmetry.

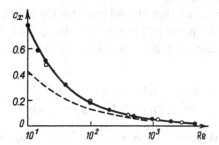

Fig. 3.5 The dependence of the drag coefficient on Reynolds number for a plate of finite length: dashed line, Blasius (1908); solid line, formula (3.1.44); filled circles, Dennis et al. (1966, 1969); open circles, Nishioka and Miyagi (1978).

$$c_x = \frac{Q}{\frac{1}{2}\rho U_\infty^2 L} = 1.328 \, \mathrm{Re}^{-1/2} + d_0 \mathrm{Re}^{-7/8} + O(\mathrm{Re}^{-1}),$$

$$d_0 = 2a_0^{-1/4} \int_{-\infty}^0 (\bar{\tau} - 1)d\bar{X}.$$

(3.1.44)

The numerical solutions of the problem (3.1.43) in the work of Jobe and Burggraf (1974), Veldman and van de Vooren (1975), and Chow and Melnik (1976) yielded, respectively, the following values for the constant d_0: 2.694; 2.651 ± 0.003; 2.668.

The experimental data of Nishioka and Miyagi (1978), as well as the results of the numerical solution of Dennis et al. (1966, 1969), give excellent agreement for the drag coefficient with the value determined from the formula (3.1.44). A comparison of these results is given in Figure 3.5 (the value for d_0 was taken equal to 2.668).

The third term in the expansion (3.1.44) is $O(\mathrm{Re}^{-1})$ and is determined by the integral of the surface friction forces over all the different regions.

First, there is a contribution from the flow region located in the vicinity of the leading edge. The value of the appropriate constant was found by Imai (1957a). Second, it is necessary to take into account the next term of the expansion in Reynolds number for the boundary-layer region along the plate. An approximate value of the corresponding constant was obtained in the work of Kuo (1953). And finally a contribution of order Re^{-1} to the drag coefficient is introduced by the second term in the expansion (3.1.29). The work of McLachlan (1991b), on the basis of numerical solutions of the Navier–Stokes equations, clarified the reason why the formula (3.1.44), disregarding the term $O(Re^{-1})$, gives good agreement with the results of the numerical solutions, even at Reynolds numbers that are not large.

To conclude, we note that the solution for the interaction region has a singularity at the point $|z^*| = 0$ (Stewartson, 1969; Messiter, 1970). The pressure gradient in regions II and III at the cross section $x^* = 0$ is discontinuous:

$$P^{*'} = O(1) \quad \text{as } x^* \to -0 \,,$$
$$P^{*'} = O(x^{*-1/3}) \quad \text{as } x^* \to +0 \,.$$

Removal of this singularity is accomplished through consideration of a region (with three-layer structure) having longitudinal scale $|x| = O(Re^{-1/2})$. Here, in the main part of the boundary layer the transverse pressure change now begins to appear in the leading terms. However, complete elimination of this singularity caused by the discontinuity in the boundary conditions is not successfully achieved; as found by Veldman (1976), for $|x| < O(Re^{-1/2})$ there appears an infinite sequence of nested subregions. This sequence converges to the region with scales $|x| = O(Re^{-3/4})$, $|y| = O(Re^{-3/4})$ where the flow is described by the full Navier–Stokes equations with the local Reynolds number equal to one. A detailed analysis and numerical solution of the boundary-value problem for this region was carried out by Dijkstra (1974) and also by Bogolepov (1985a,b). The existence of this region was first pointed out by Van Dyke (1967). It is of interest that the order of magnitude of the skin friction within the interaction region remains unchanged as the trailing edge ($x^* = -0$) is approached, up to the region where $|x| = O(Re^{-3/4})$.

3.2 The Influence of Profile Thickness

The theory of fluid flow near the trailing edge of a flat plate described in the preceding section is easily generalized to the case of flow around a thin airfoil. As was shown, an acceleration of the fluid can be observed ahead of the trailing edge of a flat plate. Passing from a flat plate to a thin airfoil, exactly the opposite effect – a flow deceleration – may occur. The intensity of this deceleration, which under certain conditions leads to boundary-layer separation near the trailing edge, depends both upon the relative profile thickness and (which is of no less importance) upon the form of the trailing edge.

The wedge-shaped edge. Let a thin symmetric airfoil placed in an incompressible fluid flow be aligned such that its chord line is parallel to the velocity vector of the undisturbed flow (Figure 3.6). We choose the origin of the Cartesian coordinate system Oxy to coincide with the airfoil trailing edge and direct the axis Ox along the flow plane of symmetry and the axis Oy normal to it. We designate the airfoil thickness ratio by ϵ and take it to be small:

$$\epsilon(\mathrm{Re}) \to 0 \quad \text{as } \mathrm{Re} \to \infty.$$

We also assume that in the vicinity of the trailing edge the upper surface of the airfoil can be represented in the form

$$y = -\theta x + \cdots \quad \text{as } x \to -0.$$

Here, θ is the wedge half-angle at the trailing edge, defined by

$$\theta = \epsilon(\mathrm{Re})\theta_0;$$

the constant θ_0 is positive and is taken to have a value of order one.

Because of the flow symmetry, we can, as in the preceding section, confine ourselves to consideration of only the upper half-plane ($y > 0$). We begin with the analysis of region 1 (Figure 3.6), where x and y have values of order one. If $\epsilon(\mathrm{Re})$ exceeds $\mathrm{Re}^{-1/2}$ in order of magnitude, the flow in this region is the same as the flow in region 1 for the problem of flow over a corner of a body contour with small turning angle (Chapter 2, Section 3). In the leading approximation the flow remains uniform, and perturbations introduced as a result of the profile thickness are of order ϵ:

$$u = 1 + \epsilon u_1(x, y) + \cdots, \quad v = \epsilon v_1(x, y) + \cdots,$$
$$p = \epsilon p_1(x, y) + \cdots. \tag{3.2.1}$$

Substituting (3.2.1) into the Navier–Stokes equations, we conclude that the complex velocity

$$w_1(z) = u_1 - iv_1$$

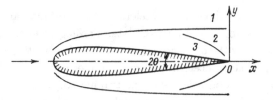

Fig. 3.6 Symmetric thin airfoil in incompressible fluid flow.

is an analytic function of the variable $z = x + iy$, and the pressure $p_1(x, y)$ satisfies Bernoulli's equation

$$p_1 = -u_1. \tag{3.2.2}$$

The boundary conditions for the function $w_1(z)$ are the tangency condition at the airfoil surface

$$v_1(x, 0) = -\theta_0 + \cdots \quad \text{as } x \to -0$$

and the condition of flow symmetry at the wake axis

$$v_1(x, 0) = 0 \quad \text{for } x > 0.$$

From these conditions it follows that the asymptotic representation of the function $w_1(z)$ as $|z| \to 0$ is

$$w_1 = \frac{\theta_0}{\pi} \ln z + d + o(1),$$

where d is a real constant. If now we use Bernoulli's equation (3.2.2), then for the pressure at the airfoil surface we obtain the following formula:

$$p_1 = -\frac{\theta_0}{\pi} \ln |x| - d + o(1) \quad \text{as } |x| \to 0. \tag{3.2.3}$$

Hence, the pressure gradient is expressed by

$$\frac{dp_1}{dx} = -\frac{\theta_0}{\pi} \frac{1}{x} + O(1) \quad \text{as } x \to -0. \tag{3.2.4}$$

On comparing (3.2.4) with formula (2.3.5) of Chapter 2, we conclude that the boundary layer in the vicinity of a wedge-shaped trailing edge of a thin airfoil is in the same state as the boundary layer ahead of a corner in a body contour. Hence, the solution of the equations for a boundary layer with a logarithmic pressure distribution at its outer edge, which was discussed in Section 3 of the preceding chapter, remains valid for this case as well. The solution, moreover, was first obtained by

Riley and Stewartson (1969) for precisely this flow in the vicinity of the trailing edge of a thin airfoil.

Thus, in the boundary layer the stream function has the following expansion (see equation (2.3.6) in Chapter 2):

$$\psi = \text{Re}^{-1/2}\Psi_0(x, Y) + \epsilon\text{Re}^{-1/2}\Psi_1(x, Y) + \cdots . \qquad (3.2.5)$$

Since the airfoil is thin, the velocity at the outer edge of the boundary layer differs only slightly from the velocity of the oncoming flow. Therefore, the leading term of the expansion (3.2.5) coincides with the Blasius solution. As for the second term, it represents the effect of the profile thickness on the flow in the boundary layer. In the viscous wall region 3 (Figure 3.6), the functions $\Psi_0(x, Y)$ and $\Psi_1(x, Y)$ are determined by the relations (2.3.12), (2.3.13) of Chapter 2. According to these relations, as $x \to -0$,

$$\Psi_0 = \frac{1}{2}(-x)^{2/3}a_0\eta^2 + \cdots , \quad \Psi_1 = f_1(\eta) + \cdots . \qquad (3.2.6)$$

Here, a_0 is a constant whose value is found from the Blasius solution: $a_0 = 0.3321$; the independent variable η, which has a value of order one in region 3, must be written taking into account the rotation of the coordinate system used here in comparison with that used in Section 3 of the preceding chapter:

$$\eta = \text{Re}^{1/2}\frac{y + x \tan\theta}{(-x)^{1/3}} .$$

The extent of the interaction region that appears in the vicinity of the trailing edge of a thin airfoil may be determined either with the help of the analysis of perturbations introduced into the flow by the viscous wake, as was done in the preceding section, or by considering the displacement effect of the wall boundary layer at the profile surface – this approach was described in Chapter 2. Both methods yield the same result:

$$\Delta x = O(\text{Re}^{-3/8}) .$$

If $\epsilon(\text{Re})$ does not exceed $\text{Re}^{-1/4}$ in order of magnitude, then everywhere beyond the interaction region the second term of the expansion (3.2.5) remains negligible in comparison with the first term. If in addition $\epsilon(\text{Re})$ is small in comparison with $\text{Re}^{-1/4}$, then the effect of the profile thickness is also insignificant within the interaction region. Therefore, in the following we shall take

$$\epsilon(\text{Re}) = \text{Re}^{-1/4} .$$

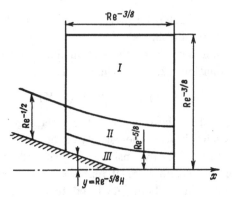

Fig. 3.7 Asymptotic structure of the flow in the interaction region.

In order to reformulate the interaction problem as applied to this case, it is necessary to revise the flow in the viscous region III (Figure 3.7). But the main part of the boundary layer (region II) and the external potential flow (region I) possess the usual features, which do not change in passing from the flow past a flat plate to the flow in the vicinity of the trailing edge of a thin airfoil.

The solution in region III is represented in the form

$$\psi = \text{Re}^{-3/4}\Psi_0^*(x^*, Y^*) + \cdots,$$

$$p = \text{Re}^{-1/4}\ln \text{Re}\frac{3\theta_0}{8\pi} + \text{Re}^{-1/4}P^*(x^*) + \cdots,$$

$$x^* = \text{Re}^{3/8}x, \quad Y^* = \text{Re}^{5/8}y.$$

Then for the function $\Psi_0^*(x^*, Y^*)$, we have the Prandtl equation

$$\frac{\partial \Psi_0^*}{\partial Y^*}\frac{\partial^2 \Psi_0^*}{\partial x^*\partial Y^*} - \frac{\partial \Psi_0^*}{\partial x^*}\frac{\partial^2 \Psi_0^*}{\partial Y^{*2}} = -\frac{dP^*}{dx^*} + \frac{\partial^3 \Psi_0^*}{\partial Y^{*3}}.$$

The boundary condition at the airfoil surface is the no-slip condition

$$\Psi_0^* = \frac{\partial \Psi_0^*}{\partial Y^*} = 0 \quad \text{at } Y^* = H(x^*), \quad x^* < 0. \qquad (3.2.7)$$

Here, $H(x^*)$ is the function characterizing the form of the airfoil trailing edge; in the present case of a wedge-shaped edge

$$H(x^*) = -\theta_0 x^*, \quad x^* < 0.$$

For the flow downstream of the trailing edge, the no-slip condition (3.2.7) must be replaced by conditions of flow symmetry

$$\Psi_0^* = \frac{\partial^2 \Psi_0^*}{\partial Y^{*2}} = 0 \quad \text{at } Y^* = 0, \quad x^* > 0.$$

It is also necessary to write out the condition for matching with the solution (3.2.5), (3.2.6) in region 3:

$$\Psi_0^* = \tfrac{1}{2}a_0(-x^*)^{2/3}\eta^2 + f_1(\eta) + \cdots \quad \text{as } x^* \to -\infty,$$

where

$$\eta = \frac{Y^* + \theta_0 x^*}{(-x^*)^{1/3}},$$

and the condition of matching with the solution (3.1.27), (3.1.28) in the main part of the boundary layer (region II, Figure 3.7):

$$\Psi_0^* = \tfrac{1}{2}a_0 Y^{*2} + a_0 A(x^*)Y^* + \cdots \quad \text{as } Y^* \to \infty.$$

The interaction problem is closed by the expression for the pressure gradient (see (2.3.24) in Chapter 2)

$$\frac{dP^*}{dx^*} = -\frac{1}{\pi}\int_{-\infty}^{\infty}\frac{A''(s)}{s - x^*}\,ds.$$

The substitution of variables

$$\Psi_0^* = a_0^{-1/2}\bar{\Psi}, \quad P^* = a_0^{1/2}\bar{P} + \frac{5\theta_0}{4\pi}\ln a_0 - d, \quad x^* = a_0^{-5/4}\bar{X},$$

$$Y^* = a_0^{-3/4}\bar{Y} + H, \quad A = a_0^{-3/4}\bar{A} - H, \quad H = a_0^{-3/4}\bar{H},$$

$$(3.2.8)$$

which include both the affine transformation of the interaction theory (see (1.7.1), Chapter 1) and the transposition theorem of Prandtl (1938), allows reduction of the interaction problem to a canonical form:

$$\frac{\partial\bar{\Psi}}{\partial\bar{Y}}\frac{\partial^2\bar{\Psi}}{\partial\bar{X}\partial\bar{Y}} - \frac{\partial\bar{\Psi}}{\partial\bar{X}}\frac{\partial^2\bar{\Psi}}{\partial\bar{Y}^2} = -\frac{d\bar{P}}{d\bar{X}} + \frac{\partial^3\bar{\Psi}}{\partial\bar{Y}^3},$$

$$\frac{d\bar{P}}{d\bar{X}} = \frac{1}{\pi}\int_{-\infty}^{\infty}\frac{\bar{H}''(s) - \bar{A}''(s)}{s - \bar{X}}\,ds,$$

$$\bar{\Psi} = \frac{\partial\bar{\Psi}}{\partial\bar{Y}} = 0 \quad \text{at } \bar{Y} = 0, \quad \bar{X} < 0, \qquad (3.2.9)$$

$$\bar{\Psi} = \frac{\partial^2\bar{\Psi}}{\partial\bar{Y}^2} = 0 \quad \text{at } \bar{Y} = 0, \quad \bar{X} > 0,$$

$$\bar{\Psi} = \tfrac{1}{2}\bar{Y}^2 + \bar{A}(\bar{X})\bar{Y} + \cdots \quad \text{as } \bar{Y} \to \infty,$$

$$\bar{\Psi} = \tfrac{1}{2}\bar{Y}^2 + \cdots \quad \text{as } \bar{X} \to -\infty.$$

Here,

$$\bar{H}(\bar{X}) = \begin{cases} -\alpha\bar{X} & \text{for } \bar{X} < 0, \\ 0 & \text{for } \bar{X} > 0. \end{cases}$$

The constant $\alpha = \theta_0 a_0^{-1/2}$ represents the similarity parameter for flows of the class considered.

The analysis of the boundary layer ahead of the interaction region, as already mentioned, was carried out by Riley and Stewartson (1969), and the formulation of the interaction problem (3.2.9) was first adduced in Stewartson's work (1970a). Before we come to the numerical solution of the problem (3.2.9), it is necessary to determine the asymptotic behavior of the function $\bar{\Psi}(\bar{X}, \bar{Y})$ at large values of the arguments. Such an analysis is carried out at any time when it becomes necessary to integrate the interaction problem. However, for brevity we shall illustrate it only for the one example of the flow near the trailing edge of a thin airfoil.

According to the condition of matching with the solution (3.2.3) outside the interaction region[3]

$$\bar{P} = -\frac{\alpha}{\pi} \ln |\bar{X}| + o(1) \quad \text{for } |\bar{X}| \to \infty. \tag{3.2.10}$$

As $\bar{X} \to -\infty$, the asymptotic expansion of the function $\bar{\Psi}(\bar{X}, \bar{Y})$ is represented in the form

$$\bar{\Psi} = (-\bar{X})^{2/3} \frac{1}{2} \bar{\eta}^2 + \alpha \bar{f}_1(\bar{\eta}) + \cdots, \quad \bar{\eta} = \frac{\bar{Y}}{(-\bar{X})^{1/3}}. \tag{3.2.11}$$

Then, substituting (3.2.10) and (3.2.11) into (3.2.9), we obtain the following equation for the function $\bar{f}_1(\bar{\eta})$:

$$\bar{f}_1''' - \frac{1}{3} \bar{\eta}^2 \bar{f}_1'' = \frac{1}{\pi}. \tag{3.2.12}$$

The solution satisfying the no-slip condition

$$\bar{f}_1(0) = \bar{f}_1'(0) = 0 \tag{3.2.13}$$

is unique under the additional condition of absence of exponentially growing terms in the expansion of the function $\bar{f}_1(\bar{\eta})$ as $\bar{\eta} \to \infty$. If this condition is met, then

$$\bar{f}_1 = \bar{A}_1 \bar{\eta} + \frac{3}{\pi} \ln \bar{\eta} + \bar{B}_1 + o(1) \quad \text{as } \bar{\eta} \to \infty, \tag{3.2.14}$$

where $\bar{A}_1$, $\bar{B}_1$ are constants whose values are found in the process of solving equation (3.2.12).

Let us substitute (3.2.14) into (3.2.11) and make the replacement

---

[3] The additive constant in the pressure transformation (3.2.8) is chosen so that the remainder term in (3.2.10) is small in comparison with one.

$\bar{\eta} = \bar{Y}(-\bar{X})^{-1/3}$. As a result we obtain the expression

$$\bar{\Psi} = \frac{1}{2}\bar{Y}^2 + \frac{\alpha\bar{A}_1}{(-\bar{X})^{1/3}}\bar{Y} + \frac{3\alpha}{\pi}\ln\bar{Y} + \alpha\bar{B}_1 - \frac{\alpha}{\pi}\ln|\bar{X}| + \cdots, \quad (3.2.15)$$

from which it follows that the asymptotic expansion of the function $\bar{\Psi}(\bar{X},\bar{Y})$ as $\bar{Y} \to \infty$ has the form

$$\bar{\Psi} = \frac{1}{2}\bar{Y}^2 + \bar{A}(\bar{X})\bar{Y} + \frac{3\alpha}{\pi}\ln\bar{Y} + \bar{B}(\bar{X}) + \cdots. \quad (3.2.16)$$

Substituting (3.2.16) into equation (3.2.9) for the function $\bar{\Psi}(\bar{X},\bar{Y})$ and making use of the expression (3.2.15), which determines the asymptotic behavior of the functions $\bar{A}(\bar{X})$ and $\bar{B}(\bar{X})$ as $\bar{X} \to -\infty$, we conclude that

$$\frac{1}{2}\bar{A}^2 - \bar{B} + \bar{P} = -\alpha\bar{B}_1. \quad (3.2.17)$$

Finally, let $\bar{X} \to \infty$; then we have the following expansion for $\bar{\Psi}(\bar{X},\bar{Y})$:

$$\bar{\Psi} = \bar{X}^{2/3}\bar{g}_0(\bar{\xi}) + \alpha\bar{g}_1(\bar{\xi}) + \cdots, \quad \bar{\xi} = \frac{\bar{Y}}{\bar{X}^{1/3}}. \quad (3.2.18)$$

The first term represents Goldstein's solution (1930) for the wake, considered in detail in the preceding section. The function $\bar{g}_0(\bar{\xi})$ appearing in the expansion (3.2.18) is related to the function $g_0(\xi)$, which enters the expansion (3.1.10), through the affine transformation

$$\bar{g}_0 = a_0^{1/3}\bar{g}_0, \quad \xi = a_0^{-1/3}\bar{\xi}.$$

Applying this transformation to the relation (3.1.13) we obtain the asymptotic expansion of the function $\bar{g}_0(\bar{\xi})$ as $\bar{\xi} \to \infty$:

$$\bar{g}_0 = \frac{1}{2}\bar{\xi}^2 + \widehat{A}_0\bar{\xi} + \frac{1}{2}\widehat{A}_0^2 + o(1), \quad (3.2.19)$$

where $\widehat{A}_0 = a_0^{1/3}A_0$.

In order to determine the second term of the expansion (3.2.18), it is necessary to substitute (3.2.18) into (3.2.9). Then we obtain the following equation for the function $\bar{g}_1(\bar{\xi})$:

$$\bar{g}_1''' + \frac{2}{3}\bar{g}_0\bar{g}_1'' = -\frac{1}{\pi}. \quad (3.2.20)$$

A specific property of equation (3.2.12) that allows unique determination of the solution with the two boundary conditions (3.2.13) is related to the fact that one of the linearly independent solutions of the homogeneous part of equation (3.2.12) grows exponentially as $\bar{\eta} \to \infty$.

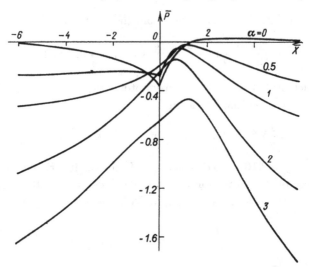

Fig. 3.8 Pressure distribution in the interaction region.

Equation (3.2.20) does not have such a solution and therefore it requires three boundary conditions. Two of them follow from flow symmetry:

$$\bar{g}_1(0) = \bar{g}_1''(0) = 0.$$

To obtain the third condition it is necessary to use the relation (3.2.17). We note that the asymptotic expansion of the function $\bar{g}_1(\bar{\xi})$ as $\bar{\xi} \to \infty$ has the form

$$\bar{g}_1 = \widehat{A}_1 \bar{\xi} + \frac{3}{\pi}\ln \bar{\xi} + \widehat{B}_1 + o(1), \qquad (3.2.21)$$

where $\widehat{A}_1$ and $\widehat{B}_1$ are unknown constants. Substituting (3.2.21) together with (3.2.19) into the expansion (3.2.18), and thereby determining the asymptotic expansion of the functions $\bar{A}(\bar{X})$ and $\bar{B}(\bar{X})$ as $\bar{X} \to \infty$, we find from (3.2.17) that

$$\widehat{B}_1 - \widehat{A}_0\widehat{A}_1 = \bar{B}_1. \qquad (3.2.22)$$

Equation (3.2.22), which relates the constants $\widehat{A}_1$ and $\widehat{B}_1$, is just the required boundary condition that completes the problem for equation (3.2.20).

A numerical integration of the interaction problem (3.2.9) was performed by Ruban (1977). For $\alpha = 0$, when the thin airfoil reduces to a flat plate, the results agree well with those of Jobe and Burggraf (1974) shown in Figures 3.3 and 3.4. At the surface of the plate ahead of the

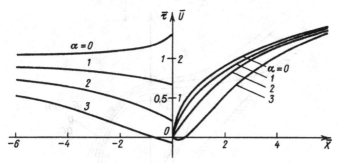

Fig. 3.9 Skin-friction distribution ($\bar{X} < 0$) and velocity on the wake axis of symmetry ($\bar{X} > 0$), in the interaction region.

trailing edge a flow acceleration is observed, which is accompanied by a decrease in pressure (Figure 3.8) and an increase in skin friction (Figure 3.9, $\bar{X} < 0$). However, at $\alpha = 0.5$ this acceleration is already offset by the effect of flow deceleration brought about through the influence of the airfoil thickness, so that the pressure ahead of the trailing edge is almost constant. A further increase in the parameter α leads to the result that the pressure gradient along the airfoil surface becomes adverse; the skin friction decreases as the trailing edge is approached, where $\bar{\tau}(\bar{X})$ reaches its minimum. The pressure rise continues also in the flow downstream from the trailing edge, and if $\bar{\tau}(-0) > 0$, then the function $\bar{P}(\bar{X})$ as $\bar{X} \to +0$ has the same kind of singularity as in the case of flow past the trailing edge of a flat plate.[4] The velocity $\bar{U}(\bar{X})$ also has singular behavior on the wake axis of symmetry (Figure 3.9, $\bar{X} > 0$).

The negative values of the skin friction $\bar{\tau}(\bar{X})$, as well as the negative values of the velocity $\bar{U}(\bar{X})$ in the wake, indicate the appearance of a region of reverse flow. A sketch of the streamlines in this region is shown in Figure 3.10 for $\alpha = 3.0$. Finally, shown in Figure 3.11 is the dependence of the skin friction $\bar{\tau}(-0)$ at the airfoil trailing edge on the parameter α. It follows from this graph that flow separation in the vicinity of the trailing edge of the thin airfoil occurs at $\alpha = 2.6$. Thus, the accelerating effect of the viscous wake results in a delay of separation near the profile trailing edge in comparison with separation in the vicinity of a corner on a body contour (Chapter 2, Section 3). Let us recall that in the latter case the region of recirculating flow appears when $\alpha = 2.0$.

[4] If $\bar{\tau}(-0)$ becomes negative, the character of the singularity change (Smith, 1983).

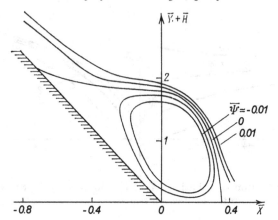

Fig. 3.10 Sketch of the streamlines near a wedge-shaped trailing edge of a profile with $\alpha = 3.0$.

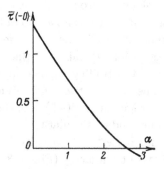

Fig. 3.11 Dependence of the skin friction at the airfoil trailing edge on the parameter α.

The elliptic profile. The theory of flow separation in the vicinity of the trailing edge of a thin airfoil has been illustrated in this section by the example of flow around a profile that has a wedge-shaped trailing edge. The theory, however, remains valid in other cases, too. In particular, Korolev (1980b) extended it to the flow near the trailing edge of a thin ellipse. As a result of his analysis, Korolev stated the following simple rule, which allows an estimate of the profile thickness at which flow separation occurs in the vicinity of the trailing edge: the thickness should be chosen such that at a distance $\Delta x = O(\mathrm{Re}^{-3/8})$ from the profile trailing edge its transverse scale reaches a value of order $\mathrm{Re}^{-5/8}$. Thus,

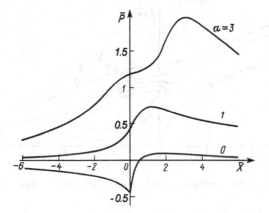

Fig. 3.12 Pressure distribution in the interaction region for an elliptic profile.

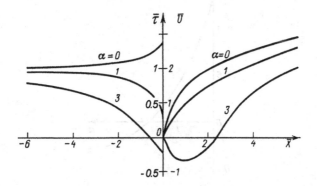

Fig. 3.13 Skin-friction distribution ($\bar{X} < 0$) and velocity on the wake axis of symmetry ($\bar{X} > 0$) near the trailing edge of an elliptic profile.

if we consider an elliptic profile

$$y = \pm 2\epsilon\sqrt{-x(1+x)}, \quad -1 \le x \le 0,$$

then according to this rule the parameter ϵ must have the form

$$\epsilon = \mathrm{Re}^{-7/16} h, \text{ where } h = O(1).$$

In the vicinity of the trailing edge of the profile, the upper surface is represented in the variables of region III (Figure 3.7) as

$$Y^* = H(x^*), \quad H(x^*) = 2h(-x^*)^{1/2}. \tag{3.2.23}$$

Applying the affine transformations (3.2.8) to equation (3.2.23), we ob-

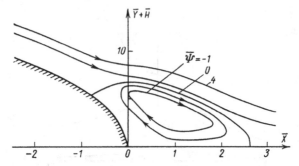

Fig. 3.14 Sketch of streamlines close to the elliptic trailing edge of a profile with $\alpha = 3.0$.

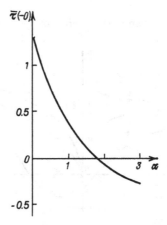

Fig. 3.15 Dependence of the skin friction at the trailing edge on the parameter α for an elliptic profile.

tain

$$\bar{H}(\bar{X}) = \begin{cases} 2\alpha(-\bar{X})^{1/2} & \text{for } \bar{X} < 0, \\ 0 & \text{for } \bar{X} > 0, \end{cases} \tag{3.2.24}$$

where $\alpha = ha_0^{1/8}$.

The solution of the interaction problem (3.2.9) with the function $\bar{H}(\bar{X})$ defined by the relation (3.2.24) has been constructed numerically (see Chapter 7). The plots in Figure 3.12 show the dependence of the pressure $\bar{P}$ on $\bar{X}$ for various values of the parameter α. The skin-friction distribution along the surface of the ellipse near its trailing edge and the velocity distribution along the wake axis of symmetry are shown in Figure 3.13. Streamlines in the region of recirculating flow calculated for $\alpha = 3$ are illustrated in Figure 3.14. Finally, from Figure 3.15 it follows that in this case flow separation first occurs at $\alpha = 1.8$. Using this result, for every elliptic profile with a given relative thickness 2ϵ, one can determine the Reynolds number starting at which the flow over the profile is accompanied by flow separation from its surface. This Reynolds number is calculated from the formula $\text{Re} = 5.25\epsilon^{-16/7}$.

3.3 The Influence of Angle of Attack

Another problem associated with fluid motion in the vicinity of an airfoil trailing edge concerns the investigation of the onset of separation resulting from angle of attack. The analysis was carried out by Brown and Stewartson (1970). This problem is of interest in particular because its solution yields a correction to the value of the lift force caused by the effect of viscosity; moreover, it makes clear the nature of the singularity that is always present in the solution of a potential-flow problem satisfying the Chaplygin (1910)–Zhukovskii (1910) (frequently written as Joukowski) condition of smooth flow leaving the trailing edge.

On the basis of the results described in the preceding section, one can assume that the appearance of separation in the vicinity of the trailing edge, that is, the process of change from attached to separated flow, takes place at small values of the relative profile thickness and angle of attack. This means that in the limit as $\text{Re} \to \infty$ the values tend to zero.[5]

In order to devote our attention exclusively to the effect of angle of attack, we assume, following the work of Brown and Stewartson (1970), that the profile thickness near the trailing edge can be neglected. (The condition under which such neglect in the leading terms in the asymptotic expansions is justified was described in the preceding section.) Furthermore, we suppose that the flow over the airfoil is unseparated, so that the dimensionless skin friction τ at the trailing edge remains finite. The classic example of a lifting profile that does not require creating special mechanisms, such as boundary-layer suction, for attached flow around the leading edge is a cambered airfoil placed at zero angle of attack. The point is that in this particular case, as has been known since the work of Kutta (1902), one can ensure attached flow around both the leading and trailing edges through the choice of the circulation. Separation will first occur on the upper surface near the trailing edge. This is

---

[5] Here and below the previous symbols are used and the airfoil chord is taken as the characteristic length L.

related to the fact that the pressure gradient here has a maximum and the skin friction has a minimum.

We denote by $\epsilon\alpha^*$ the angle that represents the slope of the trailing edge relative to the uniform oncoming flow. In accordance with what has been said above, we shall suppose that as Re $\to \infty$

$$\epsilon = \epsilon(\text{Re}) \to 0, \quad \alpha^* = O(1). \tag{3.3.1}$$

We assume that the Chaplygin–Zhukovskii condition is satisfied. Then, as is well known, the complex velocity near the trailing edge, where $x = y = 0$, has the following asymptotic representation:

$$u - iv = 1 + \epsilon[-2ic_1 z^{1/2} + O(z)] + o(\epsilon),$$
$$z = x + iy, \quad |z| \to 0. \tag{3.3.2}$$

We take the axis Ox of the rectangular Cartesian coordinate system to be in the direction of the tangent to the body contour; c_1 is an arbitrary constant whose value is found from the solution of the entire outer problem.[6]

Now let us consider the flow in the boundary layer and the wake. According to the original assumptions, the boundary layer remains attached to the surface everywhere, and the appearance of separation is possible only in some small neighborhood of the trailing edge. Consequently, as for the flows around symmetric thin bodies considered previously, and according to (3.3.1) and (3.3.2), the problem is reduced to consideration of small perturbations (of order ϵ) from the Blasius solution (3.1.3)–(3.1.5). These perturbations are of singular character. In fact, the expression (3.3.2) shows that the pressures on the upper and lower surfaces ($x < 0$) behave as follows:

$$p = \epsilon P_1(x, y) + o(\epsilon);$$
$$P_1(x, \pm 0) = p_\pm(x) = -2c_1(-x)^{1/2} \text{ sign } y + o[(-x)^{1/2}], \tag{3.3.3}$$
$$x \to -0; \quad c_1 > 0.$$

Here and below the signs $+$ and $-$ refer, respectively, to the upper ($y > 0$) and lower ($y < 0$) airfoil surfaces. The effect of the singular pressure gradient (3.3.3) must obviously lead to deceleration of the fluid along the upper surface and acceleration along the lower surface. (For simplicity, we ignore the surface curvature since it does not affect the solution for the leading terms of the asymptotic expansions given below.)

---

[6] E.g., for a circular-arc airfoil whose chord line is parallel to the uniform oncoming flow, the value $c_1 = 4h_0$, where ϵh_0 is the maximum camber.

Earlier, we have already considered problems of singular perturbations in the boundary layer. In fact, if we compare the expression (1.4.6) from Chapter 1 with (3.3.3), we see that they coincide (for $y > 0$). Hence, the solution here (for the leading terms of the expansions as $x \to -0$) will coincide with the two-layer solution that describes the flow ahead of the separation point (see Chapter 1, Section 4), if in (1.4.2) of Chapter 1 we set $p_0(x) = 0$ and for the functions $\Psi_0(x, Y)$ and $p_1(x)$ we understand the Blasius function (3.1.3)–(3.1.5) and $p_+(x)$.

On the lower airfoil surface, the solution corresponding to the function $\Psi_1(x, Y)$ from (1.4.2) of Chapter 1 is obtained from the solution for $y \geq 0$ by replacing Y with $-Y$; this follows from the fact that the expressions for the pressure gradients $p'_+(x)$ and $p'_-(x)$ differ in the leading terms of the expansions only in sign, and equation (1.4.5) of Chapter 1 is invariant with respect to the transformation

$$x \to x, \quad Y \to -Y, \quad \Psi_0 \to -\Psi_0, \quad \Psi_1 \to \Psi_1, \quad p_1 \to -p_1.$$

Along with the deceleration on the upper surface of the body caused by the angle of attack, in the vicinity of the trailing edge there is a strong acceleration of fluid elements at the "bottom" of the boundary layer resulting from the discontinuity in the boundary condition (see Section 1). As a consequence, a large pressure gradient $\partial p / \partial x = O(\mathrm{Re}^{1/8})$ is induced in the interaction region with streamwise scale $|x| = O(\mathrm{Re}^{-3/8})$. It is easy to determine the order of magnitude of the angle of attack $\epsilon(\mathrm{Re})$ at which these two effects will balance each other, and consequently for which the appearance of separation can be expected. For this, it is necessary that, when $|x| = O(\mathrm{Re}^{-3/8})$, the pressure gradient (3.3.3) have a value $O(\mathrm{Re}^{1/8})$, which is possible if

$$\epsilon = \mathrm{Re}^{-1/16}. \tag{3.3.4}$$

In the interaction region the triple-deck structure of Messiter (1970) and Stewartson (1969) will be preserved, but at the same time the effect of angle of attack will become essential in determining the leading terms in the asymptotic expansions. If we had taken $\epsilon > O(\mathrm{Re}^{-1/16})$, the value of the adverse pressure gradient (3.3.3) in the interaction region would have been found to be larger in order of magnitude than $\mathrm{Re}^{1/8}$, that is, larger than for separation with the formation of an extensive region of reverse flow (Sychev, 1972; see also Chapter 1). This means that for $\epsilon > O(\mathrm{Re}^{-1/16})$ attached flow over the trailing edge becomes impossible.

Thus, $\epsilon = \mathrm{Re}^{-1/16}$. Then, on the basis of the above, the solution for the viscous sublayer in the interaction region may be represented by the expansions (3.1.29) which, taking into account the transformation (3.1.42), assume the form

$$x = \mathrm{Re}^{-3/8} a_0^{-5/4} \bar{X}, \quad y = \mathrm{Re}^{-5/8} a_0^{-3/4} \bar{Y},$$

$$\Psi = \mathrm{Re}^{-1/4} a_0^{-1/2} \bar{\Psi}(\bar{X}, \bar{Y}) + \cdots, \quad p = \mathrm{Re}^{-1/4} a_0^{1/2} \bar{P}_{\pm}(\bar{X}) + \cdots.$$

The unknown functions satisfy the Prandtl boundary-layer equation

$$\frac{\partial \bar{\Psi}}{\partial \bar{Y}} \frac{\partial^2 \bar{\Psi}}{\partial \bar{X} \partial \bar{Y}} - \frac{\partial \bar{\Psi}}{\partial \bar{X}} \frac{\partial^2 \bar{\Psi}}{\partial \bar{Y}^2} + \frac{d\bar{P}_{\pm}}{d\bar{X}} = \frac{\partial^3 \bar{\Psi}}{\partial \bar{Y}^3}. \tag{3.3.5}$$

As usual, in the main part of the boundary layer developing along the upper surface of the body where $Y = O(1)$, the velocity profile determined by the Blasius solution is preserved; the same may be said about the flow along the lower surface. Therefore, from matching the solutions in these regions and the sublayer, it follows that

$$\frac{\partial \bar{\Psi}}{\partial \bar{Y}} \to |\bar{Y}|, \quad \bar{Y} \to \pm \infty. \tag{3.3.6}$$

From matching with the solution described above for the boundary layer ahead of the interaction region ($x \to -0$), we find that as $\bar{X} \to -\infty$

$$\frac{\partial \bar{\Psi}}{\partial \bar{Y}} \to |\bar{Y}|, \quad (-\bar{X})^{-1/2} \bar{P}_{\pm} \to -\alpha \operatorname{sign} \bar{Y}, \tag{3.3.7}$$

where

$$\alpha = 2c_1 a_0^{-9/8} \tag{3.3.8}$$

is a similarity parameter of the problem.

At the body surface the no-slip condition is satisfied:

$$\bar{\Psi} = \frac{\partial \bar{\Psi}}{\partial \bar{Y}} = 0 \text{ for } \bar{Y} = \pm 0, \quad \bar{X} < 0. \tag{3.3.9}$$

Because the pressure is constant across the boundary layer (in the approximation considered), it is necessary that for $\bar{X} \geq 0$

$$\bar{P}_{+}(\bar{X}) = \bar{P}_{-}(\bar{X}). \tag{3.3.10}$$

Finally, the last boundary condition is provided by the velocity profile as $\bar{X} \to \infty$. It has the form

$$\bar{\Psi} \to \bar{X}^{2/3} \bar{g}(\bar{\xi}), \quad \bar{\xi} = \left(\bar{Y} - \tfrac{2}{3} \alpha \bar{X}^{3/2} \right) \Big/ \bar{X}^{1/3}. \tag{3.3.11}$$

$(a_0^{1/3} \bar{g}(\bar{\xi}) = g_0(a_0^{-1/3} \bar{\xi}))$ is the solution of the problem (3.1.11)–(3.1.13)), and this means that downstream of the interaction region the solution for the sublayer approaches Goldstein's (1930) solution for the near wake

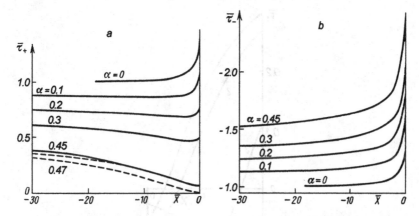

Fig. 3.16 Skin-friction distribution on the upper and lower body surfaces in the interaction region: solid line – Chow and Melnik (1976); dashed line – Korolev (1989).

(Section 1). This sublayer develops along the zero streamline, whose shape as $x \to +0$ is described according to (3.3.2) by the expression

$$y = y_S = \epsilon \left[\tfrac{4}{3} c_1 x^{3/2} + o(x^{3/2}) \right] + o(\epsilon).$$

To complete the system of relations obtained, it is necessary as usual to consider a region of external potential flow with transverse scale $|y| = O(\mathrm{Re}^{-3/8})$. Here, the solution has the form (3.1.36), (3.1.37) and then on the basis of (3.1.38) the analytic function

$$\bar{W}(\bar{Z}) = a_0^{-1/2} w_1^*(z^*), \quad \bar{Z} = a_0^{5/4} z^* \tag{3.3.12}$$

is such that on the real axis its real and imaginary parts assume the values $\bar{P}_\pm(\bar{X})$ and $\bar{G}_\pm(\bar{X})$, where $\bar{G}_\pm(\bar{X})$ is the displacement effect of the viscous sublayer:

$$\bar{G}_\pm(\bar{X}) = \lim_{\bar{Y} \to \pm\infty} \left(-\frac{\partial \bar{\Psi}}{\partial \bar{X}} \Big/ \frac{\partial \bar{\Psi}}{\partial \bar{Y}} \right). \tag{3.3.13}$$

Therefore,

$$\frac{d\bar{P}_\pm}{d\bar{X}} = \pm \frac{1}{\pi} \int_{-\infty}^{\infty} \frac{\bar{G}'_\pm(t)\, dt}{t - \bar{X}}. \tag{3.3.14}$$

The relations (3.3.13), (3.3.14) complete the boundary-value problem (3.3.5)–(3.3.11) for the region considered (Brown and Stewartson, 1970). Numerical solutions of the problem were obtained by Chow and Melnik (1976) and also by Korolev (1989, 1991b). Shown in Figures 3.16a,

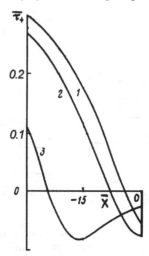

Fig. 3.17 Skin-friction distribution on the upper surface for $\alpha > \alpha_S$ (Korolev, 1989).

b is the skin-friction distribution $\bar{\tau}_{\pm}(\bar{X}) = \bar{\Psi}_{\bar{Y}\bar{Y}}|_{\bar{Y}=\pm 0}$ on the upper (a) and lower (b) body surfaces at different values of the similarity parameter α. It can be seen that as α is increased, corresponding to an increase in angle of attack, the value of the skin friction on the lower surface of the body increases and assumes a finite value at $\bar{X} = -0$. On the upper surface one can observe the appearance of a minimum in the distribution of $\bar{\tau}_+(\bar{X})$, which is caused by the accelerating effect of the wake. At the same time, as α increases, the value of this minimum tends to zero and its position is shifted toward the trailing edge. For $\alpha = \alpha_S = 0.47$ the value is $\bar{\tau}_+(-0) = 0$. In the work of Brown and Stewartson (1970), it is shown that for the existence of the solution of the boundary-value problem considered it is necessary to satisfy the condition $\dfrac{\bar{\tau}_+(-0)}{\bar{\tau}_-(-0)} \leq 0$, which excludes the possibility of fluid moving in the opposite direction at the section $\bar{X} = 0$ near the body surface. For $0 \leq \alpha \leq \alpha_S$, the condition is satisfied. The question about the existence of the solution when $\alpha > \alpha_S$ remained open until the appearance of the work of Korolev (1989), although Smith (1983) showed that the solution exists for values of α slightly larger than α_S.

Korolev's calculations showed that the solution does indeed exist for $\alpha > \alpha_S$ and that a region of reverse flow appears near the trailing edge. Shown in Figure 3.17 are plots of the function $\bar{\tau}_+(\bar{X})$ for the values

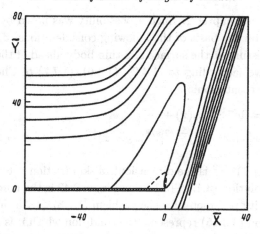

Fig. 3.18 Sketch of streamlines for $\alpha = 0.492$. Dotted line designates the first branch of the solution and solid lines the second branch (Korolev, 1989).

$\alpha = 0.48$ and $\alpha = 0.492$. (They are designated by the numerals 1 and 2, respectively.) With increasing α, the position of the separation point is quickly displaced upstream, and the size of the reverse-flow region increases. Flow reattachment occurs at some small distance ($\bar{X} = \bar{X}_R \approx -0.1$) from the trailing edge, and thus $\bar{\tau}_+(-0) > 0$.

It is most interesting, first, that a second branch of the solution is established, in agreement with the prediction of Stewartson (1980) and, second, that a solution does not exist for $\alpha > \alpha_S^* = 0.497$.

In passing to the second branch with a decrease in α, the solution also describes a motion with a region of reverse flow, but the latter is found to have a larger size in comparison with that corresponding to the solution for the first branch. This can be seen clearly in Figure 3.17, where the numeral 3 designates the second solution when $\alpha = 0.492$. Sketches of the streamlines for both branches of the solution for this value of α are shown in Figure 3.18. (Adjacent streamlines are plotted for values of $\bar{\Psi}$ at intervals equal to 125.)

This growth of the reverse-flow region makes it difficult to obtain solutions, and therefore the smallest value of α (if, of course, one moves along the second branch of the solution) at which it was possible to obtain reliable results was equal to 0.492. It is possible that, for a further decrease in α, the separation point will move upstream to infinity in $\bar{X}$, and we would thus arrive at a flow with an extensive separation region, as described in Chapter 1. Such a possibility was noted by Stewartson (1980, 1982) on the basis of the following considerations.

For the pressure at the surface of a thin body ahead of the separation point, we have, according to (1.4.2), (1.4.6), (1.7.2) of Chapter 1, the following representation:

$$p(x, 0) = \text{Re}^{-1/16} p_e(x) + \cdots ;$$

$$p_e = -\alpha_0 [a_0(x_S + 1)^{-1/2}]^{9/8}(x_S - x)^{1/2} + \cdots, \quad x \to x_S - 0,$$
$$\tag{3.3.15}$$

where $a_0(x_S + 1)^{-1/2}$ is the dimensionless skin friction determined from the Blasius solution (3.1.3)–(3.1.5) and $\alpha_0 = 0.42$ is the numerical value of the similarity parameter for the problem (1.7.3)–(1.7.5) in Chapter 1. The expression (3.3.15) represents the condition which it is necessary to impose on the pressure distribution when solving the outer problem for flow around the body at an angle of attack (or with relative thickness) of order $\text{Re}^{-1/16}$ to determine the coordinate x_S at the separation point of the free streamline.[7]

On the other hand, when approaching the trailing edge ($x \to -0$), according to (3.3.3), (3.3.4), and (3.3.8), we obtain

$$p(x, +0) = \text{Re}^{-1/16} p_+(x) + \cdots ;$$
$$p_+ = -\alpha a_0^{9/8}(-x)^{1/2} + \cdots, \quad x \to -0.$$
$$\tag{3.3.16}$$

Comparing the expressions (3.3.15) and (3.3.16), we can see that at angles of attack corresponding to values of the similarity parameter $\alpha > 0.42$ a flow with an extensive separation region can be produced. Therefore, within the range of angles of attack corresponding to values $\alpha_0 \leq \alpha \leq \alpha_S^*$, along with the solution described above, there can exist a second solution (Stewartson, 1980).

Thus, the value of the similarity parameter α_S^* determines the range of changes in the angle of attack for which the global flow can remain unseparated and the solution in the outer region satisfies the Chaplygin–Zhukovskii condition.

Now let us determine the changes in the lift force acting on the airfoil which are caused by the interaction. For this purpose we consider the asymptotic behavior of the function $\bar{W}(\bar{Z})$ as $|\bar{Z}| \to \infty$. From matching with the solution (3.3.3), it follows that

$$\bar{W} = i\alpha \bar{Z}^{1/2} + i\alpha\gamma_0 \bar{Z}^{-1/2} + O(\bar{Z}^{-2/3}), \quad |\bar{Z}| \to \infty. \tag{3.3.17}$$

---

[7] Further details concerning the condition (3.3.15) are given in Section 7 of Chapter 6.

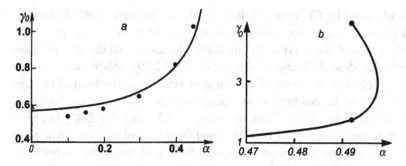

Fig. 3.19 Dependence of the parameter γ_0 on α. Filled circles – data of Chow and Melnik (1976); solid line – data of Korolev (1989).

The form of the second term and the constant $\gamma_0(\alpha)$ are found with the help of a formula of Keldysh and Sedov (1937); as a result one can obtain

$$\gamma_0(\alpha) = \frac{1}{2\pi\alpha} \int_{-\infty}^{0} \left[\frac{G_+(t) + G_-(t)}{(-t)^{1/2}} \right] dt$$

(Chow and Melnik, 1976). The third term in (3.3.17) follows from the displacement effect of the wake (see Section 1).

Turning to the expansion (3.3.17), and taking into account (3.3.12), (3.3.8), (3.3.4) and (3.1.37), (3.1.38) in the outer variables (3.1.36), we obtain

$$u - iv = 1 + \epsilon(-i2c_1 z^{1/2} + \cdots) + \cdots$$
$$+ \epsilon^7 [-i\alpha\gamma_0 a_0^{-1/8} z^{-1/2} + \cdots] + \cdots, \qquad (3.3.18)$$
$$|z| \to 0; \quad \epsilon = \mathrm{Re}^{-1/16}.$$

The coefficient of $z^{-1/2}$ in (3.3.18) characterizes the value of the change in circulation in comparison with its value found from the Chaplygin–Zhukovskii condition. Therefore, the ratio of the lift coefficient c_y to its value c_{y0} found from solving the flow problem without accounting for the boundary layer has the form

$$\frac{c_y}{c_{y0}} - 1 = b_1 \mathrm{Re}^{-3/8} + b_2 \mathrm{Re}^{-1/2}\ln \mathrm{Re} + O(\mathrm{Re}^{-1/2}),$$
$$(3.3.19)$$
$$b_1 = -n_0\gamma_0(\alpha)a_0^{-5/4},$$

where n_0 is some positive constant whose value is determined from the solution to the outer problem; for example, for a flat plate $n_0 = 2$ and for a circular-arc airfoil $n_0 = 8$. The dependence of the value of γ_0 on

α obtained by Chow and Melnik (1976) and Korolev (1989) is shown in Figures 3.19a,b. For values of α close to the critical value ($\alpha = \alpha_S^*$), the shape of the curve in Figure 3.19b illustrates well the double-valued solution described above. According to (3.3.19), this implies that the lift coefficient in terms of the angle of attack has the form of the hysteresis loop known from experimental studies (see Kuryanov, Stolyarov, and Steinberg, 1979). Thus, in reality the lift force $R = \frac{1}{2}\rho U_\infty^2 L c_y$ turns out to be smaller than that obtained from the solution of a flow problem within the framework of an ideal fluid, and, moreover, according to equation (3.3.19), this decrease results mainly from the interaction process near the trailing edge rather than from the entire boundary layer, which contributes a value of order $\mathrm{Re}^{-1/2}$. As for the second term on the right-hand side of the expression (3.3.19), it depends on the curvature of the streamlines in the near wake; without going into detail about the way it was found (see the work of Brown and Stewartson, 1975), we mention only that the constant b_2 is determined through the displacement thickness and momentum thickness of the Blasius solution.

In conclusion, we note that surveys of the results of other investigators concerning fluid motion in the vicinity of a trailing edge, and devoted also to the effect of unsteadiness, can be found in the work of Stewartson (1981, 1982), Smith (1982a), and Crighton (1985).

4

Separation at the Leading Edge of a Thin Airfoil

4.1 Experimental Observations

Boundary-layer separation at the leading edge of a thin airfoil is the principal factor that limits the lift force acting on an airfoil in a fluid stream. Jones (1934) was the first to describe this kind of separation. Since that time, many researchers have turned to the experimental study of flow around the leading edge of an airfoil. In addition to a great number of original studies, several surveys have been devoted to this theme. The reviews by Tani (1964) and Ward (1963) can be regarded as the most complete.

Experiments show that as the angle of attack increases the picture of the flow around the airfoil changes in the following way. When the angle of attack is small, the flow over the profile is attached. Then the pressure has its maximum at the stagnation point O of the flow, where the zero streamline divides into two – one branch lies along the lower surface of the airfoil, and the second one bends around the leading edge of the airfoil and then lies along its upper surface. As we move from the stagnation point along the upper branch, the pressure first falls rapidly, reaching a minimum at some point M (Figure 4.1a), and then starts to increase, so that the boundary layer downstream from point M finds itself under the influence of an adverse pressure gradient. Its magnitude increases with growth of the angle of attack, finally resulting in boundary-layer separation, which occurs earlier for smaller relative airfoil thickness.

When the boundary layer separates, one can observe the appearance of a closed region of recirculating flow on the upper surface of the airfoil (Figure 4.1a). This region is commonly referred to as a "short bubble." The length of the short bubble normally does not exceed 1% of the airfoil chord, and it has an extremely weak effect upon the flow and hence upon

135

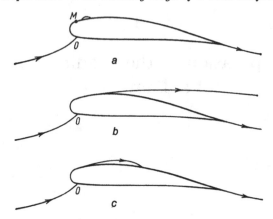

Fig. 4.1 The flow around an airfoil: (a) with a short bubble; (b) with an extended separation zone; (c) with a long bubble.

the values of the aerodynamic forces acting upon the airfoil. However, the short bubble exists only within a small range of variation of angle of attack, and then it suddenly bursts. As a result, a new flow regime is formed: this may be either flow with an extended separation zone that covers the entire upper surface of the airfoil (Figure 4.1b) or flow with a so-called "long bubble" (Figure 4.1c). In either case the change is accompanied by an abrupt decrease in the lift acting upon the airfoil and an increase in the drag.

These conditions are normal for an airfoil when the boundary layer at the leading edge remains laminar. The fact is that it is not the airfoil chord that plays the role of the characteristic geometric dimension, but rather a considerably smaller quantity – the radius of the leading edge. If we base the Reynolds number on the airfoil nose radius, we find that for most experimental data it attains a value of order 10^4–10^5 (Ely and Herring, 1978); that is, it is not high enough for the boundary layer along the surface to become turbulent.

A detached boundary layer is less stable. Therefore, transition is observed in the separation region, and the flow rejoins the profile surface being already turbulent. Transition significantly affects the behavior of the short bubble. If one induces transition by increasing the Reynolds number, the range of angle of attack where the short bubble exists expands. Then the bursting of the short bubble leads, as a rule, to the formation of an extended separation zone, shown in Figure 4.1b. Conversely, with a decrease in the Reynolds number, the short bubble bursts earlier and in its place on the upper surface of the airfoil there appears a long bubble (Figure 4.1c). Although at the moment of its appearance the long bubble may differ only slightly in size from the short bubble, it obeys other rules. The most important among them is that with increase in angle of attack, the long bubble rapidly increases in length due to displacement of the reattachment point in the direction of the airfoil trailing edge. If we recall that the Reynolds number is to be based on the nose radius of the airfoil, then it becomes clear why the second of the situations described is characteristic of thinner airfoils.

The present chapter gives a theoretical analysis of the initial stages of the phenomenon described for the short bubble. Despite the common view that transition in the separation region is a necessary condition for flow reattachment to the airfoil surface, and therefore for the very existence of a short bubble, we shall proceed from the assumption of laminar flow near the airfoil leading edge, and show that the theory constructed in this way not only describes the process of generation of a short bubble when the boundary layer on the airfoil surface is definitely laminar, but also allows us to follow its evolution right up to the moment of bursting.

4.2 Statement of the Problem. Inviscid-Flow Region

Let an airfoil placed in a two-dimensional stream of incompressible fluid have a relative thickness ϵ, so that the equation of the surface can be written in the form

$$y' = \epsilon F_\pm(x'), \quad 0 \le x' \le 1,$$

where the plus sign refers to the upper surface of the profile and the minus sign to the lower. A rectangular Cartesian coordinate system $O'x'y'$ is introduced such that its origin O' coincides with the leading edge and the axis $O'x'$ is directed along the tangent to the mean line of the profile (Figure 4.2). Henceforth, we use dimensionless variables: distances along the axes $O'x'$ and $O'y'$ are referred to the airfoil chord L, components of the velocity vector are referred to the free-stream speed U_∞, and the increment in pressure, relative to its value at infinity, is referred to twice the dynamic pressure ρU_∞^2.

Let us assume for simplicity that the nose of the airfoil has a parabolic shape. In this case the parameter ϵ can be chosen so that

$$F_\pm = \pm\sqrt{2x'} + \cdots \quad \text{as } x' \to 0.$$

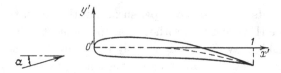

Fig. 4.2 Thin airfoil in a fluid flow.

Then the radius of curvature of the leading edge is $r = L\epsilon^2$.

We shall construct a theory of the fluid flow around the profile on the basis of an asymptotic analysis of the Navier–Stokes equations, using a limit process where the Reynolds number $\mathrm{Re} = U_\infty r/\nu$, based on the radius of the leading edge of the profile, tends to infinity and the relative thickness ϵ tends to zero.[1]

Following the method of matched asymptotic expansions, we consider first the basic inviscid-flow region with characteristic dimensions of the order of the airfoil chord: $x' = O(1)$, $y' = O(1)$. If the angle of attack α' has a value of order ϵ, that is,

$$\alpha' = \epsilon\alpha_*, \quad \text{where } \alpha_* = O(1),$$

then for this region one can apply the classical thin-airfoil theory.[2] According to this theory the pressure gradient on the airfoil surface is of order ϵ; it cannot produce separation of the boundary layer.

At the same time we know that thin-airfoil theory becomes invalid in the vicinity of the leading edge; one should analyze separately the region where the variables $X' = \epsilon^{-2}x'$ and $Y' = \epsilon^{-2}y'$ are of order unity and the airfoil contour is represented by the infinite parabola $Y' = \pm\sqrt{2X'}$ (Figure 4.3). In this region, the perturbations in the flow variables are no longer small. In particular, the tangential component of the velocity vector on the airfoil surface (referred to the free-stream flow speed) is expressed, as demonstrated by Van Dyke (1956), in the form

$$U_e = \frac{Y' + k}{\sqrt{(Y')^2 + 1}}. \tag{4.2.1}$$

Here, Y' represents the distance from the axis of the parabola to the surface at a point where the value of the speed is to be calculated.

---

[1] The second of these restrictions is used henceforth only in determining the numerical values of certain constants arising in the solution of the problem. In other respects, the theory remains quite universal and is valid not only in the case of flow around an airfoil of finite thickness, but also in a number of other cases. Some of them will be considered at the end of the chapter.

[2] See, for instance, Sedov's (1966) book.

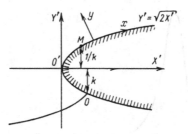

Fig. 4.3 Flow near the leading edge of a thin airfoil.

The parameter k characterizing the degree of flow asymmetry near the parabola is determined from the condition of matching with the solution for the outer flow region, and is expressed in terms of the angle of attack and the slope of the mean line of the profile as follows:

$$k = \sqrt{2}\left(\alpha_* + \frac{1}{\pi}\int_0^1 \frac{G(x')}{\sqrt{x'(1-x')}}dx'\right), \quad G = -\frac{1}{2}\left(\frac{dF_+}{dx'} + \frac{dF_-}{dx'}\right).$$

(4.2.2)

The stagnation point is determined by $Y' = -k$; for $k = 0$ it coincides with the vertex of the parabola, and for $k > 0$ it moves to the lower surface.

4.3 The Boundary Layer

In examining the boundary layer on the parabolic surface of the airfoil in the vicinity of its leading edge, it is convenient to use the orthogonal curvilinear coordinate system Oxy, where x is measured from the stagnation point O along the upper surface of the profile and y along the normal to it (Figure 4.3). We take the radius r of the leading edge of the profile as the unit of length. The components of the velocity vector in the chosen coordinate system, referred to U_∞, will be denoted by u and v, respectively. We also introduce the stream function ψ, referred to rU_∞, and note that it is related to the components of the velocity vector by the equations

$$\frac{\partial\psi}{\partial x} = -(1 + \kappa y)v, \quad \frac{\partial\psi}{\partial y} = u,$$

where κ is the dimensionless curvature of the surface of the profile.

The leading term in the expansion of the stream function in the boundary layer

$$\psi = \mathrm{Re}^{-1/2}\Psi(x, Y) + \cdots, \quad \text{where } Y = \mathrm{Re}^{1/2}y,$$

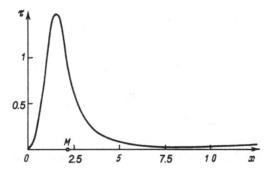

Fig. 4.4 Distribution of friction along the upper surface of the airfoil near the leading edge for $k = 1.15$.

satisfies the classical equation of Prandtl

$$\frac{\partial \Psi}{\partial Y} \frac{\partial^2 \Psi}{\partial x \partial Y} - \frac{\partial \Psi}{\partial x} \frac{\partial^2 \Psi}{\partial Y^2} = -\frac{dp_e}{dx} + \frac{\partial^3 \Psi}{\partial Y^3}, \qquad (4.3.1)$$

which must be integrated with the no-slip condition at the airfoil surface

$$\Psi = \frac{\partial \Psi}{\partial Y} = 0 \quad \text{at } Y = 0, \qquad (4.3.2)$$

with the condition of matching with the solution in the outer inviscid-flow region

$$\frac{\partial \Psi}{\partial Y} = U_e(x) \quad \text{at } Y = \infty, \qquad (4.3.3)$$

and also with an initial condition at the flow stagnation point O:

$$\Psi = 0 \quad \text{at } x = 0. \qquad (4.3.4)$$

The pressure gradient dp_e/dx appearing in equation (4.3.1) does not change across the boundary layer and can be calculated with the help of Bernoulli's equation

$$\frac{dp_e}{dx} = -U_e \frac{dU_e}{dx}.$$

The velocity $U_e(x)$ at the outer edge of the boundary layer is determined by equation (4.2.1). As has already been mentioned, $U_e(x)$ vanishes at the stagnation point O. Away from that point $U_e(x)$ increases, reaching its maximum value $\sqrt{1 + k^2}$ at the point M, for which $Y' = 1/k$ (see Figure 4.3). Downstream from the point M the speed $U_e(x)$ drops monotonically, tending to unity.

The first calculation of the boundary layer at the leading edge of a thin airfoil was carried out by Ermak (1969). Somewhat later a more

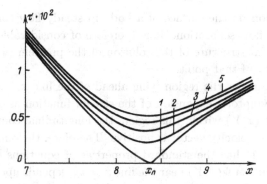

Fig. 4.5 Distribution of skin friction near the point $x = x_0$: (1) for $k = k_0$; (2) for $k = k_0 - 0.0003$; (3) for $k = k_0 - 0.0006$; (4) for $k = k_0 - 0.0009$; (5) for $k = k_0 - 0.0012$.

detailed numerical solution of the problem (4.3.1)–(4.3.4) was given by Werle and Davis (1972) who used the Crank–Nicolson (1947) method in their calculations. It was found that the properties of the boundary layer under consideration depend significantly on the value of the parameter k. If $k < k_0$, where $k_0 = 1.1556$, the skin friction

$$\tau = \left. \frac{\partial^2 \Psi}{\partial Y^2} \right|_{Y=0}$$

remains positive for all $x > 0$ and has a minimum beyond the point M (Figure 4.4), where the pressure gradient is adverse. The appearance of a minimum of the friction against a background of monotonically growing pressure is explained by diffusion of vorticity to the wall. If $dp_e/dx > 0$, then farther from the wall the shear stress increases, and consequently friction forces tend to accelerate the fluid layer next to the wall. Therefore, when the adverse pressure gradient becomes weak enough, the shear stress at the profile surface begins to increase.

As the parameter k increases, the minimum value of the skin friction decreases (Figure 4.5) and vanishes for $k = k_0$. We denote by x_0 the coordinate of the point where that takes place; for a parabola, $x_0 = 8.37$. If $k > k_0$ the point of zero friction $x = x_S$ shifts to the left of x_0. The magnitude of this displacement is greater the more the parameter k differs from its critical value k_0. Werle and Davis (1972) noted that for all $k > k_0$, a singularity appears in the solution of the boundary-layer equations ahead of the point of zero friction. We shall discuss this later.

The solution ahead of the point of zero friction. According to the traditional view dating back to the work of Prandtl (1904), the point of zero friction on the surface of a body in steady flow coincides with the point of flow separation. It is therefore of considerable interest to study the local structure of the solution of the problem (4.3.1)–(4.3.4) in the vicinity of that point.

We first consider the region lying ahead of the line $x = x_S$ and construct an asymptotic expansion of the stream function as $x \to x_S - 0$. At any point (x, Y) within this region, the longitudinal component $u = \partial\Psi/\partial Y$ of the velocity vector is positive. Therefore, the boundary-layer equation (4.3.1) has the standard properties of equations of parabolic type – its solution $\Psi(x, Y)$ near the line $x = x_S$ depends upon the velocity distribution $U_e(x)$ at the outer edge of the boundary layer throughout the whole range of values of x from the stagnation point O to the point $x = x_S$ of zero friction. On the other hand, the procedure utilized below for constructing the asymptotic expansion of the function $\Psi(x, Y)$ as $x \to x_S - 0$ is restricted to consideration only of the vicinity of the line $x = x_S$, and thus it does not take into account all the boundary conditions affecting the function $\Psi(x, Y)$. Therefore, the desired asymptotic expansion must be constructed as an expansion of eigensolutions of the local problem. The coefficients multiplying these eigenfunctions remain arbitrary in the local analysis. At the same time, they are determined uniquely in the solution of the problem (4.3.1)–(4.3.4) as a whole.

We introduce the variable $s = x - x_S$ and write the pressure gradient in the vicinity of the point of zero friction in the form of a Taylor series expansion:

$$\frac{dp_e}{dx} = \lambda_0 + \lambda_1 s + \cdots \quad \text{as } s \to 0. \qquad (4.3.5)$$

The leading term λ_0 of this expansion coincides with the pressure gradient at the point $x = x_S$. For all $k \geq k_0$, when a point of zero friction exists, $\lambda_0 > 0$.

From the boundary-layer equation (4.3.1) and the boundary conditions at $Y = 0$, it follows that at any point on the surface

$$\left.\frac{\partial^3\Psi}{\partial Y^3}\right|_{Y=0} = \frac{dp_e}{dx}.$$

At the point of zero friction we also know the second derivative of the stream function: $\partial^2\Psi/\partial Y^2 = 0$. Hence, the asymptotic representation of the stream function in the vicinity of this point is

$$\Psi = \frac{1}{6}\lambda_0 Y^3 + \cdots. \qquad (4.3.6)$$

We can now determine the width of the viscous flow region, that is, of the region where the viscous term of equation (4.3.1) agrees with the inertia term in order of magnitude:

$$\frac{\partial \Psi}{\partial Y} \frac{\partial^2 \Psi}{\partial x \partial Y} \sim \frac{\partial^3 \Psi}{\partial Y^3}.$$

The quantity $\partial \Psi / \partial Y$ on the left-hand side of this relation is estimated according to (4.3.6) as $O(Y^2)$. Therefore, the width of the viscous region decreases when $s \to -0$ as $Y = O[(-s)^{1/4}]$. The main part of the boundary layer, where $Y = O(1)$, represents a region of locally inviscid flow.

The asymptotic expansion of the stream function within the viscous region will be sought in the following form:

$$\Psi = (-s)^{3/4} \tfrac{1}{6} \lambda_0 \eta^3 + (-s)^{\alpha} f_1(\eta) + (-s)^{2\alpha - 3/4} f_2(\eta) + \cdots . \qquad (4.3.7)$$

The first term of this expansion agrees with the expression (4.3.6) written in terms of the variable $\eta = Y(-s)^{-1/4}$, which is a quantity of order unity in the region under consideration. The second term represents an eigenfunction with an unknown eigenvalue α. The third term is a forced term, arising from the square of the second term, and appears in the expansion (4.3.7) due to the nonlinearity of the boundary-layer equation.

Substituting the expansion (4.3.7) into equation (4.3.1) and taking the limit as $s \to -0$, we obtain differential equations for the functions $f_1(\eta)$ and $f_2(\eta)$. Let us analyze the first of them:

$$f_1''' - \tfrac{1}{8} \lambda_0 \eta^3 f_1'' + \tfrac{1}{2} \lambda_0 (\alpha + \tfrac{1}{4}) \eta^2 f_1' - \lambda_0 \alpha \eta f_1 = 0. \qquad (4.3.8)$$

Because the region considered adjoins the solid boundary, we must impose the no-slip condition on the function $f_1(\eta)$,

$$f_1(0) = f_1'(0) = 0. \qquad (4.3.9)$$

Three linearly independent solutions of equation (4.3.8) can be chosen in such a way that their Taylor series expansions about the point $\eta = 0$ start with 1, η, η^2, respectively. The first two solutions do not satisfy the boundary conditions (4.3.9) and must be rejected. As for the third solution, it can easily be seen to be identically equal to η^2. Hence, a nontrivial solution of equation (4.3.8) with boundary conditions (4.3.9) exists for all α:

$$f_1 = \frac{1}{2} a_0 \eta^2 ; \qquad (4.3.10)$$

here, a_0 is an arbitrary constant.

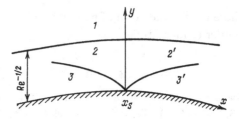

Fig. 4.6 Asymptotic structure of characteristic flow regions near the point of zero skin friction.

In order to determine the eigenvalue α, it is necessary to consider the third term of the expansion (4.3.7). It satisfies the equation

$$f_2''' - \tfrac{1}{8}\lambda_0\eta^3 f_2'' + \lambda_0(\alpha - \tfrac{1}{4})\eta^2 f_2' - \lambda_0(2\alpha - \tfrac{3}{4})\eta f_2 = \tfrac{1}{4}(1 - 2\alpha)a_0^2\eta^2\,,$$

and the solution that satisfies the no-slip condition $f_2(0) = f_2'(0) = 0$ can be written in terms of a confluent hypergeometric function $F(\kappa, \gamma, z)$:[3]

$$f_2 = \frac{a_0^2}{2\lambda_0}(\eta - h_\alpha) + \frac{b_0}{2}\eta^2\,,$$

$$h_\alpha(\eta) = \eta - \eta^2 \int_0^\eta \eta^{-2}\left[F\left(\kappa, \gamma, \frac{\lambda_0}{32}\eta^4\right) - 1\right]d\eta\,,$$

(4.3.11)

where b_0 is an arbitrary constant and the parameters of the function $F(\kappa, \gamma, z)$ are determined as $\kappa = 1 - 2\alpha$, $\gamma = \frac{5}{4}$.

It is known that the confluent hypergeometric function is defined throughout the entire complex z plane and grows according to an exponential law if z tends to infinity along a ray that lies in the right half-plane Re $z > 0$. Therefore, the function $h_\alpha(\eta)$, together with $f_2(\eta)$, also grows exponentially as $\eta \to \infty$. An exception occurs if

$$\kappa = -m, \text{ where } m = 0,\ 1,\ 2,\ldots, \tag{4.3.12}$$

for which the confluent hypergeometric function reduces to a polynomial of degree m and grows as $O(z^m)$ as $z \to \infty$.

We now note that, regardless of how the parameter κ is chosen, the solutions (4.3.7), (4.3.10), (4.3.11) for the viscous region do not satisfy the boundary conditions at the outer edge of the boundary layer. Therefore, in addition to the viscous region (region 3 in Figure 4.6), it is also necessary to analyze a locally inviscid region 2 covering the major part of the boundary layer: $Y = O(1)$. In this region the longitudinal velocity component $u = \partial\Psi/\partial Y$ is of the order of the velocity at the outer

[3] For more detail, the reader may refer to the paper of Ruban (1981).

edge of the boundary layer: $u = O(1)$, so that matching the solutions in regions 2 and 3 is possible only under the condition (4.3.12), when $f_2(\eta)$ does not grow exponentially. It is just this condition that determines the eigenvalues

$$\alpha = \frac{m+1}{2}, \text{ where } m = 0, 1, 2, \ldots.$$

Goldstein's singularity. Returning to the expansion (4.3.7), we note that it makes sense only for $\alpha > 3/4$, when each successive term in the expansion is of smaller order of magnitude than the preceding one. Therefore, the first eigenvalue is $\alpha = 1$. The coefficient a_0 multiplying the first eigenvalue depends on $U_e(x)$, or (what is the same thing) the pressure distribution $p_e(x)$ along the boundary layer. In the general case, when the distribution of pressure is not prescribed in a special way, a_0 differs from zero and the solution (4.3.7), (4.3.10), (4.3.11) for region 3 is represented in the form[4]

$$\Psi = (-s)^{3/4}\frac{1}{6}\lambda_0\eta^3 + (-s)f_1(\eta) + (-s)^{5/4}f_2(\eta) + \cdots,$$
$$f_1 = \frac{a_0}{2}\eta^2, \quad f_2 = \frac{b_0}{2}\eta^2 - \frac{a_0^2}{2\cdot 5!}\eta^5. \tag{4.3.13}$$

In addition to the terms shown, the expansion of the function $\Psi(x, Y)$ contains the sum of an infinite number of successive eigenfunctions, and also includes additional terms related to subsequent terms in the expansion of the pressure gradient (4.3.5). All of them, however, are of very small order in comparison with $(-s)^{5/4}$.

We now consider region 2, where $Y = O(1)$. The form of the asymptotic expansion of the stream function $\Psi(x, Y)$ in this region may be determined with the help of a standard procedure based on the principle of matched asymptotic expansions. We make the change of variables $\eta = Y(-s)^{-1/4}$ in (4.3.13) and, for $Y = O(1)$, collect terms of the same order of smallness as $s \to -0$. As a result, we find that in region 2

$$\Psi = \Psi_{00}(Y) + (-s)^{1/2}\Psi_{01}(Y) + \cdots \tag{4.3.14}$$

and that as $Y \to 0$

$$\Psi_{00} = \frac{\lambda_0}{6}Y^3 - \frac{a_0^2}{2\cdot 5!}Y^5 + \cdots, \quad \Psi_{01} = \frac{a_0}{2}Y^2 + \cdots. \tag{4.3.15}$$

Substituting the expansion (4.3.14) into the boundary-layer equation

[4] Subsequent terms in the expansion (4.3.13) are given in the work of Stewartson (1958).

(4.3.1), and taking into account the conditions (4.3.15), we obtain

$$\Psi_{01} = \frac{a_0}{\lambda_0}\Psi'_{00}(Y).$$

If we now calculate the transverse component $V = -\partial\Psi/\partial x$ of the velocity vector, it turns out to have the singularity

$$V = \mathrm{Re}^{1/2}v = (-s)^{-1/2}\frac{a_0}{2\lambda_0}\Psi'_{00}(Y) + \cdots \quad \text{as } s \to -0. \qquad (4.3.16)$$

It follows from (4.3.13) that the skin friction also behaves in a singular way:

$$\tau = (-s)^{1/2}a_0 + \cdots \quad \text{as } s \to -0.$$

Because the friction is positive for $s < 0$, we must consider $a_0 > 0$.

The above analysis of the boundary layer ahead of the point of zero skin friction was first published in the work of Goldstein (1948) that was already mentioned in Chapter 1, with the difference, however, that Goldstein, knowing the numerical results of Hartree (1939), set $\alpha = 1$ in the expansion (4.3.7) from the very beginning. Goldstein also showed – and this result is especially important – that for $a_0 \neq 0$ the solution of the boundary-layer equation cannot be extended beyond the point of zero friction.

In fact, suppose that the solution is extended continuously through the line $x = x_S$. Then, using the continuity equation and the expression (4.3.14), we come to the conclusion that the first term of the asymptotic expansion of the stream function in region $2'$ (Figure 4.6) coincides with $\Psi_{00}(Y)$:

$$\Psi = \Psi_{00}(Y) + \cdots \quad \text{as } s \to +0, \quad Y = O(1). \qquad (4.3.17)$$

This means that the longitudinal component $u = \partial\Psi/\partial Y$ of the velocity vector decreases as $O(Y^2)$ when $Y \to 0$, and hence the thickness of the viscous region $3'$ is estimated as $Y = O(s^{1/4})$. Thus, the asymptotic analysis of the boundary-layer equation (4.3.1) in region $3'$ is associated with the limit process

$$s \to +0, \quad \xi = Ys^{-1/4} = O(1).$$

We represent the stream function in region $3'$ in the form of the asymptotic expansion

$$\Psi = s^{3/4}\frac{1}{6}\lambda_0\xi^3 + s^\mu\widehat{f}_1(\xi) + s^{2\mu-3/4}\widehat{f}_2(\xi) + \cdots \qquad (4.3.18)$$

with the exponent μ unknown in advance. The functions $\widehat{f}_1(\xi)$ and $\widehat{f}_2(\xi)$

are determined in the same way as the functions $f_1(\eta)$ and $f_2(\eta)$ that appear in the expansion (4.3.7). As a result we obtain the following expressions for $\widehat{f}_1(\xi)$ and $\widehat{f}_2(\xi)$:

$$\widehat{f}_1 = \frac{1}{2}\widehat{a}_0\xi^2, \quad \widehat{f}_2 = \frac{\widehat{a}_0^2}{2\lambda_0}(\xi - h_\mu) + \frac{1}{2}\widehat{b}_0\xi^2,$$

$$h_\mu(\xi) = \xi - \xi^2 \int_0^\xi \xi^{-2}\left[F\left(\kappa, \gamma, -\frac{\lambda_0}{32}\xi^4\right) - 1\right]d\xi,$$

where $\widehat{a}_0$ and $\widehat{b}_0$ are arbitrary constants and the parameters of the confluent hypergeometric function $F(\kappa, \gamma, z)$ are determined as $\kappa = 1 - 2\mu$, $\gamma = 5/4$.

It remains to carry out the matching of the expansions of the stream function in regions 2' and 3'. In order to perform this procedure it is necessary to know the asymptotic expansion of the function $\widehat{f}_2(\xi)$ as $\xi \to \infty$. As we know, the asymptotic behavior of a confluent hypergeometric function for large values of z depends on the way that z tends to infinity. If the passage to the limit is performed along a ray lying in the left half-plane (Re $z < 0$), then

$$F(\kappa, \gamma, z) = \frac{\Gamma(\gamma)}{\Gamma(\gamma - \kappa)}(-z)^{-\kappa}\left[1 + O\left(\frac{1}{z}\right)\right],$$

where Γ is the Euler gamma function.

Therefore, the leading term of the asymptotic expansion of $\widehat{f}_2(\xi)$ as $\xi \to \infty$ is

$$\widehat{f}_2 = \frac{\widehat{a}_0^2}{2\lambda_0}\frac{\Gamma(\gamma)}{\Gamma(\gamma - \kappa)}\left(\frac{\lambda_0}{32}\right)^{2\mu-1}\frac{\xi^{8\mu-3}}{8\mu - 5} + \cdots.$$

Substituting into (4.3.18) and making the change of variable $\xi = Ys^{-1/4}$, we find that at the outer edge of region 3'

$$\Psi = \frac{1}{6}\lambda_0 Y^3 + \frac{\widehat{a}_0^2}{2\lambda_0}\frac{\Gamma(\gamma)}{\Gamma(\gamma - \kappa)}\left(\frac{\lambda_0}{32}\right)^{2\mu-1}\frac{Y^{8\mu-3}}{8\mu - 5} + \cdots. \qquad (4.3.19)$$

On the other hand, we know the one-term asymptotic representation (4.3.17) of the stream function in region 2'. Using the expansion (4.3.15) for $\Psi_{00}(Y)$ as $Y \to 0$, we find that near the inner edge of region 2'

$$\Psi = \frac{1}{6}\lambda_0 Y^3 - \frac{\widehat{a}_0^2}{2 \cdot 5!}Y^5 + \cdots.$$

According to the principle of matched asymptotic expansions, this expression must agree with the expression (4.3.19). Therefore, $\mu = 1$ and

$$\widehat{a}_0^2 = -a_0^2. \tag{4.3.20}$$

If the constant a_0 is found in the process of solving the boundary-layer equation from the initial cross section $x = 0$ up to the point of zero skin friction, then the equation (4.3.20) serves to determine the constant $\widehat{a}_0$. This attests to the impossibility of finding a real solution downstream of the point of zero friction. An exception is the case when the first eigenfunction in the solution ahead of the point of zero friction is absent: $a_0 = 0$.

The above analysis also shows that the hypotheses of the boundary-layer theory cease to be valid in the neighborhood of the point of zero friction. Indeed, for all $a_0 \neq 0$ the transverse component of the velocity vector (4.3.16) increases without limit on the approach to this point. Hence, there exists a neighborhood where the effect of the boundary layer upon the outer flow due to the displacement of streamlines from the surface becomes considerable and leads to redistribution of the pressure along the boundary layer. The question arises whether the displacement effect of the boundary layer causes such a change of the pressure gradient as to smooth the singularity and create the opportunity for a smooth transition of the boundary layer into the separation region.

To answer this question, Stewartson (1970b) investigated the vicinity of the singular point on the basis of the Navier–Stokes equations and came to the following conclusion. If Goldstein's solution holds for the boundary layer ahead of the point of zero friction, then the interaction region generated in the vicinity of this point has a normal triple-deck structure similar to that considered in the previous chapters. This structure has already been anticipated by Goldstein's solution: the two characteristic regions 3 and 2 have developed in the boundary layer ahead of the point of zero friction. The first of them forms a viscous wall layer in the interaction region, and the second one a middle layer. To these must still be added an outer region of potential flow where the perturbations in the slope of the streamlines caused by the boundary layer are transformed into pressure perturbations. As usual, the interaction occurs between the wall layer and the potential part of the flow while the intermediate layer plays the role of a passive transmission link between them. However, the distinctive conditions for which the region of interaction under consideration appears (in the present case, the boundary layer ahead of this region has a preseparation velocity profile with vanishingly small friction at the surface) lead to an essential change in the formulation of the boundary-value problem for it. The consequences of these changes turn out to be catastrophic – the solution of the interaction problem ceases to exist. Thus, Goldstein's singularity proves to be unavoidable.

This result obtained by Stewartson on the basis of direct mathematical analysis of the interaction problem has a simple explanation. Recall that the impossibility of continuing the boundary-layer equations through the line $x = x_S$ is related to the form of the initial distribution (4.3.17) of the stream function in region 2' or, to be more precise, to the minus sign in the second term of the expansion of $\Psi_{00}(Y)$ for $Y \to 0$ in (4.3.15). Its appearance in the right-hand side of the relation (4.3.20) leads directly to a contradiction. The formation of the interaction region results in the fact that now the initial velocity profile in the region 2' must be found not from the condition of its continuity on the line $x = x_S$, but rather from the condition of matching with the solution in the middle layer of the interaction region. It is known, however, that the change Δu in the longitudinal component of the velocity vector is small in this case:

$$\Delta u \to 0 \quad \text{as Re} \to \infty .$$

Therefore, an interaction region does not alter the initial condition for region 2', and hence leads to a contradiction in the procedure for constructing the solution of the boundary-layer equation (4.3.1) for $x > x_S$.

A singular solution of the boundary-layer equation extended continuously through the zero-friction point. As already mentioned, the behavior of the boundary layer at the leading edge of the airfoil depends on the parameter k, which is determined by the angle of attack and the profile curvature according to the relation (4.2.2). If the parameter k is less than its critical value k_0, the skin friction $\tau(x)$ remains positive, and the solution of the problem (4.3.1)–(4.3.4) exists for all values of x. The function $\tau(x)$ has a minimum whose value decreases as the parameter k is increased, as shown in Figure 4.5. At $k = k_0$ the value of the minimum friction vanishes and for the first time a point of zero friction $x_S = x_0$ appears on the profile surface. We consider this case separately.

We denote the solution of the problem (4.3.1)–(4.3.4) for $k = k_0$ by $\Psi_0(x, Y)$, and study its behavior near the point of zero friction $x = x_0$. First of all we note that the procedure described above for constructing a solution in the vicinity of this point is valid in the present case as well. One need only bear in mind that the function $\Psi_0(x, Y)$ may be considered as the limit of a solution of the boundary-layer equation as $k \to k_0 - 0$. At the same time, for all $k < k_0$ the solution of this equation can be continued across the line $x = x_0$. Hence, the limiting solution $\Psi_0(x, Y)$ can also be continued. The latter is possible only under the condition that the coefficient a_0 multiplying the first eigenfunction in the expansion (4.3.13) is equal to zero.

The same conclusion results from the calculation of Werle and Davis (1972). They discovered that for all $k > k_0$, Goldstein's singularity develops in the solution of the boundary-layer equation ahead of the point of zero skin friction. However, as the parameter k is decreased, the point of zero friction x_S is displaced downstream, approaching the point x_0, and the coefficient a_0 multiplying the first eigenfunction in the expansion (4.3.13) decreases. At $k = k_0$ it becomes equal to zero, and the first eigenfunction disappears from the expansion (4.3.7).

The second eigenvalue is $\alpha = 3/2$. In this case we obtain for region 3

$$
\begin{aligned}
\Psi_0 &= (-s)^{3/4} \frac{\lambda_0}{6} \eta^3 + (-s)^{3/2} f_1(\eta) \\
&\quad + (-s)^{7/4} F_1(\eta) + (-s)^{9/4} f_2(\eta) + \cdots, \\
f_1 &= \frac{a_0}{2} \eta^2, \quad F_1 = -\frac{\lambda_1}{6} \eta^3 + \frac{2\lambda_0 \lambda_1}{7!} \eta^7, \\
f_2 &= \frac{b_0}{2} \eta^2 - \frac{a_0^2}{5!} \eta^5 + \frac{\lambda_0 a_0^2}{8!} \eta^9.
\end{aligned}
\tag{4.3.21}
$$

To maintain uniformity of writing we shall denote the coefficient multiplying the eigenfunction by a_0 as before; however, it is necessary to keep in mind that now this is the coefficient multiplying the second eigenfunction. The appearance of an additional term $(-s)^{7/4} F_1(\eta)$ in the expansion for $\Psi_0(x, Y)$ is related to the linear term in the expansion for the pressure coefficient (4.3.5).

The solution in region 2 (Figure 4.6) is written using the expansion

$$
\Psi_0 = \Psi_{00}(Y) + (-s)\Psi_{01}(Y) + \cdots \quad \text{as } s \to -0.
\tag{4.3.22}
$$

Substituting (4.3.22) into the boundary-layer equation (4.3.1) we obtain

$$
\Psi_{01} = \Psi_{00}' \left[c_0 + \int_0^Y \frac{\lambda_0 - \Psi_{00}'''}{(\Psi_{00}')^2} dY \right].
$$

From matching the expansions (4.3.21) and (4.3.22), it follows that $c_0 =$

a_0/λ_0 and that as $Y \to 0$,

$$\Psi_{00} = \frac{1}{6}\lambda_0 Y^3 + \frac{2\lambda_0\lambda_1}{7!}Y^7 + \frac{\lambda_0 a_0^2}{8!}Y^9 + \cdots. \qquad (4.3.23)$$

The solution ahead of the point of zero friction has been constructed. We show that it can be extended continuously downstream from this point. In region $2'$ we seek it in the form

$$\Psi_0 = \Psi_{00}(Y) + s\widehat{\Psi}_{01}(Y) + \cdots \quad \text{as } s \to +0. \qquad (4.3.24)$$

The boundary-layer equation is satisfied if

$$\widehat{\Psi}_{01} = \Psi'_{00}\left[\widehat{c}_0 + \int_0^Y \frac{\Psi'''_{00} - \lambda_0}{(\Psi'_{00})^2}dY\right]. \qquad (4.3.25)$$

We represent the solution in the region $3'$ (Figure 4.6) in the following form:

$$\Psi_0 = s^{3/4}\frac{1}{6}\lambda_0\xi^3 + s^{3/2}\widehat{f}_1(\xi) + s^{7/4}\widehat{F}_1(\xi) + s^{9/4}\widehat{f}_2(\xi) + \cdots. \qquad (4.3.26)$$

Substituting (4.3.26) into the boundary-layer equation and taking into account the no-slip condition at the surface, we obtain

$$\widehat{f}_1 = \frac{1}{2}\widehat{a}_0\xi^2, \quad \widehat{F}_1 = \frac{1}{6}\lambda_1\xi^3 + \frac{2\lambda_0\lambda_1}{7!}\xi^7,$$

$$\widehat{f}_2 = \frac{1}{2}\widehat{b}_0\xi^2 + \frac{\widehat{a}_0^2}{5!}\xi^5 + \frac{\lambda_0\widehat{a}_0^2}{8!}\xi^9.$$

It remains to match the expansions (4.3.24) and (4.3.26). We recall that it is just this procedure that was found impossible to carry out when an attempt was made to extend the solution with Goldstein's singularity downstream from the point of zero friction. The result was equation (4.3.20), which could not be solved for $\widehat{a}_0$ in terms of real numbers. Now instead of (4.3.20) we obtain

$$\widehat{a}_0^2 = a_0^2. \qquad (4.3.27)$$

Matching the expansions (4.3.24) and (4.3.26) also allows finding the constant $\widehat{c}_0$ in the expression (4.3.25) for $\widehat{\Psi}_{01}(Y)$: $\widehat{c}_0 = \widehat{a}_0/\lambda_0$.

From the relation (4.3.27) it follows that the solution of the boundary-layer equation that is unique ahead of the point of zero friction can be continued downstream from this point in two ways. The first one ($\widehat{a}_0 = -a_0$) provides smooth behavior of the solution – the stream function in this case is represented with the help of the expansion

$$\Psi_0 = \Psi_{00}(Y) + s\Psi_{00}'(Y) \left[\int_0^Y \frac{\Psi_{00}''' - \lambda_0}{(\Psi_{00}')^2} dY - \frac{a_0}{\lambda_0} \right] + O(s^2), \quad (4.3.28)$$

which remains uniformly valid in the entire vicinity of the line $x = x_0$. Now if we calculate the skin friction

$$\tau = \frac{\partial^2 \Psi_0}{\partial Y^2}\bigg|_{Y=0} = -a_0 s + O(s^2),$$

it turns out to change sign at the point $x = x_0$, which is evidence of the formation of a region of reverse flow downstream of this point. The previous chapters contain several examples of such a solution. It is realized every time that there appears in the interaction region a point of zero skin friction about which the solution of Prandtl's equation for the viscous wall layer remains smooth.

However, there exists also a second way of extending the solution, which is realized with $\hat{a}_0 = a_0$. In this case the solution

$$\Psi_0 = \Psi_{00}(Y) + \Psi_{00}'(Y) \left[\frac{a_0}{\lambda_0}|s| + s \int_0^Y \frac{\Psi_{00}''' - \lambda_0}{(\Psi_{00}')^2} dY \right] + O(s^2) \quad (4.3.29)$$

is singular. In particular, the angle of inclination ϑ of the streamlines in the boundary layer undergoes a discontinuity when passing through the line $x = x_0$:

$$\text{Re}^{1/2}\vartheta = -\frac{\partial \Psi_0}{\partial x} \bigg/ \frac{\partial \Psi_0}{\partial Y} = -\frac{a_0}{\lambda_0} \text{sign}(s) + \int_0^Y \frac{\lambda_0 - \Psi_{00}'''}{(\Psi_{00}')^2} dY + O(s),$$

$$(4.3.30)$$

and the skin friction behaves as

$$\tau = a_0|s| + O(s^2) \quad \text{as } s \to 0. \quad (4.3.31)$$

The possibility of constructing a solution of the boundary-layer equations with smooth transition through the point of zero friction was mentioned earlier by Goldstein (1948). The existence of a special branch of the solution was discovered by Ruban (1981). It is just this special solution of the problem (4.3.1)–(4.3.4) with a characteristic minimum in the distribution of skin friction that is the limiting solution as $k \to k_0 - 0$. To convince ourselves of the validity of this statement, let us study the behavior of the boundary layer for values of the parameter k close to the critical value k_0.

The appearance of a singularity in the boundary layer. If the parameter k differs only slightly from its critical value k_0, then the following Taylor expansion is valid for the velocity (4.2.1) at the outer edge of the boundary layer:

$$U_e(x, k) = U_e(x, k_0) + \Delta k \frac{\partial U_e}{\partial k} + O[(\Delta k)^2],$$

where $\Delta k = k - k_0$ and the derivative $\partial U_e/\partial k$ is calculated at $k = k_0$. It follows from Bernoulli's equation that a similar expansion holds for the pressure as well:[5]

$$p_e = p_0(x) + \Delta k p_1(x) + \cdots.$$

Therefore, we represent the solution of the problem (4.3.1)–(4.3.4) in the form

$$\Psi = \Psi_0(x, Y) + \Delta k \Psi_1(x, Y) + O[(\Delta k)^2]. \tag{4.3.32}$$

The properties of the function $\Psi_0(x, Y)$ have been thoroughly studied in the preceding part of this section. Now let us consider the function $\Psi_1(x, Y)$. Substituting (4.3.32) into (4.3.1)–(4.3.4) yields

$$\frac{\partial \Psi_0}{\partial Y} \frac{\partial^2 \Psi_1}{\partial x \partial Y} + \frac{\partial^2 \Psi_0}{\partial x \partial Y} \frac{\partial \Psi_1}{\partial Y} - \frac{\partial \Psi_0}{\partial x} \frac{\partial^2 \Psi_1}{\partial Y^2} - \frac{\partial^2 \Psi_0}{\partial Y^2} \frac{\partial \Psi_1}{\partial x} = -\frac{dp_1}{dx} + \frac{\partial^3 \Psi_1}{\partial Y^3},$$

$$\Psi_1 = \frac{\partial \Psi_1}{\partial Y} = 0 \quad \text{at } Y = 0. \tag{4.3.33}$$

For the boundary conditions we have written here only the no-slip condition at the surface. Strictly speaking, it is also necessary to impose the boundary conditions at the outer edge of the boundary layer and at the initial cross section. However, their specific form has no importance for the subsequent analysis and therefore is not shown.

The solution of equation (4.3.33) ahead of the point of zero friction $x = x_0$ is constructed in the same way as the solution of equation (4.3.1). We consider first the viscous wall region 3. Here, we seek the function $\Psi_1(x, Y)$ in the form

$$\Psi_1 = (-s)^\beta g_1(\eta) + (-s)^{\beta+3/4} g_2(\eta) + \cdots \quad \text{as } s \to -0, \tag{4.3.34}$$

where β is the eigenvalue of the problem.

---

[5] In the following, λ_0 and λ_1 will denote Taylor series coefficients for the leading terms in the expansion of the pressure gradient:

$$\frac{dp_0}{dx} = \lambda_0 + \lambda_1 s + \cdots \quad \text{as } s = x - x_0 \to 0.$$

Substituting the expansion (4.3.34) into (4.3.33), we obtain

$$g_1 = \frac{1}{2}a_1\eta^2, \quad g_2 = \frac{a_0a_1}{\lambda_0}(\eta - h_\beta) + \frac{1}{2}b_1\eta^2,$$

$$h_\beta(\eta) = \eta - \eta^2 \int_0^\eta \eta^{-2}\left[F\left(\kappa, \gamma, \frac{\lambda_0}{32}\eta^4\right) - 1\right]d\eta.$$

Now $\kappa = -\left(\beta + \frac{1}{2}\right)$ and γ is still equal to 5/4.

The function $h_\beta(\eta)$ does not grow exponentially as $\eta \to \infty$ if

$$\beta = m - \frac{1}{2}, \quad \text{where } m = 0, 1, 2, \ldots,$$

with $h_\beta = \eta$ corresponding to the minimum eigenvalue $\beta = -1/2$. Therefore, the solution (4.3.34) for region 3 assumes the form

$$\Psi_1 = (-s)^{-1/2}\frac{1}{2}a_1\eta^2 + (-s)^{1/4}\frac{1}{2}b_1\eta^2 + \cdots. \tag{4.3.35}$$

The constants a_1 and b_1 appearing in this expression remain arbitrary. To determine them it is necessary to carry out a complete solution of equation (4.3.33) that takes into account the specific form of the boundary conditions at the outer edge of the boundary layer and at the initial cross section.

Relying on the asymptotic matching principle and taking into account the solution (4.3.35) for region 3, one can obtain for region 2 (Figure 4.6)

$$\Psi_1 = (-s)^{-1}\frac{a_1}{\lambda_0}\Psi_{00}'(Y) + O[(-s)^{-1/4}] \quad \text{as } s \to -0. \tag{4.3.36}$$

Let us now return to the expansion (4.3.32) for the stream function in the boundary layer. According to the definition of an asymptotic expansion, each successive term must be of smaller order than its predecessor. The assumption that the expansion (4.3.32) satisfies this condition was used in deriving equation (4.3.33), and thus it became the basis of the entire procedure for determining the function $\Psi_1(x, Y)$. It is actually realized due to the smallness of Δk, for each fixed point of space located upstream of the point $x = x_0$. However, as $x \to x_0 - 0$ the function $\Psi_1(x, Y)$ increases without bound, leading to violation of the supposed relationship between the orders of magnitude in the expansion (4.3.32). Therefore, we must examine separately the neighborhood of the point $x = x_0$. Because the next section will describe the flow near the point $x = x_0$ on a more general basis, including the effect of interaction of the boundary layer with the outer inviscid flow, we omit here the description of the procedure for constructing the corresponding solution and give it without derivation:

$$\Psi = \Psi_{00} + \Psi_{00}' \left[\frac{a_0}{\lambda_0} \sqrt{s^2 + 2\frac{a_1}{a_0}\Delta k} + s \int_0^Y \frac{\Psi_{00}''' - \lambda_0}{(\Psi_{00}')^2} dY \right] + \cdots . \quad (4.3.37)$$

Equation (4.3.37) represents an asymptotic solution of the boundary-layer equation (4.3.1). It holds for any relationship between the orders of magnitude of s and Δk. It is only required that both of them be small compared with unity.

The domain of existence of the solution is determined by the sign of the expression under the radical in (4.3.37). If $a_1 \Delta k > 0$, the solution exists for all s. The skin friction

$$\tau = \frac{\partial^2 \Psi}{\partial Y^2}\bigg|_{Y=0} = a_0 \sqrt{s^2 + 2\frac{a_1}{a_0}\Delta k}$$

remains positive everywhere and at the point $s = 0$ it reaches its minimum, the value of which decreases as $\Delta k \to 0$ according to the rule

$$\tau_{\min} = \sqrt{2a_0 a_1 \Delta k}. \quad (4.3.38)$$

If, however, $a_1 \Delta k < 0$, the solution exists only up to the point of zero friction,

$$x_S = x_0 - \sqrt{2|a_1 \Delta k|/a_0};$$

downstream of it, the expression in the radicand in (4.3.37) becomes negative. As one might expect, Goldstein's singularity appears ahead of the point $x = x_S$. In this case, however, it is weak. As $x \to x_S - 0$, the skin friction behaves as

$$\tau = (8a_0^3 |a_1 \Delta k|)^{1/4}(x_S - x)^{1/2} + \cdots$$

and for the transverse component of the velocity vector we have

$$V = \frac{(8a_0^3 |a_1 \Delta k|)^{1/4}}{2\lambda_0} \Psi_{00}'(Y)(x_S - x)^{-1/2} + \cdots$$

According to the numerical results of Werle and Davis (1972), the first of the situations described is realized for $\Delta k < 0$ and the second for $\Delta k > 0$. This means that the constant a_1 is negative.

In the intermediate case when $\Delta k = 0$, the boundary-layer equation (4.3.1) admits two solutions, as mentioned earlier. One of them (4.3.28) is smooth, while the other (4.3.29) has a singularity at the line $x = x_0$. Comparing equations (4.3.29) and (4.3.37), it is easy to satisfy oneself

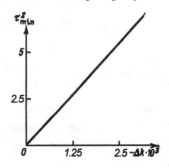

Fig. 4.7 Dependence of the square of the minimum skin friction on $\Delta k = k - k_0$.

that it is just the singular solution of the boundary-layer equation that is its limiting solution as $k \to k_0 - 0$.

To determine the constants a_0 and a_1 it is necessary to apply the results of a numerical solution of the problem (4.3.1)–(4.3.4). We first consider the curve 1 in Figure 4.5. It corresponds to the critical value of the parameter $k = k_0$, where equation (4.3.31) is valid. Using this formula we find that $a_0 = 0.0085$. The constant a_1 may be found with the help of equation (4.3.38). From that it follows that the square of the minimum value of the skin friction is proportional to Δk. Plotting, on the basis of Figure 4.5, the graph of $\tau^2_{\min}$ versus Δk (Figure 4.7) and determining from it the coefficient of proportionality between $\tau^2_{\min}$ and Δk, we find[6] $a_1 = -1.24$.

4.4 The Interaction Region

Up to this point the analysis of the flow around the airfoil leading edge has been carried out within the framework of the boundary-layer theory. For the investigation of the inviscid-flow region we did not take into account the existence of the boundary layer at the wall, and instead of the conditions of matching with the solution in this layer we employed the condition of no fluid flow across the profile surface. When the parameter k is smaller than its critical value k_0, the influence of the boundary layer on the external flow is indeed weak. It is mainly revealed in the pushing of the streamlines away from the surface. One can take this effect into account with the help of Prandtl's hierarchical procedure mentioned in

---

[6] One should take note of the fact that the assumption of a parabolic leading edge of the airfoil is in essence used only in calculating the constants a_0, a_1, and others. Otherwise, the theory remains quite universal.

Chapter 1. Fluid flow around a parabola at subcritical values of the parameter k represents a classic example of a flow for which the solution in the boundary layer remains regular along the entire surface, and application of the hierarchical procedure does not lead to a contradiction. The situation changes when the parameter k approaches its critical value k_0 and a singularity appears in the boundary layer. If $k = k_0$, the streamlines within the boundary layer and at its outer edge develop a sharp bend at $x = x_0$. The slope of the streamlines (4.3.30) changes when passing through this point by the angle

$$\theta = -\mathrm{Re}^{-1/2}\frac{2a_0}{\lambda_0}.$$ (4.4.1)

The minus sign is introduced into equation (4.4.1) by analogy with Chapter 2, Section 3, where the fluid flow around a corner on a body contour was described. If one uses this analogy, it is easy to find that the pressure perturbation induced due to the displacement effect of the boundary layer is estimated as $x \to x_0$ as

$$p_i \sim \mathrm{Re}^{-1/2}\left\{\frac{2a_0 U_0^2}{\pi\lambda_0}\ln|x - x_0| + O(1)\right\}.$$

Here, U_0 is the value of U_e, the longitudinal component of the velocity vector at the outer edge of the boundary layer, evaluated at the point $x = x_0$ for $k = k_0$.[7]

At a sufficient distance from the point $x = x_0$, the induced pressure gradient is small and does not exert a noticeable influence on the flow in the boundary layer. At the same time its unbounded growth

$$\frac{dp_i}{dx} = O\left(\frac{\mathrm{Re}^{-1/2}}{x - x_0}\right)$$ (4.4.2)

with the approach to $x = x_0$ indicates the inevitable appearance of the interaction region, where the pressure perturbations induced by the displacement effect of the boundary layer now influence the flow within the boundary layer itself in the first approximation.

In order to estimate the extent of the interaction region, it is necessary to reconsider the procedure of constructing the solution of the boundary-layer equation in the vicinity of the point $x = x_0$, and determine when it becomes invalid. This procedure includes as a main element the solution of the eigenvalue problem for the function $f_2(\eta)$. It is from just this problem that we determine the dependence of the angle of inclination of

---

[7] For a profile with a nose of parabolic shape $U_0 = 1.286$ and $\lambda_0 = 0.024$.

the streamlines in the boundary layer upon the longitudinal coordinate x. Hence, if in a certain neighborhood of the point $x = x_0$ the induced pressure gradient becomes so large that it must be taken into account in the equation for $f_2(\eta)$, this neighborhood is the interaction region.

The induced pressure gradient can be compared with any term of the first equation of the Navier–Stokes system of equations (see Chapter 1, (1.1.2)) that contributes to the equation for $f_2(\eta)$. In particular, the leading viscous term in region 3 where the expansion (4.3.21) is valid is represented in the form

$$\frac{\partial^3 \Psi_0}{\partial Y^3} = \lambda_0 + (-s)^{3/4} f_1''' + (-s) F_1''' + (-s)^{3/2} f_2''' + \cdots .$$

The fourth term of this expansion is a part of the equation for the function $f_2(\eta)$. It is to be compared with the induced pressure gradient (4.4.2). As a result we find that the extent of the interaction zone is estimated as

$$|x - x_0| = O(\mathrm{Re}^{-1/5}). \qquad (4.4.3)$$

The reasoning given above is based upon the properties of the solution of the boundary-layer equation in the special case when $k = k_0$. One may expect that it will also remain valid when the parameter k differs slightly from k_0. The final part of the preceding section was devoted to the analysis of the boundary-layer equation (4.3.1) under these conditions. On applying the limit process

$$\Delta k \to 0, \quad x = O(1), \quad Y = O(1), \qquad (4.4.4)$$

to (4.3.1) we found the Taylor series (4.3.32) for the stream function. This expansion, as was mentioned, is the asymptotic solution of equation (4.3.1) in the entire boundary layer, except for a small neighborhood of the point $x = x_0$.

The behavior of the solution of the boundary-layer equation near the point $x = x_0$ is governed by eigenfunctions contained in the stream function expansion in region 3. This refers both to the leading approximation $\Psi_0(x, Y)$ and to the perturbation of the solution $\Delta k \Psi_1(x, Y)$. In the first case (4.3.21) the eigenfunction is estimated as $O[(-s)^{3/2}]$, and in the second case (4.3.35) as $O[\Delta k(-s)^{-1/2}]$. When

$$|x - x_0| = O(|\Delta k|^{1/2}), \qquad (4.4.5)$$

the second eigenfunction is comparable to the first in order of magnitude, and the solution (4.3.32), constructed with the help of the limit process (4.4.4), becomes invalid.

Depending on the values of the angle of attack and the Reynolds number, different relationships between the scales (4.4.3) and (4.4.5) are possible. If the scale (4.4.5) is smaller than the scale (4.4.3) or agrees with it in order of magnitude, the solution (4.3.32) remains valid up to the vicinity (4.4.3) of the point $x = x_0$ and the estimate of the extent of the interaction region also holds. But if the scale (4.4.5) is wider than the scale (4.4.3), then everything depends on the sign of $\Delta k = k - k_0$. When $\Delta k < 0$, the singularity contained in the solution (4.3.32) is smoothed out within the region (4.4.5) and the interaction effect upon the flow in the boundary layer is insignificant. In this case, the stream function near the point $x = x_0$ is determined by equation (4.3.37). However, when $\Delta k > 0$, the solution (4.3.37) of the boundary-layer equation for the region with the scale (4.4.5) indicates the appearance of a point of zero skin friction $x = x_S$, with Goldstein's singularity upstream from it. Taking a formal approach, we would have introduced an interaction region in the vicinity of the point $x = x_S$, as was done by Stewartson (1970b) This path, however, leads to a contradiction because in this case the solution of the boundary-value problem for the interaction region does not exist.

Thus, the most interesting case is when $\Delta k = O(\mathrm{Re}^{-2/5})$, and the scales (4.4.3) and (4.4.5) coincide. This flow regime covers the whole range of angle of attack associated with the evolution of a short bubble from the moment of its generation until its destruction. Therefore, we shall take

$$k = k_0 + \mathrm{Re}^{-2/5} k_1 \,, \qquad (4.4.6)$$

and we shall show that the theory constructed on the basis of a limit process where the constant k_1 remains of order unity as $\mathrm{Re} \to \infty$ not only describes the appearance of separation at the leading edge but also allows (in the limit as $k_1 \to -\infty$) transition to the conventional theory of the boundary layer.

Proceeding to the solution of this problem, we first of all mention that the induced pressure gradient is essential only within the interaction region, which covers a neighborhood of the point $x = x_0$ of extent $\Delta x = O(\mathrm{Re}^{-1/5})$. Outside this neighborhood the boundary-layer theory remains valid. The solutions obtained in Section 3 for regions 2 and 3 lying ahead of the interaction region also remain valid. As for the interaction region itself, it is easy to see that it has a triple-deck structure (Figure 4.8). Region 3 transforms into region III, the lower layer of the interaction region, when extended into the vicinity of the point

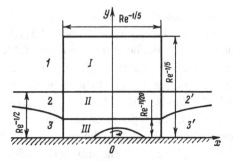

Fig. 4.8 Asymptotic structure of the interaction region.

$x = x_0$ that is of interest to us. The continuation of region 2 is region II, the middle layer of the interaction region. To these must still be added region I, which lies outside the boundary layer.

The middle layer. The asymptotic analysis of the Navier–Stokes equations in region II is related to the limit process

$$x^* = \mathrm{Re}^{1/5}(x - x_0) = O(1), \quad Y = O(1), \quad \mathrm{Re} \to \infty.$$

The asymptotic expansion of the stream function in this region has the form

$$\psi = \mathrm{Re}^{-1/2}\Psi_{00}(Y) + \mathrm{Re}^{-7/10}\widetilde{\Psi}_1(x^*, Y) + \cdots . \tag{4.4.7}$$

One can confirm this as usual by writing the solution (4.3.32), (4.4.6), (4.3.22), (4.3.36) for region 2 in the variables of region II.

The expansion for the pressure gradient must represent the superposition of the original Taylor expansion (4.3.5) and the induced pressure gradient, whose value is estimated in equation (4.4.2). Therefore,

$$\frac{\partial p}{\partial x} = \lambda_0 + \mathrm{Re}^{-1/5}\lambda_1 x^* + \mathrm{Re}^{-3/10}\frac{\partial P^*}{\partial x^*} + \cdots . \tag{4.4.8}$$

Substituting (4.4.7), (4.4.8) into the first Navier–Stokes equation, we obtain

$$\widetilde{\Psi}_1 = \Psi_{00}'\left[\frac{A_1(x^*)}{\lambda_0} + x^*\int_0^Y \frac{\Psi_{00}''' - \lambda_0}{(\Psi_{00}')^2}dY\right]. \tag{4.4.9}$$

The function $A_1(x^*)$ is arbitrary. From the condition of matching with the solution in region 2 it is known only that

$$A_1 = a_0(-x^*) + k_1 a_1(-x^*)^{-1} + \cdots \quad \text{as } x^* \to -\infty. \tag{4.4.10}$$

Substituting (4.4.7), (4.4.8) into the second momentum equation, we find that the induced pressure gradient remains unchanged across region II:

$$\frac{\partial}{\partial Y}\left(\frac{\partial P^*}{\partial x^*}\right) = 0.$$

The lower layer. We turn now to consideration of region III. This is the continuation of the viscous wall region 3, whose thickness decreases as $s \to -0$ according to the rule $Y = O[(-s)^{1/4}]$. Since the length of region III has magnitude of order $\mathrm{Re}^{-1/5}$, the estimate $Y = O(\mathrm{Re}^{-1/20})$ holds for its thickness. Thus, the asymptotic analysis of the Navier–Stokes equations in region III must be based upon the limit process

$$x^* = O(1), \quad Y^* = \mathrm{Re}^{11/20}y = O(1), \quad \mathrm{Re} \to \infty.$$

Writing the solution (4.3.32), (4.4.6), (4.3.21), (4.3.35) for region 3 in the variables of region III, we reach the conclusion that the asymptotic expansion of the stream function in region III is

$$\psi = \mathrm{Re}^{-13/20}\frac{1}{6}\lambda_0 Y^{*3} + \mathrm{Re}^{-16/20}\Psi_1^*(x^*, Y^*) \tag{4.4.11}$$

$$+ \mathrm{Re}^{-17/20}\left(\frac{2\lambda_0\lambda_1}{7!}Y^{*7} + \frac{\lambda_1}{6}x^*Y^{*3}\right) + \mathrm{Re}^{-19/20}\Psi_2^*(x^*, Y^*) + \cdots.$$

The asymptotic expansion of the pressure gradient retains the previous form (4.4.8).

We substitute (4.4.11) into the first momentum equation and match with the solution for region 3 and the solution for region II.[8] As a result we obtain the following boundary-value problem for Ψ_1^*:

$$\frac{1}{2}\lambda_0 Y^{*2}\frac{\partial^2\Psi_1^*}{\partial x^*\partial Y^*} - \lambda_0 Y^*\frac{\partial\Psi_1^*}{\partial x^*} = \frac{\partial^3\Psi_1^*}{\partial Y^{*3}},$$

$$\Psi_1^* = \frac{\partial\Psi_1^*}{\partial Y^*} = 0 \quad \text{at } Y^* = 0, \tag{4.4.12}$$

$$\Psi_1^* = \frac{1}{2}A_1(x^*)Y^{*2} + \cdots \quad \text{as } Y^* \to \infty, \quad x^* \to -\infty.$$

Through direct substitution it is easy to verify that the solution of the problem (4.4.12) is

$$\Psi_1^* = \frac{1}{2}A_1(x^*)Y^{*2}. \tag{4.4.13}$$

The function $A_1(x^*)$ remains arbitrary. To determine it, one must also

---

[8] Here, one must take into account the asymptotic representation (4.3.23) of the function $\Psi_{00}(Y)$ as $Y \to 0$.

consider the problem for the coefficient $\Psi_2^*(x^*, Y^*)$ in the fourth term of the expansion (4.4.11):

$$\frac{\lambda_0}{2} Y^{*2} \frac{\partial^2 \Psi_2^*}{\partial x^* \partial Y^*} - \lambda_0 Y^* \frac{\partial \Psi_2^*}{\partial x^*} = \frac{\partial^3 \Psi_2^*}{\partial Y^{*3}} - \frac{1}{2} A_1 A_1' Y^{*2} - \frac{\partial P^*}{\partial x^*},$$

$$\Psi_2^* = \frac{\partial \Psi_2^*}{\partial Y^*} = 0 \quad \text{at } Y^* = 0, \tag{4.4.14}$$

$$\Psi_2^* = \frac{\lambda_0 a_0^2}{8!} Y^{*9} + \frac{a_0^2}{5!} x^* Y^{*5} + \frac{1}{2} A_2 Y^{*2}$$

$$+ \left[A_1^2(x^*) - a_0^2 x^{*2} - 2k_1 a_0 a_1 \right] \frac{Y^*}{2\lambda_0} + o(1)$$

$$\text{as } Y^* \to \infty, \ x^* \to -\infty.$$

Here, $A_2(x^*)$ is an unknown function which plays the same role in the problem (4.4.14) as the function $A_1(x^*)$ in the problem (4.4.12). Matching the expansion (4.4.11) with the solution in region 3, one can obtain

$$A_2 = b_0(-x^*)^{7/4} + k_1 b_1(-x^*)^{-1/4} + \cdots \quad \text{as } x^* \to -\infty.$$

In contrast to the problem (4.4.12), the boundary-value problem (4.4.14) is not closed, because the pressure gradient $\partial P^*/\partial x^*$ appearing in the right-hand side of the equation for $\Psi_2^*(x^*, Y^*)$ remains undetermined. To close the problem it is necessary to consider region I (Figure 4.8), in which there occurs the development of a pressure gradient induced by the displacement effect of the boundary layer. Before proceeding to an analysis of the flow within this region, we recall that the derivative $\partial P^*/\partial x^*$ does not change across region II. Substituting the expansions (4.4.8), (4.4.11) into the second momentum equation of the Navier–Stokes system, one can see that also in region III

$$\frac{\partial}{\partial Y^*} \left(\frac{\partial P^*}{\partial x^*} \right) = 0.$$

Thus, $\partial P^*/\partial x^*$ maintains its value across the entire boundary layer.

The upper layer of the interaction region. Just as in the cases described previously, the slope of the streamlines within region I is small. Therefore, the linear theory of potential fluid flow is valid here. According to this theory the induced pressure gradient is determined by the curvature of the streamlines at the bottom of region I, which can be found from matching with the solution in region II. Using the relations (4.4.7) and (4.4.9), we find that in region II

$$u = \Psi_{00}'(Y) + \cdots,$$

$$v = -\mathrm{Re}^{-1/2}\Psi'_{00}(Y)\left[\frac{A'_1(x^*)}{\lambda_0} + \int_0^Y \frac{\Psi'''_{00} - \lambda_0}{(\Psi'_{00})^2}dY\right] + \cdots.$$

Therefore, the angle of inclination of the velocity vector is determined as

$$\vartheta = -\mathrm{Re}^{-1/2}\left[\frac{A'_1(x^*)}{\lambda_0} + \int_0^Y \frac{\Psi'''_{00} - \lambda_0}{(\Psi'_{00})^2}dY\right],$$

and for the curvature of the streamlines we obtain the expression

$$\frac{\partial\vartheta}{\partial x} = -\mathrm{Re}^{-3/10}\frac{A''_1(x^*)}{\lambda_0}. \tag{4.4.15}$$

It remains to substitute (4.4.15) into the integral of small-perturbation theory. As a result we find that

$$\frac{\partial P^*}{\partial x^*} = -\frac{U_0^2}{\pi\lambda_0}\int_{-\infty}^{\infty} \frac{A''_1(s)}{s - x^*}ds.$$

Solution of the interaction problem. We introduce the change of variables:

$$\Psi_2^* = \Psi_2 + \frac{\lambda_0 a_0^2}{8!}Y^{*9} + \frac{a_0^2}{5!}x^*Y^{*5} + \frac{A_1^2 - a_0^2x^{*2} - 2k_1a_0a_1}{2\lambda_0}Y^*$$

and the affine transformations

$$\Psi_2 = \frac{a_0^{11/10}U_0^{9/5}}{\lambda_0^{17/10}}\bar{\Psi}, \quad A_1 = \frac{a_0^{3/5}U_0^{4/5}}{\lambda_0^{1/5}}A, \quad \frac{\partial P^*}{\partial x^*} = \frac{a_0^{7/5}U_0^{6/5}}{\lambda_0^{4/5}}r(X),$$

$$x^* = \frac{U_0^{4/5}}{a_0^{2/5}\lambda_0^{1/5}}X, \quad Y^* = \frac{U_0^{1/5}}{a_0^{1/10}\lambda_0^{3/10}}\bar{Y}. \tag{4.4.16}$$

Then the interaction problem assumes the form

$$\frac{1}{2}\bar{Y}^2\frac{\partial^2\bar{\Psi}}{\partial X\partial\bar{Y}} - \bar{Y}\frac{\partial\bar{\Psi}}{\partial X} = \frac{\partial^3\bar{\Psi}}{\partial\bar{Y}^3} - r(X), \quad r(X) = -\frac{1}{\pi}\int_{-\infty}^{\infty}\frac{A''(s)}{s - X}ds,$$

$$\bar{\Psi} = 0, \quad \frac{\partial\bar{\Psi}}{\partial\bar{Y}} = g(X) \quad \text{at } \bar{Y} = 0, \tag{4.4.17}$$

$$\bar{\Psi} = O(\bar{Y}^2) \quad \text{as } \bar{Y} \to \infty, \quad \bar{\Psi} = o(1) \quad \text{as } X \to -\infty.$$

Here, $g(X)$ denotes the combination

$$g(X) = -\frac{1}{2}(A^2 - X^2 + 2a), \tag{4.4.18}$$

whose value, according to the condition (4.4.10), tends to zero as $X \to -\infty$. The parameter a is determined by the equation

$$a = k_1 \frac{(-a_1)\lambda_0^{2/5}}{a_0^{1/5}U_0^{8/5}}$$

and characterizes the deviation of the parameter k from its critical value k_0, with growth of the angle of attack leading to increase in the parameter a.

Equation (4.4.17) for the function $\bar{\Psi}(X,\bar{Y})$ can be solved by employing the classical method of Fourier transformation.[9] It turns out that the solution of this equation satisfying the conditions at $\bar{Y} = 0$ does not grow exponentially as $\bar{Y} \to \infty$ only in the case when

$$g(X) = \frac{\lambda}{2} \int_{-\infty}^{X} \frac{r(\xi)}{(X-\xi)^{1/2}} d\xi, \quad \lambda = \frac{\Gamma(3/4)}{\sqrt{2}\Gamma(5/4)}. \qquad (4.4.19)$$

If the relation (4.4.19) is satisfied, a solution exists and satisfies the remaining boundary conditions. At the same time it proves not to be unique, which we can verify with the help of the following reasoning. Suppose we know some solution $\bar{\Psi}(X,\bar{Y})$ of the given boundary-value problem. Then by direct substitution it is easy to verify that a solution is also

$$\bar{\Psi} + \tfrac{1}{2}B(X)\bar{Y}^2$$

with an arbitrary function $B(X)$. This arbitrariness attests to the impossibility of determining the function $A_2(x^*)$, appearing in the statement of the boundary-value problem (4.4.14), from the solution of just this problem. In order to find $A_2(x^*)$ it is necessary to consider the next term in the expansion (4.4.11) of the stream function in region III.

The solvability condition (4.4.19) for the boundary-value problem (4.4.17), written taking into account the expression (4.4.18) for $g(X)$,

$$A^2 - X^2 + 2a = -\lambda \int_{-\infty}^{X} \frac{r(\xi)}{(X-\xi)^{1/2}} d\xi, \qquad (4.4.20)$$

yields a first connection between the function $A(X)$ and the pressure gradient. The second one is expressed by the Cauchy integral

$$r(\xi) = -\frac{1}{\pi} \int_{-\infty}^{\infty} \frac{A''(s)}{s-\xi} ds. \qquad (4.4.21)$$

---

[9] A detailed description of this procedure is contained in the appendix of Stewartson's paper (1970b) and in the work of Ruban (1982a) as well.

Substituting this into the right-hand side of the relation (4.4.20) and interchanging the order of integration, we obtain an integro-differential equation

$$A^2 - X^2 + 2a = \lambda \int_X^\infty \frac{A''(s)}{(s-X)^{1/2}} ds \qquad (4.4.22)$$

for the function $A(X)$.

We now summarize the analysis that has been performed. We have seen that leading-edge separation on a thin airfoil, as well as separation from a smooth surface (accompanied by formation of an extensive zone of recirculating flow), originates within the interaction region. If, however, separation from a smooth surface arises from self-induced growth of the pressure gradient applied to the boundary layer, then separation from the leading edge of a thin airfoil differs in that the boundary layer entering the interaction region already has a separation velocity profile with vanishingly small shear at the surface. It is developed under the action of an adverse pressure gradient distributed over some portion of the profile surface ahead of the interaction region. The pressure gradient remains bounded in this region, and the appearance of separation is explained only by the singular behavior of the solution of the boundary-layer equations at a point with zero skin friction.

The observed difference leads to an essential change in the mathematical theory of the fluid flow within the interaction region. In particular, the formulation of the interaction problem is changed. Whereas for separation from a smooth surface it was written with the help of the relations (1.7.3)–(1.7.4) of Chapter 1, now this problem is reduced to the integro-differential equation (4.4.22).

To emphasize the distinctiveness of the theory of short separation zones at the leading edge of a thin airfoil, this is commonly referred to as the theory of marginal separation, and the equation (4.4.22) for $A(X)$ is called the fundamental equation of marginal-separation theory. We recall that the function $A(X)$ is related by the affine transformation (4.4.16) to the function $A_1(x^*)$, which appears in the expression (4.4.13) for Ψ_1^* and hence determines the two-term expansion (4.4.11) of the stream function within the viscous wall layer of the interaction region. Differentiating (4.4.11), we conclude that the skin friction is proportional to $A(X)$:

$$\tau = \left.\frac{\partial u}{\partial Y}\right|_{Y=0} = \mathrm{Re}^{-1/5} \frac{a_0^{3/5} U_0^{4/5}}{\lambda_0^{1/5}} A(X) + \cdots.$$

The theory of marginal separation was developed independently in two studies: by Ruban (1982a) and Stewartson, Smith, and Kaups (1982). In describing the theory in this chapter we have relied upon Ruban's paper and also his earlier work (Ruban, 1981) devoted to the analysis of the boundary layer near the point of zero skin friction (see Section 3 of this chapter).

4.5 Numerical Solution of the Fundamental Equation

The boundary conditions for equation (4.4.22) are the conditions

$$A(X) = |X| + \cdots \quad \text{as } |X| \to \infty, \tag{4.5.1}$$

which follow from matching with the singular branch of the solution of the boundary-layer equations.[10]

It is easy to construct an asymptotic solution of the fundamental equation for large negative values of the parameter a. If $a \to -\infty$, the required function $A(X)$ is a quantity of order $|a|^{1/2}$. The region where the fundamental equation is being solved expands according to the same law: $X = O(|a|^{1/2})$. Therefore, the integral on the right-hand side of equation (4.4.22) is estimated as $O(|a|^{-1/4})$. At the same time the left-hand side of this equation grows and is proportional to $|a|$. Hence, the leading term of the asymptotic expansion of the function $A(X)$ as $a \to -\infty$ will be

$$A(X) = \sqrt{X^2 - 2a} + \cdots . \tag{4.5.2}$$

Considering that the right-hand side of equation (4.4.22), connected with the induced pressure gradient, becomes negligibly small as $a \to -\infty$, we come to the conclusion that the asymptotic solution (4.5.2) of the fundamental equation is essentially a solution of the boundary-layer equation (4.3.1). If we rewrite (4.5.2) taking account of the transformations (4.4.16) in the form

$$A_1(x^*) = a_0 \sqrt{x^{*2} + 2k_1 \frac{a_1}{a_0}} + \cdots ,$$

substitute the expressions (4.4.7) and (4.4.9), and then transform to the original boundary-layer variables, we obtain as a result the above formula (4.3.37) for the stream function in the vicinity of the point

---

[10] If the conditions (4.5.1) are not imposed the solution of equation (4.4.22) is not unique. In particular, for $a = 0$, this equation allows at least two more solutions $A = X$ and $A = -X$ besides the one that meets the condition (4.5.1).

$x = x_0$. A detailed derivation of this formula based on the solution of the boundary-layer equation is given in the paper of Ruban (1981).

To solve the fundamental equation Stewartson, Smith, and Kaups (1982) applied two numerical methods. One of these coincides with the method used by Ruban (1982a). In developing this method the fundamental equation is first transformed using a procedure employed in solving Abel's integral equation, and then is integrated with respect to X, taking into account the relationship[11]

$$A^2 - X^2 + 2a = -\frac{3}{4}a\lambda\pi X^{-5/2} + \cdots \quad \text{as } X \to \infty, \qquad (4.5.3)$$

which results in

$$A'(X) = 1 - \frac{1}{\pi\lambda}\int_X^\infty \frac{A^2(s) - s^2 + 2a}{(s - X)^{1/2}}ds. \qquad (4.5.4)$$

Then a sufficiently large value of X_∞ is chosen, and the infinite region $-\infty < X < \infty$, where the solution of the fundamental equation is sought, is replaced by the segment $[-X_\infty, X_\infty]$. This segment is divided into intervals of length ΔX. The unknowns are the values of $A(X)$ at the end points of the intervals.

An iterative process is constructed such that each subsequent approximation to the function $A(X)$ results from substituting the previous approximation into the right-hand side of equation (4.5.4). Here, the integral over the segment $[X + \Delta X, X_\infty]$ is calculated according to the trapezoidal rule; for $s > X_\infty$, the integration is extended using the asymptotic formula (4.5.3), and across the segment $[X, X + \Delta X]$ the numerator of the integrand is approximated by a linear function and the integration is performed analytically. These calculations result in the values of the derivative $A'(X)$ at the nodes of the computational grid. The function $A(X)$ is recovered from $A'(X)$ using the asymptotic formula

$$A^2 - X^2 + 2a = 2\lambda(-X)^{-1/2} + \cdots \quad \text{as } X \to -\infty, \qquad (4.5.5)$$

which serves to determine the values of the function $A(X)$ at the point $X = -X_\infty$. Convergence of the iteration process is achieved by the use of underrelaxation.

Still more effective was the second numerical method, where the equation (4.4.22) was approximated by a system of nonlinear algebraic equations. Their derivation was based on application of the trapezoidal rule

---

[11] This relationship as well as the formula (4.5.5) were obtained from equation (4.4.22) by using the boundary conditions (4.5.1).

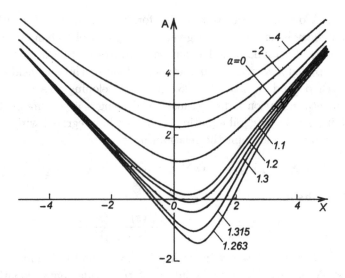

Fig. 4.9 Graphs of the function $A(X)$ constructed for various values of the parameter a.

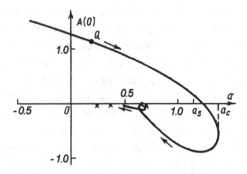

Fig. 4.10 The fundamental curve.

to the integral on the right side of equation (4.4.22). The unknowns are the values of the function $A(X)$ at the interior points of the segment $[-X_\infty, X_\infty]$, and its values on the boundaries of this segment are given according to the formulas (4.5.3) and (4.5.5). The solution of the resulting system of equations is found with the help of Newton's method.

The results of the computation are shown in Figure 4.9. To illustrate these results it is convenient to use the plot of $A(0)$ versus a (Figure 4.10). Each solution of the fundamental equation corresponds to a

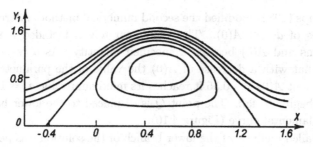

Fig. 4.11 Sketch of the streamlines within the interaction region for $a = 1.3$.

certain point on such a graph, and as a whole it gives the set of admissible solutions. Henceforth, we call it the fundamental curve.

Let the point Q move along the fundamental curve in the direction of the arrows. Its motion is accompanied by a continuous variation of the solution. For large negative values of the parameter a, the graph of the function $A(X)$ agrees well with the asymptotic formula (4.5.2). With increasing a, the skin friction, proportional to $A(X)$, decreases (Figure 4.9). Simultaneously, there occurs a downstream displacement of the minimum skin friction. Separation occurs at $a = a_S = 1.14$, when the minimum value of the skin friction vanishes; at that moment $A(0)$ still remains positive (Figure 4.10). For $a > a_S$ a zone of recirculating flow appears at the profile surface. As an example, Figure 4.11 (Ruban, 1982a) shows the picture of the streamlines within this zone for $a = 1.3$. This was constructed with the help of an expansion

$$\psi_1 = \frac{1}{6}Y_1^3 + \frac{1}{2}A(X)Y_1^2,$$

which can be obtained as a result of the change of variables

$$\psi = \mathrm{Re}^{-11/10}\frac{U_0^{12/5}a_0^{9/5}}{\lambda_0^{13/5}}\psi_1, \quad Y^* = \mathrm{Re}^{-3/20}\frac{U_0^{4/5}a_0^{3/5}}{\lambda_0^{6/5}}Y_1$$

in the two-term expansion of the stream function (4.4.11), (4.4.13) for the viscous wall layer of the interaction region. Neighboring streamlines are shown in Figure 4.11 at intervals of 0.02 in ψ_1.

Up to this point the calculation was carried out such that during the iterative process the parameter a was maintained fixed, and after the solution was obtained the value of the parameter was increased and the calculations repeated. Along this path, however, an obstacle arose: for all $a > 1.33$ the iterative process proved to be divergent, and no solution was obtained. To extend the calculations Stewartson, Smith, and Kaups (1982) modified the second numerical method, interchanging the roles of a and $A(0)$. The parameter a was included among the unknowns and $A(0)$ became a specified quantity. As a result it was shown that with a decrease of $A(0)$ the value of the parameter a first increases, at $A(0) = -0.55$ it reaches its maximum value $a_c = 1.33$, and then it begins to fall. The point Q is displaced to the lower branch of the fundamental curve (Figure 4.10).

A detailed analysis of the lower branch of the solution was performed by Brown and Stewartson (1983). They discovered the existence of a loop on the fundamental curve, and they also showed that the point $a = 0$ is a singular point of the lower branch. A study of the features of the solution near this point was conducted with the help of an asymptotic analysis of equation (4.4.22) as $a \to +0$. As it turned out, as the parameter a approaches its limiting value $(a = 0)$, the extent of the separation region increases without bound. Here, the separation point approaches $X = 0$ and the reattachment point recedes to infinity.

Thus, for all $a < 0$ the solution of the fundamental equation exists and is unique. This shows that within the corresponding range of angle of attack the flow past the leading edge is realized without separation. For $0 < a < a_c$ the fluid motion within the boundary layer may assume at least two forms. If $0 < a < a_S$, then one of the solutions of the fundamental equation corresponds to unseparated flow and the other to a flow with a local zone of separation. If $a_S < a < a_c$, then both solutions indicate the existence of a local zone of separation, this zone being larger on the lower branch of the fundamental curve. Finally, for $0.60 < a < 0.68$ the number of solutions increases up to four. One of them corresponds to unseparated flow, and the rest to flows with local zones of separation.

The above analysis also shows (and this result is especially important) that for any $a > a_c$ the fundamental equation does not have a solution.[12] Thus, passage through the critical value of the parameter a cannot be realized within the framework of the marginal-separation theory. We recall that this theory is based upon the assumption of a local region of recirculating flow. For $a > a_c$ it seems better to abandon this limitation and to seek the solution within the class of flows with an extended zone of separation (Chapter 1). This jump transition from one regime of flow around a profile to another one is well known from experiment as a process that accompanies the phenomenon of short-bubble bursting.

---

[12] To be more precise, the solution becomes complex (Chernyshenko, 1985).

The fluid flow near the leading edge of an airfoil was the first for which nonuniqueness of a solution of the equations of hydrodynamics was established within the framework of the asymptotic theory of boundary-layer separation. Commenting on this result, Stewartson, Smith, and Kaups (1982) drew attention to the considerable difference of this nonuniqueness from that encountered in self-similar flows, when an appropriate solution can be chosen out of a number of admissible ones resulting from analysis of the whole flow picture – as a rule, only one of the solutions matches the actual boundary conditions. In the present case, however, a nonunique one proves to be the solution of a problem for which, in accordance with the formalism of the method of matched asymptotic expansions, all the required boundary conditions have been formulated.

It is known from experimental observations that the picture of flow past a solid body under the same conditions may assume various configurations. This phenomenon, called quasi-steady hysteresis, is also observed in flow past an airfoil (e.g., see the paper by Kuryanov, Stolyarov, and Steinberg, 1979). If the angle of attack is at first increased and then decreased (so slowly that at every instant of time the flow is nearly steady), then it may happen that the moment of appearance of flow separation will not coincide with the moment of return to unseparated flow. The graph of the dependence of lift upon angle of attack assumes the shape of a hysteresis curve, and for one value of the angle of attack two flow states become possible, and the choice between them must be made taking into account the history of the formation of the flow. It appears that we deal with nonuniqueness of this kind in the theory of marginal separation.

Using the results of the above analysis we can, for any profile with a given contour, reduce the problem of determining the angle of attack at which a short bubble is generated, and the angle of attack at which this bubble is destroyed, to integration of the classical equation of Prandtl (1904) for the boundary layer on the profile surface. We denote by α'_0 the value of the angle of attack corresponding to the moment that the minimum skin friction vanishes. For a thin airfoil with a parabolic leading edge, such a condition of the boundary layer is observed at $k = k_0$. If one uses the relation (4.2.2) and for simplicity considers a symmetric profile ($F_+ = -F_-$), then it is easy to find that the leading term in the

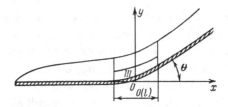

Fig. 4.12 The boundary layer on a hyperbolic surface.

expansion for α'_0 as $\epsilon \to 0$ will be

$$\alpha'_0 = \epsilon \frac{k_0}{\sqrt{2}} + O(\epsilon^2) = 0.8172\epsilon + O(\epsilon^2).$$

Interaction of the boundary layer with the outer inviscid part of the flow delays the separation. According to the results of the numerical solution of the fundamental equation (4.4.22), the appearance of a short bubble can be expected at

$$\alpha' = \alpha'_0(1 + 2.037 \, \mathrm{Re}^{-2/5} + \cdots)$$

and its bursting at

$$\alpha' = \alpha'_0(1 + 2.376 \, \mathrm{Re}^{-2/5} + \cdots).$$

In the present chapter the theory of marginal separation was for concreteness described by using the example of flow near the leading edge of a thin airfoil. However, the range of its application is far wider than this example. In particular, Fomina (1983) showed that the marginal-separation theory remains valid even when one deals with the boundary layer on the surface of a solid body (Figure 4.12) whose contour has a hyperbolic shape

$$y = \frac{\theta}{2}(x + \sqrt{x^2 + l^2}). \tag{4.5.6}$$

Here, one uses a Cartesian coordinate system Oxy, where the leading edge of the body is placed at the point $x = -1$. The parameter l defining the length of the curved section of the surface is considered small. If l is identically equal to zero we recover the problem of flow around a corner of a body contour analyzed in Chapter 2, Section 3. In this case, the self-similar solution of Blasius is realized ahead of the corner. The Blasius solution becomes invalid within a small neighborhood of the corner where the interaction region appears. Its length is $\Delta x = O(\mathrm{Re}^{-3/8})$. We recall that it is in just this region that boundary-layer separation originates.

The theory described in Chapter 2, Section 3 remains valid also in the case when l differs from zero but does not exceed $Re^{-3/8}$ in order of magnitude. But if this condition is violated, the interaction effect disappears and the flow within the viscous wall layer (region III in Figure 4.12) obeys the classical theory of Prandtl. As the calculations showed, this flow possesses the same properties as the boundary layer at the leading edge of an airfoil: the distribution of skin friction over the surface has a minimum, its magnitude depending upon the angle θ, and it vanishes at a certain value of the angle. Whenever such behavior of the boundary layer is encountered, one can be assured of the applicability of the marginal-separation theory.

We mention also the work by Hackmüller and Kluwick (1989) devoted to analysis of the flow within the interaction region considered above in the presence of unevenness of the surface – a crest or a trough.

Another interesting example was examined by Zametaev (1986). He investigated the separation of a viscous wall jet flowing past the surface of a hyperbolic body (4.5.6). Under certain limitations on the parameter l it was possible to show that this flow is also described by marginal-separation theory, but now it appears in a modified form. The situation is that the induced pressure for a jet is proportional to the curvature of the streamlines at the outer edge of the viscous sublayer: $r = -A'''(X)$.[13] Therefore, instead of (4.4.22) we obtain from (4.4.20) the following equation:

$$A^2 - X^2 + 2a = \lambda \int_{-\infty}^{X} \frac{A'''(\xi)}{(X - \xi)^{1/2}} d\xi .$$

Its solution shows that the corresponding fundamental curve has the form of a logarithmic spiral (Figure 4.13) so that as $|a| \to 0$ the number of solutions becomes infinite (Zametaev, 1987).

Ruban (1982b) and independently Smith (1982b) extended the marginal-separation theory to the case of unsteady fluid flows. The boundary layer at the leading edge of a thin airfoil turns out to be highly sensitive to the effect of unsteadiness, which may account for the experimentally observed dynamic separation from the leading edges of thin airfoils that perform relatively slow oscillations in the flow. The

---

[13] The phenomenon of interaction and separation of the viscous wall jet was investigated by Messiter and Liñán (1976) and Shidlovsky (1977) and also by Smith and Duck (1977).

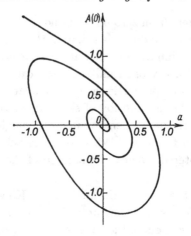

Fig. 4.13 The fundamental curve for a viscous wall jet (calculations of Zametaev, 1986).

nonsteady analog of equation (4.4.22) has the form

$$\frac{\partial A}{\partial T} = -\frac{\gamma}{2}\frac{\partial}{\partial X}\int_{-\infty}^{X}\left[A^2(\xi, T) - \xi^2 + 2a\right.$$
$$\left. - \lambda\int_{\xi}^{\infty}\frac{A''(s, T)}{(s-\xi)^{1/2}}ds\right]\frac{d\xi}{(X-\xi)^{3/4}},$$
(4.5.7)

where T is a dimensionless time referred to $\frac{r}{U_\infty}\left(\frac{\lambda_0^{3/10}}{a_0^{9/10}U_0^{1/5}}\right)\mathrm{Re}^{1/20}$, and γ is a constant: $\gamma = \Gamma(5/4)/2^{1/4}\pi$.

Ruban (1982b) used equation (4.5.7) to study the stability of the flow at the leading edge of a thin airfoil, a nonlinear theory of stability being constructed as well as a linear one. The oscillation in the flow under consideration was found to branch off from the steady solution in the domain of subcritical values of the Reynolds number. Typical of such flows is a subcritical type of transition to turbulence. In our opinion it is just this circumstance that explains the fact that the transition zone in the region of separation at the leading edge of a thin airfoil is usually extremely short.

The work of Smith (1982b), his later study performed in cooperation with Ryzhov (Ryzhov and Smith, 1984), and also the studies of Smith and Elliott (1985) and of Elliott and Smith (1987) were devoted to the solution of the Cauchy problem for the unsteady equation of marginal-separation theory. The main result of these studies consists in the fact that perturbations of the steady solution are not damped in the course of time. On the contrary, a singularity appears in the solution and the function $A(T, X)$ increases without bound at a certain point $X = X_S$ on the surface of the body, this taking place during either a finite or even an infinitesimal interval of time.

Finally, the marginal-separation theory was applied to the analysis of pulsatile flows in the boundary layer. This was done by Timoshin (1988a, 1988b, 1988c). First, he studied the effect of pulsations of the oncoming flow upon the angle of attack, for which a short zone of separation is formed at the leading edge of a thin airfoil (Timoshin, 1988a); and second, he performed an analysis of the pulsations appearing in the boundary layer on the surface of a flat plate under the conditions that harmonic oscillations of the longitudinal component of the velocity vector occur at the outer edge of the boundary layer (Timoshin, 1988b, 1988c).

Further development of the marginal-separation theory, including, in particular, the analysis of three-dimensional flows, is found in the survey of Ruban (1991).

5

The Theory of Unsteady Separation

5.1 The Analogy between Unsteady Separation and Separation from a Moving Surface

For the steady fluid motions considered in the preceding chapters, flow separation leads to a change in the flow structure as a whole. The limiting state when the Reynolds number tends to infinity is defined by the Helmholtz–Kirchhoff theory for ideal fluid flows with free streamlines, and the location of the separation point, in accordance with the criterion of Prandtl, coincides with the point where the shear stress at the surface of the body vanishes (see Chapter 1).

The situation is somewhat different if the flow is unsteady. To illustrate this fact we consider the example of flow past a circular cylinder that is set into motion instantaneously from rest. The starting motion of a body in a viscous fluid can be likened to the introduction of the no-slip condition at the surface of a body as it moves through a fluid that has no internal friction. Therefore, at the first instant the flow is potential and is described by the well-known solution for unseparated steady flow past a cylinder with zero circulation. At the body surface, where the no-slip condition is imposed, there arises the process of vorticity diffusion and convection, which leads to the formation of a boundary layer.

Blasius (1908) formulated this problem and gave an approximate solution. It turns out that at a certain instant the point of zero skin friction starts moving upstream along the body surface from the rear stagnation point. At the same time a region of reverse flow appears within the boundary layer (Figure 5.1). Although the pressure distribution is given, and a zero-friction point is located on the body surface, the solution of the unsteady boundary-layer equations is regular. Thus, in contrast to steady-flow problems, the appearance of reverse flow does not lead to violation of Prandtl's hierarchical concept (see Chapter 1). Therefore,

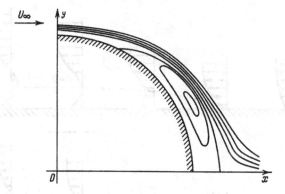

Fig. 5.1 An instantaneous picture of the streamlines within the boundary layer on an impulsively started cylinder; calculation of Blasius (1908).

the appearance of a point of zero skin friction in an unsteady boundary layer is not to be identified with the occurrence of flow separation.

This circumstance was first noted by Rott (1956), Sears (1956), and Moore (1958). The same authors gave another criterion for separation, different from that of Prandtl. The most fruitful and descriptive approach to the analysis proves to be an approach based upon the analogy between separation of an unsteady boundary layer and steady boundary-layer separation from the surface of a moving body. Let us describe this analogy and define the criterion for unsteady separation, taking the work of Moore (1958) as a basis.

We consider a steady boundary-layer flow, assuming that the body surface moves downstream with constant speed. If the boundary layer develops under the action of pressure that is increasing downstream, then the fluid in this layer will be decelerated. Since the body surface moves, entraining the nearest portion of the fluid, the boundary-layer velocity profile will, beginning at a certain section, have a minimum point that evidently coincides with the point of zero shear stress ($\tau = 0$) (Figure 5.2a). The totality of these points represents a line that divides the boundary layer into zones in one of which the shear stress is positive and in the other negative. (This line is shown dashed in Figure 5.2a.) If at some point the minimum of the longitudinal velocity component vanishes, then beyond this point lies a region of reverse flow. Thus, the shear stress and the longitudinal velocity component should vanish simultaneously at the separation point for a surface that is moving downstream. However, in contrast to steady flow past a fixed surface, this point lies within the boundary layer rather than on the body. Fig-

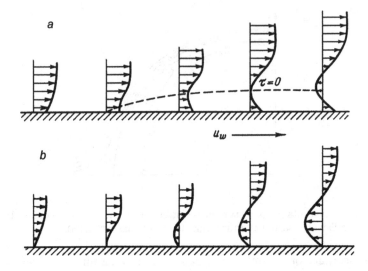

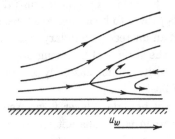

Fig. 5.2 Velocity profiles in a boundary layer near the separation point on a surface that is moving downstream, and corresponding velocity profiles in an unsteady boundary layer.

Fig. 5.3 Streamlines in the neighborhood of the separation point for a surface moving downstream.

ure 5.3 shows a qualitative picture of the streamlines. It is important to emphasize that the appearance of reverse flow will change the structure of the external flow, as in the case of separation from a fixed surface.

If we now transform to a system of coordinates attached to the moving surface, then within this new system the separation point will move upstream relative to the fixed surface with a constant speed equal to that of the surface in the original steady flow. The point of zero surface friction is located at some finite distance from the separation point. Velocity profiles corresponding to such an unsteady flow are shown in Figure 5.2b; they are obtained from the velocity profiles shown in Fig-

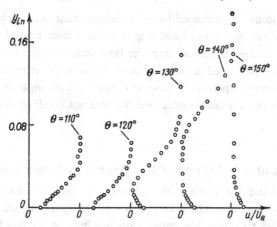

Fig. 5.4 Velocity profiles in the boundary layer on the surface of a rotating cylinder. Experimental data are from Ludwig (1964). (θ is the polar angle measured from the forward stagnation point.)

ure 5.2a with the help of the transformation to a new coordinate system mentioned above. If the separation point moves with variable speed, its acceleration being not very large, then this evidently should not lead to qualitative changes in the flow structure in comparison with that considered.

Thus, according to the Moore–Rott–Sears criterion, which generalizes the criterion of Prandtl (1904), unsteady flow separation occurs inside the boundary layer at a point of simultaneous vanishing of the shear stress and the longitudinal velocity component in the coordinate system fixed with respect to that point. Henceforth, this will be referred to as the Moore–Rott–Sears point; sometimes it is called the boundary-layer breakaway point.

The flows near separation points described earlier were studied experimentally by Ludwig (1964). Figure 5.4 shows the velocity profiles u/U_e obtained for the boundary layer on a surface moving downstream. (The dimensionless value of the surface speed is 0.3; U_e is the speed at the outer edge of the boundary layer.) These results are seen to agree qualitatively with the character of the flow as described earlier. In other experimental work, Koromilas and Telionis (1980) carried out studies of velocity profiles for unsteady motion, and also confirmed the Moore–Rott–Sears concept.

We remark that in the case where the body surface moves in the direction opposite to the outer flow, the formation of the flow near the separation point is influenced by disturbances that come both from the region where the boundary layer originates and from that of the reverse flow, which makes understanding the flow structure far more difficult. Moore (1958) proposed using the same separation criterion when the surface is moving upstream. However, since an asymptotic theory for this case has not yet been developed, we will not dwell on consideration of such flows.

5.2 Singular Solutions of the Boundary-Layer Equations

After describing the qualitative behavior of the flow near separation points on a moving surface and in an unsteady boundary layer, one could proceed to study the separation mechanism on the basis of an asymptotic analysis of the Navier–Stokes equations. However, prior to that let us analyze solutions of the Prandtl boundary-layer equations near Moore–Rott–Sears points in the classical way, that is, with a given regular pressure distribution. In advance one might mention that in this case the separation is self-induced, just as for steady flows over a fixed surface, whereas the classical approach leads to the appearance of non-removable singularities. It is of interest to analyze these singularities in the solution of the system of boundary-layer equations, because nonremovable singularities indicate the impossibility of unseparated flow and the necessity of reconsidering the solution of the outer problem.

Statement of the problem. The appearance of a third independent variable makes the investigation of the unsteady boundary-layer equations more complicated in comparison with the corresponding steady-flow equations; therefore, we first analyze steady solutions of the boundary-layer equations on a surface that moves with constant speed, taking into consideration the analogy described in the preceding section.

We denote by Ox° and Oy° the axes of an orthogonal curvilinear coordinate system that are directed, respectively, along the surface of the body and normal to it; by u° and v° the velocity components along these axes; by ψ° the stream function; and by p° the pressure. We introduce the dimensionless variables:

$$x = \frac{x^\circ}{L}, \quad y = \frac{y^\circ}{L}, \quad u = \frac{u^\circ}{U_\infty},$$

$$v = \frac{v^\circ}{U_\infty}, \quad p = \frac{p^\circ - p_\infty}{\rho U_\infty^2}, \quad \psi = \frac{\psi^\circ}{LU_\infty},$$

where L is a typical dimension of the body, U_∞ and p_∞ are parameters of the uniform oncoming flow, and ρ is the density. The Reynolds number is $\mathrm{Re} = U_\infty L/\nu$, where ν is the coefficient of kinematic viscosity.

We shall make use of the Prandtl boundary-layer equations in both the usual variables x, y and the von Mises (1927) variables x, ψ. In the latter case, together with the corresponding boundary conditions, they have the form

$$u\frac{\partial u}{\partial x} + \frac{dp_e}{dx} = u\frac{\partial}{\partial \Psi}\left(u\frac{\partial u}{\partial \Psi}\right), \quad \frac{\partial Y}{\partial \Psi} = \frac{1}{u},$$

$$V = u\frac{\partial Y}{\partial x}; \qquad\qquad\qquad (5.2.1)$$

$$u = u_w, \quad Y = V = 0 \quad \text{at } \Psi = 0;$$

$$u \to U_e(x) \text{ as } \Psi \to \infty; \quad u = U^*(\Psi) \quad \text{at } x = x_0.$$

Here, as usual, $\Psi = \mathrm{Re}^{1/2}\psi$, $Y = \mathrm{Re}^{1/2}y$, $V = \mathrm{Re}^{1/2}v$; the functions $U_e(x)$ and $p_e(x)$ define the distributions of velocity and pressure at the outer edge of the boundary layer, and $U^*(\Psi)$ is the initial velocity profile. Here, u_w denotes the speed of the body surface, which we take as a constant of order unity.

The boundary layer on a surface moving downstream ($u_w > 0$). As a result of the numerical solution of the problem (5.2.1) (for different $U_e(x)$) by Telionis and Werle (1973), and also by Williams and Johnson (1974a), it was shown for the first time that, in agreement with the ideas of Moore (1958), for an adverse pressure gradient at a point where the values of shear stress and longitudinal velocity vanish simultaneously, there arises a singularity that is characterized by unbounded growth of the vertical velocity component; at the point of zero shear stress on the moving surface, the solutions remain regular.

We turn now to an analysis of the structure of the solution in the neighborhood of this point (Sychev, 1980).

For convenience we take the longitudinal coordinate of the Moore–Rott–Sears point to be zero. The functions $U_e(x)$ and $p_e(x)$ in (5.2.1) are considered to be known from the solution of the outer problem and are assumed to be regular, that is, as $x \to 0$,

$$U_e = U_{00} - \lambda_0 U_{00}^{-1}x + O(x^2), \quad \frac{dp_e}{dx} = -U_e\frac{dU_e}{dx} = \lambda_0 + O(x). \quad (5.2.2)$$

Here, U_{00} and λ_0 are positive constants.

We denote the value of the stream function Ψ at the Moore–Rott–Sears point by Ψ_S. In the main part of the boundary layer, where $\Psi > \Psi_S$, and in the part adjacent to the moving body surface ($\Psi < \Psi_S$), the

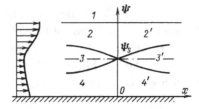

Fig. 5.5 Asymptotic structure of the solution in the vicinity of the singular point where $u = \partial u / \partial y = 0$.

longitudinal velocity component is of order unity as $x \to -0$. Therefore, according to (5.2.2), as $x \to -0$ the solution in these regions (regions 2 and 4 in Figure 5.5) can be written in the form

$$
\begin{aligned}
u &= U_0(\Psi) + (-x)U_1(\Psi) + O[(-x)^2], \quad \Psi > \Psi_S, \\
u &= \overline{U}_0(\Psi) + (-x)\overline{U}_1(\Psi) + O[(-x)^2], \quad \Psi < \Psi_S.
\end{aligned}
\tag{5.2.3}
$$

Substituting these expansions into (5.2.1) and (5.2.2), we find that

$$
U_1 = \lambda_0 U_0^{-1} - (U_0 U_0')', \quad \overline{U}_1 = \lambda_0 \overline{U}_0^{-1} - (\overline{U}_0 \overline{U}_0')'.
\tag{5.2.4}
$$

From the condition at the outer edge of the boundary layer and the no-slip condition at the body surface, it follows that $U_0(\infty) = U_{00}$ and $\overline{U}_0 = u_w + c_0 \Psi + (\lambda_0 - c_0^2 u_w)(2u_w^2)^{-1}\Psi^2 + O(\Psi^3)$ as $\Psi \to 0$; c_0 is an arbitrary constant that characterizes the value of the shear stress at the body surface.

Now substituting the solutions (5.2.3) and (5.2.4) into the second equation of the system (5.2.1), we find that

$$
\begin{aligned}
Y &= G(x) + Y_0(\Psi) + O[(-x)], \quad U_0 Y_0' = 1, \\
Y &= \overline{G}(x) + \overline{Y}_0(\Psi) + O[(-x)], \quad \overline{U}_0 \overline{Y}_0' = 1.
\end{aligned}
\tag{5.2.5}
$$

Satisfying the boundary condition at the body surface, we obtain

$$
\overline{Y}_0 = \int_0^\Psi \overline{U}_0^{-1} d\overline{\Psi}, \quad \overline{G}(x) = 0.
$$

The function $G(x)$ in (5.2.5) remains arbitrary. (Expansions for the vertical velocity component are found from the third equation in (5.2.1), and will not be written out in the following.)

The qualitative analysis presented in the preceding section shows that as the Moore–Rott–Sears point is approached, under the influence of the adverse pressure gradient there appears a minimum point in the boundary-layer velocity profile (Figure 5.2a), with the minimum value

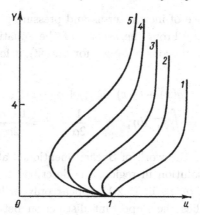

Fig. 5.6 Velocity profiles in the boundary layer. Numerical solution of Zubarev (1983); curves 1–5 correspond to the values $X = 2.0, 2.2, 2.3, 2.4, 2.45$.

tending to zero as $x \to -0$. This is also indicated by the results of numerical solutions. Figure 5.6 shows velocity profiles in the boundary layer for various values of x, obtained by Zubarev (1983).[1] (The outer boundary conditions here are determined by the well-known solution for unseparated flow past a circular cylinder with zero circulation, and therefore $U_e = 2 \sin(x + X_S)$, $x = X - X_S$, $0 \leq X \leq \pi$. The longitudinal coordinate of the singular point is $X = X_S = 2.479$ and $u_w = 1$.)

The appearance of a minimum in the velocity profile at the section $x = -0$ means that near the streamline $\Psi = \Psi_S$ the value of the longitudinal velocity component $u(x, \Psi)$ tends to zero. Therefore, we may assume that as $\Psi \to \Psi_S + 0$,

$$U_0 = a_0(\Psi - \Psi_S)^\alpha + o[(\Psi - \Psi_S)^\alpha], \qquad (5.2.6)$$

where a_0 and α are positive constants. Substituting this expression into (5.2.4) and (5.2.3), we observe that as $x \to -0$ and $\Psi - \Psi_S \to +0$ the first and second terms of the expansion (5.2.3) become quantities of the same order when $|\Psi - \Psi_S| = O[(-x)^{1/2\alpha}]$. Consequently, a new characteristic region 3 lies in the vicinity of the streamline $\Psi = \Psi_S$.

Since $u(x, \Psi)$ should tend to zero in this region as $x \to -0$, we represent the solution in the following form:

$$u = (-x)^\beta f_0(\eta) + o[(-x)^\beta], \quad \eta = (\Psi - \Psi_S)/(-x)^{1/2\alpha}. \qquad (5.2.7)$$

Substituting this expression into (5.2.1), and considering in the first

---

[1] See also the work of Zubarev (1984).

equation the balance of inertia terms and pressure gradient (5.2.2), we find that $\beta = 1/2$. From the second of the equations (5.2.1), after substituting the expressions obtained for $u(x, \Psi)$, it follows that as $x \to -0$,

$$Y = g(x) + (-x)^{\gamma} h_0(\eta) + o[(-x)^{\gamma}],$$

$$h_0 = \int f_0^{-1} d\eta, \quad \gamma = \frac{1}{2\alpha} - \beta = \frac{1-\alpha}{2\alpha}. \tag{5.2.8}$$

As was shown in the work of Sychev mentioned above (see also his 1983 paper), the solution in region 3 is described by the inviscid-flow equations; viscous terms in (5.2.1) appear only in later terms of the expansion. Here lies the important distinction between the singular solution under study and the solution in the vicinity of the point of zero skin friction on a fixed surface (see Chapter 4, Section 3). The exponent α in (5.2.6), which also determines the characteristic dimension of region 3, can assume the following discrete values:

$$\alpha = \frac{n+1}{2}, \quad n = 1, 2, 3, \ldots . \tag{5.2.9}$$

This follows from solution of the eigenvalue problem for a "viscous sublayer," where $|\Psi - \Psi_S| = O[(-x)^{3/4}]$; the details are not cited here, since similar considerations will appear in Section 3 of this chapter for the study of separation with a self-induced pressure gradient.

We see that the most general type of singularity corresponds to the smallest value $\alpha = 1$. According to (5.2.7) and (5.2.8), the solution for the first two terms of the expansion in region 3 then has the form

$$u = (-x)^{1/2} f_0(\eta) + (-x) f_1(\eta) + O[(-x)^{3/2}],$$

$$Y = g(x) + h_0(\eta) + (-x)^{1/2} h_1(\eta) + O[(-x)], \tag{5.2.10}$$

$$\eta = (\Psi - \Psi_S)/(-x)^{1/2},$$

where

$$f_0 = (a_0^2 |\eta|^2 + 2\lambda_0)^{1/2}, \quad f_1 = a_1 \eta^3 f_0^{-1} - a_0^2 f_0^2/(3\lambda_0),$$

$$h_0 = \frac{1}{2a_0} \ln \left[\frac{f_0 + a_0 \eta}{f_0 - a_0 \eta} \right], \quad h_1 = \frac{a_0^2}{3\lambda_0} \eta - \frac{a_1 (f_0^2 + 2\lambda_0)}{a_0^4 f_0}. \tag{5.2.11}$$

The constant a_1 characterizes the asymmetry of the solution with respect to the streamline $\Psi = \Psi_S$.

Matching this solution with the solution (5.2.3)–(5.2.5) for regions 2 and 4, we find that the functions $G(x)$ and $g(x)$ in (5.2.5) and (5.2.10)

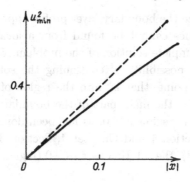

Fig. 5.7 Minimum values of speed in the boundary layer on a moving surface: dashed line – equation (5.2.14); solid line – numerical solution of Zubarev (1983).

can be written as

$$G = -a_0^{-1} \ln[(-x)\lambda_0/(2a_0^2)] + b_0 + o(1),$$
$$g = -(2a_0)^{-1} \ln[(-x)\lambda_0/(2a_0^2)] + b_0 + o(1),$$

(5.2.12)

and as $\Psi \to \Psi_S - 0$,

$$\overline{U}_0 = a_0(\Psi_S - \Psi) + O[(\Psi_S - \Psi)^2],$$
$$\overline{Y}_0 = -a_0^{-1} \ln(\Psi_S - \Psi) + b_0 + O[(\Psi_S - \Psi)], \quad b_0 = \text{const}.$$

(5.2.13)

Thus, ahead of the Moore–Rott–Sears point the boundary layer has a three-layer structure (Figure 5.5). Adjacent to the surface is region 4, which maintains the thickness of the boundary layer, and above it lies the nonlinear region 3, where the velocity component $u(x, \Psi)$ approaches zero. One can see from (5.2.5), (5.2.10), and (5.2.12) that the thickness of this region, as well as that of the main part of the boundary layer, increases without limit according to a logarithmic law as $x \to -0$, and, therefore, the vertical velocity component at the outer edge of the boundary layer behaves as $\text{Re}^{-1/2}|x|^{-1}$; that is, the solution here has a stronger singularity than in the vicinity of the point of zero skin friction on a fixed surface.

According to (5.2.10) and (5.2.11), the minimum speed behaves in the following way as $x \to -0$:

$$u_{\min}(x) = (2\lambda_0)^{1/2}|x|^{1/2} + O(|x|).$$

(5.2.14)

This dependence agrees with the results of the numerical solutions of Williams and Johnson (1974a) and also of Zubarev (1983). Figure 5.7 shows the comparison with the solution of Zubarev for $u_w = 1$.

The solution derived above contains arbitrary constants a_0, a_1, b_0, c_0,

Ψ_S that characterize the boundary-layer profile approaching the section $x = -0$; their values cannot be found from a local analysis and are determined by a complete solution of the problem (5.2.1).

We consider the possibility of continuing the solution through the Moore–Rott–Sears point, that is, into the region of positive values of x. If the solution for the main part of the boundary layer is extended continuously into the region $x > 0$, as has been done repeatedly before (see Chapter 3, Section 1 and Chapter 4, Section 3), then, according to (5.2.2) and (5.2.3), the solution as $x \to +0$ (region $2'$, Figure 5.5) will have the form

$$u = U_0(\Psi) + xU_1^\circ(\Psi) + O(x^2), \quad U_1^\circ = -\lambda_0 U_0^{-1} + (U_0 U_0')'. \quad (5.2.15)$$

For the case under consideration, when $\alpha = 1$, according to (5.2.6), $U_0 = a_0(\Psi - \Psi_S) + O[(\Psi - \Psi_S)^2]$ as $\Psi \to \Psi_S + 0$. Then, just as in region 2 for $x < 0$, the expansion (5.2.15) becomes invalid for $|\Psi - \Psi_S| = O(x^{1/2})$ (region $3'$). In this region the solution is analogous to (5.2.10)):

$$u = x^{1/2}\varphi_0(\xi) + o(x^{1/2}), \quad \xi = (\Psi - \Psi_S)/x^{1/2}. \quad (5.2.16)$$

Substituting this expansion into (5.2.1) and (5.2.2), we find that the solution of the equation for $\varphi_0(\xi)$ that satisfies the conditions of matching with (5.2.15) has the form

$$\varphi_0 = (a_0^2|\xi|^2 - 2\lambda_0)^{1/2}. \quad (5.2.17)$$

For $\xi^2 < 2\lambda_0 a_0^{-2}$ this expression becomes imaginary. Thus, just as for the singular solution for the boundary layer on a fixed surface, the solution obtained cannot be extended continuously through the singular point.

Among all possible singularities we have considered only the weakest one, which corresponds to the smallest value $\alpha = 1$ in (5.2.9). Now let us briefly describe other situations.

For integer values $\alpha > 1$ the solution is similar to the one analyzed, but it has a stronger nonremovable singularity. According to (5.2.8) an increase in the value of α corresponds to a decrease in γ, which leads to unbounded growth of the boundary-layer thickness, but now according to a power law; thus, for $\alpha = 2$ the value of γ is $-1/4$, and therefore in (5.2.5) $G(x) = O(|x|^{-1/4})$ as $x \to -0$. However, such a singularity can be realized only under the condition that the velocity profile approaching the section $x = -0$ is less arbitrary than in the case when $\alpha = 1$. Indeed, if $\alpha = 2$, this implies that in the expansions (5.2.6) and (5.2.13) the constant a_0 in front of $|\Psi - \Psi_S|$ must be taken equal to zero; if $\alpha = 3$ it is also necessary that the constant in front of $|\Psi - \Psi_S|^2$ vanish; etc. The values of the constants appearing in the expansion in the vicinity of the section $x = -0$ are determined by the solution of the problem (5.2.1) as a whole, and the vanishing of any of them seems exceptional. Such a situation is in principle possible only for special choices of the functions that define the outer boundary conditions in (5.2.1).

Half-integer values of α correspond to the boundary-layer solution for a surface that is moving upstream ($u_w < 0$). (A detailed analysis of singularities of this type was given by Williams and Stewartson (1983) and also by Elliott et al. (1983).) Even the smallest ($\alpha = 3/2$) gives a solution in which the boundary-layer thickness, according to (5.2.8), increases as $|x|^{-1/6}$ when $x \to -0$. However, van Dommelen and Shen (1983a) showed that, in addition to this kind of singularity for $u_w < 0$ at the Moore–Rott–Sears point, a weaker nonremovable singularity may arise. It seems that it is this weaker singularity that should be realized in the general case, just as for the boundary layer on a surface moving downstream.

We shall not describe these problems in detail, since no numerical solutions of the boundary-layer equations with $u_w = $ const. < 0 have been derived to date in which such singularities would occur for a given regular pressure distribution. The main difficulties result from the fact that the solution in the region under consideration is determined by "initial" conditions imposed downstream as well as upstream. A more elaborate discussion is given in the work of Chernyi (1974), devoted to the study of a boundary layer with constant speed at its outer edge, as well as in the survey of Williams (1977).

We note that, in addition to singular solutions, the boundary-layer equations also admit a regular representation in the vicinity of the Moore–Rott–Sears point. However, this is less general than the solutions described above (Sychev, 1983).

A singularity in the unsteady boundary layer. The analogy between local properties of the flow in the boundary layer on a moving surface and in an unsteady boundary layer indicates that the Moore–Rott–Sears point is singular in both cases. This fact is also confirmed by results of numerical solutions first obtained by Telionis, Tsahalis, and Werle (1973) and by Williams and Johnson (1974b).[2] Indeed, if the longitudinal coordinate of the Moore–Rott–Sears point is denoted by $x =$

---

[2] See also the surveys by Sears and Telionis (1975), Williams (1977), and Shen (1978) devoted to this topic.

$x_0(t)$, then using the standard variables x and y in a coordinate system fixed to that point we arrive at an unsteady flow in the boundary layer on a surface that is moving downstream with speed $u_w(t) = -\dot{x}_0(t) > 0$. (Here, the dot denotes differentiation with respect to t, where $LU_\infty^{-1}t$ is the time.) It is easy to see that in this case as $\tilde{x} = x - x_0(t) \to -0$, the leading terms of the solution have just the same form in the vicinity of the Moore–Rott–Sears point as for the steady case considered above. The only difference is that the constants λ_0, Ψ_S, a_0, b_0, and others appearing in the solution are now functions of t. This is related to the fact that in region 3 (Figure 5.5) the inertia terms and pressure gradient are of order unity, while the unsteady term tends to zero as $\tilde{x} \to -0$.

The work of Williams and Stewartson (1983), which was a continuation of the work of Williams (1982a,b), studied the initial stage of the flow in the vicinity of the wedge-shaped trailing edge of an airfoil that is instantaneously set into motion from rest. The results of this analysis showed that in similarity variables the solution has a singularity analogous to that considered above. In the general case, which as before corresponds to the weakest singularity, a logarithmic thickening of the self-similar boundary layer takes place. The numerical solutions obtained are in good agreement with the asymptotic expansions in the vicinity of the singular point.

Removable singularity at the Moore–Rott–Sears point. As we have seen in Chapter 4 while analyzing the steady boundary layer on a fixed surface, there exist conditions under which the solution of Prandtl's equations, for a given regular pressure distribution $p_e(x)$, is singular at the point of zero skin friction but can be continued smoothly through it; that is, the singularity is removable. A similar situation also arises when the boundary layer develops on a surface that is moving downstream.

A removable singularity may appear at the Moore–Rott–Sears point in the case when a given pressure distribution $p_e(x)$ reaches a maximum value at this point ($x = 0$), so that the flow is decelerated ahead of this point and accelerated behind it. In contrast to (5.2.14), now $u_{\min}(x) = O(|x|)$, and, in the representation (5.2.5) for $Y(x,\Psi)$, the function $G(x) = O(|x|^{-1/2})$ as $|x| \to 0$. Thus, the removable singularity in this case is stronger than the nonremovable one. This singularity appears, for example, in the solution that describes the flow in the periodic boundary layer on the surface of a circular cylinder that is rotating in a uniform oncoming stream at a certain quite definite value of its angular velocity (Nikolaev, 1982; Lam, 1988; Chipman and Duck, 1993).

Unlimited growth of the boundary-layer thickness as $|x| \to 0$ leads to a pressure change in the external-flow region. As a result, there occurs an interaction of the flow in the boundary layer with the outer potential flow in the vicinity of the Moore–Rott–Sears point. The longitudinal dimension of the interaction region is $|x| = O(\mathrm{Re}^{-1/7})$, and the boundary-layer thickness here is $O(\mathrm{Re}^{-3/7})$. Unlike the corresponding flows on a fixed surface (Chapter 4), the flow in the interaction region is inviscid, in the leading terms, despite the fact that the pressure gradient is small (of order $\mathrm{Re}^{-1/7}$).

A detailed description of the structure of the removable singularity and of the flow within the interaction region was given in the work of Sychev (1987b) and of Negoda and Sychev (1987).

5.3 Self-Induced Separation on a Moving Surface

We now study the flow near the separation point of the boundary layer that develops on a surface moving downstream at constant speed. Solving this local problem will permit us to approach the analysis of a more complicated problem – the unsteady flow near a moving separation point – on the basis of the analogy described above.

For flows over a surface moving upstream, it has been already mentioned that until now no suitable theory has been developed, and for this reason the question of the validity of the Moore–Rott–Sears criterion for such flows remains open.

Statement of the problem. Thus far we have, for the most part, discussed only the formal properties of solutions of the boundary-layer equations on a moving surface, in order to demonstrate that the singularities are nonremovable and therefore nonphysical. The appearance of nonremovable singularities always attests to the incompatibility of the solution of the boundary-layer equations with a prescribed external pressure distribution, and, as was shown in the preceding chapters, the contradiction that arises is resolved by reconsidering the solution of the outer problem.

We consider plane steady flow past a body of finite dimensions in a uniform oncoming stream. Following the main ideas of the theory of laminar separation described in Chapter 1, we seek the local limiting state of the flow field as Re $\to \infty$ in the class of separated flows of an ideal fluid with free streamlines.

The pressure distribution along the zero streamline in the vicinity of the separation point thus has the form (see Chapter 2, Section 1 and Chapter 1, Section 3):

$$p_e(x) = \begin{cases} p_{00} - 2k(-x)^{1/2} - \frac{16}{3}k^2(-x) + O[(-x)^{3/2}], & x \to -0, \\ p_{00}, & x > 0, \end{cases} \tag{5.3.1}$$

and the form of the free streamline as $x \to +0$ is

$$y = y_S(x) = \frac{4}{3}\frac{k}{U_{00}^2}x^{3/2} + O(x^{5/2}). \tag{5.3.2}$$

Here and in the following the previous notation is used, and the origin of the coordinate system is placed at the point of departure of the free streamline from the body surface; the pressure and the speed on the free streamline are denoted by p_{00} and $U_{00} = (1 - 2p_{00})^{1/2}$. To simplify the presentation, we neglect the curvature of the body surface, since it has no effect whatever upon the solution for the leading terms of the asymptotic expansions.

For laminar separation from a smooth fixed surface the value of the constant k in (5.3.1) and (5.3.2) (see Chapter 1) vanishes as Re $\to \infty$, and therefore the solution for the limiting state satisfies the Brillouin–Villat condition of finite curvature of the free streamline at the separation point. As is well known from numerous experimental data, downstream motion of the body surface leads to a displacement of the separation point, which corresponds to an increase in the value of the constant k (Figure 1.4). This is related to the fact that in the boundary layer on a moving surface, the longitudinal velocity component $u(x, \Psi)$ assumes finite positive values (Figure 5.2a). Consequently, for separation to occur, that is, for a point where $u(x, \Psi)$ vanishes to appear in the boundary layer, a finite pressure change is necessary. Therefore, the constant k in (5.3.1) and (5.3.2) must remain finite rather than tending to zero as Re $\to \infty$. This initial hypothesis became the basis of the asymptotic theory (Sychev, 1979), which will now be described.

Flow in the boundary layer ahead of the separation point. Near the body surface ahead of the separation point ($x < 0$), the flow in a first approximation as Re $\to \infty$ is described by the Prandtl boundary-layer equations which, together with the boundary conditions corresponding to the flow under consideration, have the form (5.2.1).

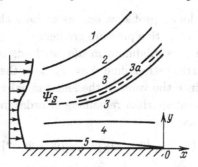

Fig. 5.8 Asymptotic structure of the flow in the boundary layer ahead of the separation point.

The pressure distribution as $x \to -0$ is given by the expression (5.3.1), and then, for the speed at the outer edge of the boundary layer, we have

$$U_e(x) = U_{00} + \frac{2k}{U_{00}}(-x)^{1/2} + O[(-x)], \quad k > 0. \qquad (5.3.3)$$

The boundary layer ahead of the separation point is acted upon by a strong adverse pressure gradient (5.3.1). The qualitative behavior of the flow in the boundary layer will here be similar to that ahead of the Moore–Rott–Sears point with a regular pressure distribution (Section 1); therefore, as $x \to -0$, the structure of the solution of the problem (5.2.1), (5.3.1), and (5.3.3) will be quite similar to that studied in Section 2.

In the main part of the boundary layer (region 2, Figure 5.8), the longitudinal velocity component is of order unity, and as $x \to -0$, the solution according to (5.3.1) and (5.3.3) may be written in the form

$$u = U_0(\Psi) + (-x)^{1/2}U_1(\Psi) + O[(-x)], \quad Y = G(x) + Y_0(\Psi) + O[(-x)^{1/2}]. \qquad (5.3.4)$$

Substituting these expansions into (5.2.1) and (5.3.1), we find that

$$U_1 = 2kU_0^{-1}, \quad Y_0 = \int U_0^{-1}d\Psi; \qquad (5.3.5)$$

the function $G(x)$ still remains arbitrary. From matching with the outer potential-flow region, it follows that $U_0(\infty) = U_{00}$. The viscous terms in the boundary-layer equation have in this region a very high order of smallness in comparison with the inertia terms and pressure gradient; that is, the flow here is locally inviscid.

A key role in the examination of the flow ahead of the separation point is played by the analysis of a region located in the vicinity of the minimum in the velocity profile, which, as we have already seen, results from the action of a positive pressure gradient.

The appearance of a minimum in $u(x, \Psi)$ leads to the fact that as $x \to -0$ near a certain streamline $\Psi = \Psi_S > 0$, the expansion (5.3.4) becomes invalid, since the value of the longitudinal velocity component here tends to zero rather than remaining of order unity. Indeed, if we assume as before that

$$U_0 = a_0(\Psi - \Psi_S)^\alpha + o[(\Psi - \Psi_S)^\alpha] \qquad (5.3.6)$$

as $\Psi \to \Psi_S + 0$, then the expansion (5.3.4), (5.3.5) ceases to be uniformly valid in region 3 where $|\Psi - \Psi_S| = O[(-x)^{1/4\alpha}]$. (Here a_0 and α are positive constants.) We therefore introduce a new independent variable $\eta = (\Psi - \Psi_S)/(-x)^{1/4\alpha}$, and taking into account, as usual, the balance of the inertia terms and the pressure gradient (5.3.1), we seek a solution in this region in the form

$$u = (-x)^{1/4} f_0(\eta) + o[(-x)^{1/4}], \quad \eta = (\Psi - \Psi_S)/(-x)^{1/4\alpha} = O(1),$$
$$(5.3.7)$$

taking $\eta \geq 0$ for the present. As a result of substituting this expansion into (5.2.1) and (5.3.1), we find that the viscous terms in the boundary-layer equation are of order $(-x)^\lambda$, $\lambda = (3\alpha - 2)/4\alpha$, while the inertia terms and pressure gradient are of order $(-x)^{-1/2}$; therefore, for $\lambda > -1/2$ ($\alpha > 2/5$) they are not essential for determining the coefficient of the leading term in the expansion (5.3.7); that is, the flow here is locally inviscid.

We take $\alpha > 2/5$ and integrate the equation obtained by substituting the expansion (5.3.7) into (5.2.1) and (5.3.1). As a result,

$$f_0 = (a^* \eta^{2\alpha} + 4k)^{1/2}. \qquad (5.3.8)$$

By matching the solution with (5.3.4)–(5.3.6), we find that $a^* = a_0^2$. The value of the constant α remains arbitrary and hence as $\eta \to 0$ the solution (5.3.8), in general, is not regular.

The singularity, as often happens, is due to the loss of the highest derivative in the equation. Therefore, to eliminate the nonuniformity we introduce the region 3a, where now all terms of the boundary-layer equation are important in the zeroth approximation. From the remarks above, it follows that in this region $\lambda = -1/2$ and $|\Psi - \Psi_S| = O[(-x)^{5/8}]$; then the solution can be represented in the form

$$u = (-x)^{1/4} g_0(\xi) + (-x)^{3/4} g_1(\xi) + (-x)^\beta g_\beta(\xi) + o[(-x)^{3/4}],$$
$$\xi = (\Psi - \Psi_S)/(-x)^{5/8}, \quad \beta > 1/4. \qquad (5.3.9)$$

The first two terms of this expansion are determined by the corresponding terms of the expansion (5.3.1) for the pressure, and the third is an eigenfunction with as yet unknown exponent β.

Substituting (5.3.9) into (5.2.1) and (5.3.1), we obtain for $g_0(\xi)$ the equation

$$g_0(g_0 g_0')' - \frac{5}{8}\xi g_0 g_0' + \frac{1}{4}g_0^2 = k. \tag{5.3.10}$$

For matching with the solutions in adjacent regions, it is necessary that as $|\xi| \to \infty$ the function $g_0(\xi)$ grow no faster than some power of ξ. The solution of equation (5.3.10) that meets these conditions is $g_0 = 2k^{1/2}$, and it is apparently unique. This is first of all confirmed by the results of numerical analysis of this equation. Furthermore, it follows immediately from (5.3.9) that, if $g_0' \neq 0$, then the value of the shear stress in region $3a$ increases without limit as $|x|^{-1/8}$ for $x \to -0$, but this contradicts the original ideas on the character of the flow.

Using this solution, we now obtain the following equations for the coefficients of other terms in the expansion (5.3.9):

$$g_1'' - \frac{5}{16k^{1/2}}\xi g_1' + \frac{1}{2k^{1/2}}g_1 = \frac{4k}{3},$$

$$g_\beta'' - \frac{5}{16k^{1/2}}\xi g_\beta' + \frac{4\beta+1}{8k^{1/2}}g_\beta = 0. \tag{5.3.11}$$

The solutions of these equations should satisfy the condition of the absence of exponential growth as $|\xi| \to \infty$. In this case a nontrivial solution of the second equation (5.3.11) (Weber's equation) exists only if $\beta = (5n-2)/8$, $n = 0, 1, 2, \ldots$, and is represented by a Chebyshev–Hermite polynomial:

$$g_\beta(\xi) = c_\beta H_n(\tau), \quad H_n = (-1)^n e^{\tau^2}\frac{d^n}{d\tau^n}e^{-\tau^2},$$

$$\xi = 4k^{1/4}(2/5)^{1/2}\tau, \quad c_\beta = \text{const}.$$

Since the homogeneous part of the first equation (5.3.11) is also Weber's equation with $\beta = 3/4$, its solution has the form $g_1 = 8k^{3/2}/3$. Now carrying out the matching of the solutions in regions 3 and $3a$ as $\eta \to 0$ and $\xi \to \infty$, we find that $\alpha = (4\beta+1)/5$; finally, recalling the expression for β derived above, as well as the fact that $\beta > 1/4$, we obtain

$$\alpha = n/2, \quad n = 1, 2, 3, \ldots. \tag{5.3.12}$$

Thus, we have found a discrete set of values for the constant α.

In the analysis of a boundary layer with a regular pressure gradient, which we studied in Section 2 of this chapter, we pointed out the expression (5.2.9). It is obtained in the same way as the expression (5.3.12) and has the same meaning. Therefore, the flow in the boundary layer on a surface moving downstream, which is considered here, corresponds, as before, only to integer values of α. However, for $\alpha = m$, where $m = 1, 2, 3, \ldots$, the solution (5.3.8) is regular as $\eta \to 0$ and is meaningful for all values of η. The region $3a$ is found to be degenerate, since the solution there is represented by the expansion of the solution (5.3.7) and (5.3.8) for $\eta \to 0$, rewritten in the inner variable $\xi = \eta/(-x)^q$, $q = (5\alpha - 2)/8\alpha$. The introduction of this region is necessary only for formal determination of the set of values for α.

We note that if from the very beginning we had assumed that the effect of viscous stresses was essential in region 3, that is, $\alpha = 2/5$, then the solution here would have had the form (5.3.9). Since $g_0 = 2k^{1/2}$, it would have been impossible to match with (5.3.4) and (5.3.5) for the main part of the boundary layer.

Thus, in the solution obtained for region 3 the constant α may take positive integer values, and the flow here is locally inviscid.

As before (see Section 2 of this chapter), the most general case corresponds to the smallest value $\alpha = 1$. It is this solution that is realized as $x \to -0$ unless some special choice is made for the functions defining the outer boundary conditions for the problem (5.2.1).

Thus, as $x \to -0$ in the vicinity of the streamline $\Psi = \Psi_S$ (region 3), where the minimum of the velocity profile is located, on the basis of (5.3.7), (5.3.8), and everything that has been said above, the solution has the form

$$u = (-x)^{1/4} f_0(\eta) + O[(-x)^{1/2}], \quad \eta = (\Psi - \Psi_S)/(-x)^{1/4},$$
$$f_0 = (a_0^2 |\eta|^2 + 4k)^{1/2}. \tag{5.3.13}$$

Substituting the solution into the second equation (5.2.1), we obtain

$$Y = g(x) + h_0(\eta) + O[(-x)^{1/4}], \quad h_0 = \frac{1}{2a_0} \ln \left[\frac{f_0 + a_0 \eta}{f_0 - a_0 \eta} \right]. \tag{5.3.14}$$

Now, matching the solutions for regions 2 and 3, we have

$$U_0 = a_0(\Psi - \Psi_S) + O[(\Psi - \Psi_S)^2],$$
$$Y_0 = a_0^{-1} \ln(\Psi - \Psi_S) + O[(\Psi - \Psi_S)] \text{ as } \Psi \to \Psi_S + 0; \tag{5.3.15}$$
$$g = G(x) + (4a_0)^{-1} \ln(-x) - (2a_0)^{-1} \ln(a_0^2/k),$$

From the expression (5.3.13), it follows that there is a line within the boundary layer along which the shear stress vanishes:

$$u\frac{\partial u}{\partial \Psi} = (-x)^{1/4}\tau_0(\eta) + O[(-x)^{1/2}], \quad \tau_0 = f_0 f_0' = a_0^2 \eta.$$

For the leading terms of the expansion this line coincides with the streamline $\Psi = \Psi_S$.

Beneath region 3 lies a part 4 of the boundary layer, adjacent to the moving body surface. Here, as in the main part of the boundary layer, the longitudinal velocity component is of order unity. The solution in this region that satisfies the condition of no flow through the body surface ($\Psi = 0$) is quite similar to (5.3.4), (5.3.5):

$$u = \overline{U}_0(\Psi) + (-x)^{1/2}2k\overline{U}_0^{-1}(\Psi) + O[(-x)],$$
$$Y = \overline{Y}_0(\Psi) + O[(-x)^{1/2}], \quad \overline{Y}_0 = \int_0^\Psi \overline{U}_0^{-1} d\overline{\Psi}. \qquad (5.3.16)$$

From matching with the solution (5.3.13), (5.3.14) for region 3 it follows that

$$\overline{U}_0 = a_0(\Psi_S - \Psi) + O[(\Psi_S - \Psi)^2],$$
$$\overline{Y}_0 = -a_0^{-1}\ln(\Psi_S - \Psi) + b_0 + O[(\Psi_S - \Psi)]$$

as $\Psi \to \Psi_S - 0$, $b_0 = \text{const.}$;

$$g = -(4a_0)^{-1}\ln(-x) - (2a_0)^{-1}\ln(k/a_0^2) + b_0 + O[(-x)^{1/2}\ln(-x)]. \qquad (5.3.17)$$

The last expression, together with (5.3.15), determines the form of the required function $G(x)$:

$$G = -(2a_0)^{-1}\ln(-x) - a_0^{-1}\ln(k/a_0^2) + b_0 + O[(-x)^{1/2}\ln(-x)]. \quad (5.3.18)$$

The effect of the singular pressure gradient (5.3.1) as $x \to -0$ is that the flow in regions 2–4 becomes locally inviscid, and therefore, in order to satisfy the no-slip condition on the body surface, it is necessary to introduce the wall sublayer 5, where the effect of internal friction forces is essential. Taking into account the balance of inertia and viscous terms in the boundary-layer equation with the pressure gradient, as well as the condition that on the body surface $u(x,0) = u_w$, we represent the solution in the form

$$u = u_w + (-x)^{1/2}\ln(-x)F_1(\zeta) + (-x)^{1/2}F_2(\zeta) + O[(-x)\ln^2(-x)],$$
$$\zeta = \Psi/(-x)^{1/2}. \qquad (5.3.19)$$

We will see that the appearance of the intermediate term with $F_1(\zeta)$ in this expansion is necessary to satisfy the boundary conditions at $\zeta = 0$ and as $\zeta \to \infty$.

Substituting (5.3.19) into (5.2.1) and (5.3.1), we obtain

$$2u_w F_1'' - \zeta F_1' + F_1 = 0, \quad 2u_w F_2'' - \zeta F_2' + F_2 = 2ku_w^{-1} - 2F_1. \quad (5.3.20)$$

The solution of the first equation (Weber's equation) which satisfies the condition $F_1(0) = 0$ has the form

$$F_1 = c_1 \zeta, \quad c_1 = \text{const.} \quad (5.3.21)$$

On the solution of the second equation in (5.3.20) there should be imposed, in addition to the condition $F_2(0) = 0$, the requirement that exponential growth be absent as $\zeta \to \infty$, which is necessary for matching with (5.3.16). Such a solution exists for a definite value of c_1, and may be written in explicit form. As a result, we obtain

$$c_1 = k/(\pi u_w^3)^{1/2}, \; F_2 = 2c_1 \zeta \ln \zeta + c^* \zeta + O(1) \text{ as } \zeta \to \infty, \quad c^* = \text{const.} \quad (5.3.22)$$

Now, after substituting the solution for $u(x, \Psi)$ into the second equation (5.2.1), we find that in region 5,

$$Y = (-x)^{1/2} H_0(\zeta) + (-x)\ln(-x)H_1(\zeta) + (-x)H_2(\zeta)$$
$$+ O[(-x)^{3/2}\ln^2(-x)], \quad (5.3.23)$$

$$H_0 = u_w^{-1}\zeta, \quad H_1 = -c_1\zeta^2/(2u_w^2), \quad H_2 = -u_w^{-2}\int_0^\zeta F_2 d\bar{\zeta}.$$

From matching the solutions in regions 4 and 5 it follows that $\overline{U}_0 = u_w + 2c_1\Psi\ln\Psi + O(\Psi)$ as $\Psi \to 0$.

It is of interest to note that the action of the adverse pressure gradient (5.3.1) leads to the appearance of a singularity in the skin-friction distribution; from the expressions (5.3.19) and (5.3.21) it follows that $\tau = u \frac{\partial u}{\partial \Psi}\big|_{\Psi=0} = c_1 u_w \ln(-x) + O(1)$ as $x \to -0$.

Thus, the analysis carried out shows that as $x \to -0$ one can distinguish four characteristic regions in the boundary layer (Figure 5.8). There are the viscous wall sublayer and the adjacent region 4, which maintains the thickness of the original boundary layer. Above this region lies region 3, where the longitudinal velocity component and the shear stress decrease as $|x|^{1/4}$; the flow here, being locally inviscid, is described by a nonlinear equation. Fluid deceleration in this region, due to the effect of a positive pressure gradient (5.3.1), leads to a growth in thickness, which is transmitted through the main part of the boundary layer to its outer edge. This implies that the main contribution to the displacement effect of the boundary layer is created by the flow in region 3, which we will call "nonlinear locally inviscid." Since, according to (5.3.4) and (5.3.18), the growth of the boundary-layer thickness as $x \to -0$ follows a logarithmic law, at the outer edge of the boundary layer the vertical velocity component and the slope of the streamlines increase without limit as $\mathrm{Re}^{-1/2}|x|^{-1}$ when $x \to -0$; hence, the solution of the boundary-layer equations has a stronger singularity at $x = -0$ in comparison with the corresponding solution for the flow over a fixed surface (Sychev, 1972; see also Chapter 1, Section 4).

The most important distinction between the flows near the separation points on moving and fixed surfaces lies in the fact that in the first case the main contribution to the displacement effect of the boundary layer is made by the flow in the nonlinear locally inviscid region rather than by the viscous wall sublayer, which here plays a secondary role.

With the displacement effect of the boundary layer taken into account, the expansion for the complex velocity in the outer potential-flow region 1 near the separation point has the form

$$u - iv = \left[U_{00} - i2kU_{00}^{-1}z^{1/2} + O(z) \right] + \dots$$
$$+ \mathrm{Re}^{-1/2} \left[(iU_{00}/2a_0)z^{-1} + O(z^{-1/2}\ln z) \right] + o(\mathrm{Re}^{-1/2}),$$
$$z = x + iy, \qquad |z| \to 0. \tag{5.3.24}$$

Here, the terms of order unity represent the solution for the limiting state of the flow field, and so as $\arg z \to \pi$ and $\arg z \to 0$, there follow the corresponding expressions (5.3.3), (5.3.1), and (5.3.2). The terms of order $\mathrm{Re}^{-1/2}$ are due to the displacement effect of the boundary layer and consequently as $\arg z \to \pi$, matching is achieved with the solution obtained for the boundary-layer equations as $\Psi \to \infty$. (The dots in (5.3.24) mean that terms may appear here whose form is found from the solution of the problem for the interaction region.)

The vicinity of the point $z = 0$ in which terms $O(\mathrm{Re}^{-1/2}/z)$ and $O(z^{1/2})$ in (5.3.24) become of the same order, that is, $|z| = O(\mathrm{Re}^{-1/3})$, is the interaction region. The streamline slopes resulting from the displacement effect of the boundary layer have the same order here as the streamline slopes in the outer potential flow. In other words, for $|z| = O(\mathrm{Re}^{-1/3})$ the expansion (5.3.24) loses its uniform validity, and now in the first approximation the pressure distribution on the body surface is not given by the expression (5.3.1), but must be determined

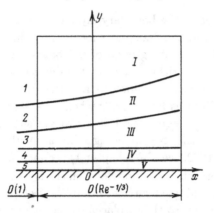

Fig. 5.9 Asymptotic structure of the interaction region.

in the process of concurrent analysis of the flows within and outside the boundary layer.

The flow in the interaction region. Having determined the characteristic dimension of the interaction region, we introduce an inner longitudinal variable $x^* = \mathrm{Re}^{1/3}x = O(1)$. Considering the solution (5.3.4), (5.3.5), (5.3.18), (5.3.1) of the boundary-layer equations for $\Psi = O(1)$ and $x \to -0$ as the inner limit of the outer asymptotic expansion, we represent the solution for the main part of the boundary layer (region II, Figure 5.9) in the interaction region in the following form:

$$x = \mathrm{Re}^{-1/3}x^*, \quad \psi = \mathrm{Re}^{-1/2}\Psi,$$
$$u = U_0(\Psi) + \mathrm{Re}^{-1/6}U_1^*(x^*, \Psi) + O(\mathrm{Re}^{-1/3}\ln \mathrm{Re}), \tag{5.3.25}$$
$$y = (6a_0)^{-1}\mathrm{Re}^{-1/2}\ln \mathrm{Re} + \mathrm{Re}^{-1/2}y_0(x^*, \Psi) + O(\mathrm{Re}^{-2/3}\ln \mathrm{Re}),$$
$$p = p_{00} + \mathrm{Re}^{-1/6}p_0(x^*, \Psi) + O(\mathrm{Re}^{-1/3}\ln \mathrm{Re}).$$

Substituting these asymptotic expansions into the original Navier–Stokes equations, we obtain

$$\frac{\partial p_0}{\partial \Psi} = 0, \quad p_0 = p^*(x^*), U_1^* = -p^*(x^*)U_0^{-1}(\Psi),$$
$$y_0 = G^*(x^*) + Y_0(\Psi), \quad Y_0 = \int U_0^{-1}d\Psi, \tag{5.3.26}$$

where $G^*(x^*)$ is some function that is so far arbitrary. Matching with the solution in region 2 shows that as $x^* \to -\infty$,

$$p^* \to -2k(-x^*)^{1/2}, \quad G^* = -(2a_0)^{-1}\ln[k^2(-x^*)/a_0^4] + b_0 + o(1).$$

The increase in boundary-layer thickness as $x \to -0$, determined by the expressions (5.3.4) and (5.3.18), leads to the fact that it becomes of order $\mathrm{Re}^{-1/2}\ln \mathrm{Re}$ in the interaction region.

In the leading term of the expansion, the transverse pressure change in the main part of the boundary layer is absent. It is not difficult to convince oneself that the situation is the same in the other regions of the boundary layer, and therefore that no further expansions for the pressure need be written. The transverse pressure variation appears for the first time in the third term of the expansion (5.3.25), which is of order $\mathrm{Re}^{-1/3}$.

In the lower part of the boundary layer (region IV), which is a continuation of region 4 and maintains the thickness of the original boundary layer ahead of the interaction region, the solution is analogous to (5.3.25) and (5.3.26):

$$u = \overline{U}_0(\Psi) + \mathrm{Re}^{-1/6}\overline{U}_1^*(x^*,\, \Psi) + O(\mathrm{Re}^{-1/3}\ln \mathrm{Re})\,,$$
$$y = \mathrm{Re}^{-1/2}\overline{Y}_0(\Psi) + O(\mathrm{Re}^{-2/3})\,, \tag{5.3.27}$$
$$\overline{U}_1^* = -p^*(x^*)\overline{U}_0^{-1}(\Psi)\,, \quad \overline{Y}_0 = \int_0^{\Psi} \overline{U}_0^{-1} d\overline{\Psi}\,.$$

As $x^* \to -\infty$, this solution matches with (5.3.16), and at $\Psi = 0$ it satisfies the condition of zero flow through the body surface.

The expressions (5.3.15) and (5.3.17) for the velocity profile as $|\Psi - \Psi_S| \to 0$ show that the first and second terms of the expansions (5.3.25) and (5.3.27) for the longitudinal velocity component become quantities of equal order when $|\Psi - \Psi_S| = O(\mathrm{Re}^{-1/12})$. This means that near the streamline $\Psi = \Psi_S$ there lies a new characteristic region of slow flow (region III), which obviously represents the continuation of region 3. In fact, transforming expressions (5.3.13), (5.3.14), and (5.3.17) to the inner variables

$$x^* = \mathrm{Re}^{1/3}x\,, \quad \psi^* = \mathrm{Re}^{1/12}(\Psi - \Psi_S)\,, \tag{5.3.28}$$

it is easy to see that in region III the solution can be represented in the following form:

$$u = \mathrm{Re}^{-1/12}u_0^*(x^*,\, \psi^*) + O(\mathrm{Re}^{-1/6})\,, \tag{5.3.29}$$
$$y = (12a_0)^{-1}\mathrm{Re}^{-1/2}\ln \mathrm{Re} + \mathrm{Re}^{-1/2}y_0^*(x^*,\, \psi^*) + O(\mathrm{Re}^{-7/12})\,.$$

Substituting these expansions together with (5.3.28) into the original

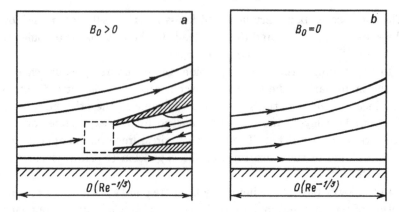

Fig. 5.10 Schematic representation of the flow in the interaction region for $B_0 > 0$ and $B_0 = 0$.

equations, we find that

$$\frac{u_0^{*2}}{2} + p^* = H^*(\psi^*), \quad u_0^* \frac{\partial y_0^*}{\partial \psi^*} = 1. \qquad (5.3.30)$$

Thus, the flow in this region is locally inviscid and Bernoulli's integral is valid here. This is related to the fact that the pressure gradient (5.3.1) grows without bound as $|x|^{-1/2}$ for $x \to -0$ and, according to (5.3.25), becomes of order $\mathrm{Re}^{1/6}$ in the interaction region, while the viscous terms of the equation are here of order $\mathrm{Re}^{-1/12}$.

An expression for the function $H^*(\psi^*)$ in (5.3.30) is obtained from matching the expansions (5.3.29) and (5.3.13):

$$H^*(\psi^*) = a_0^2 |\psi^*|^2/2 + B_0, \qquad (5.3.31)$$

where B_0 is an additive constant that can always be added to the Bernoulli function.

The structure of the solution of the problem within the interaction region depends in an essential way on whether or not B_0 is equal to zero, and therefore it is necessary to examine both possible situations.

The nonuniqueness arising in the solution is related to the fact that the flow in region III is locally inviscid, and consequently the question of the value of B_0 is essentially the question of choosing a flow "model."

Initially, in the work of Sychev (1979) mentioned above, the solution was derived with $B_0 > 0$ (with $B_0 < 0$ the solution becomes imaginary). It turned out that in this case at a certain point within region III the streamline $\psi^* = 0$ bifurcates (Figure 5.10a). Beyond this point lies a "stagnation region" where $p^*(x^*) = 0$ and the streamlines $\psi^* = 0$ enclosing this region are dividing streamlines along which $u_0^*(x^*, 0) = (2B_0)^{1/2}$, according to (5.3.30) and (5.3.31). In other words, for $B_0 > 0$ there appears in region III a flow with free streamlines similar to the flow in the Helmholtz–Kirchhoff model. The minimum value assumed here by the longitudinal component $u_0^*(x^*, \psi^*)$ is equal to $(2B_0)^{1/2}$, and therefore there is no point within the region under consideration where $u_0^*(x^*, \psi^*)$ vanishes. It was supposed (Sychev, 1983) that vanishing of the longitudinal velocity component, and the consequent appearance of reverse flow, occurs in an inner region that lies in the vicinity of the bifurcation point of the streamline $\psi^* = 0$. (In Figure 5.10a this inner region is shown as a square.) However, attempts to construct an appropriate solution proved unsuccessful. The reason for this lies in the following. Free streamlines really represent thin mixing layers (the shaded regions in Figure 5.10a), whose entraining effect causes a slow reverse flow within the "stagnation zone," so that in it $u = O(\mathrm{Re}^{-5/24})$ and $p - p_{00} = O(\mathrm{Re}^{-5/12})$. Since the mixing layers are layers of very large vorticity ($\omega = \psi_{yy} = O(\mathrm{Re}^{13/24})$), then in the vicinity of the bifurcation point of the streamline $\psi^* = 0$, where these layers originate, there must lie a region that is the source of this vorticity. The unreality of this flow pattern follows from the impossibility of explaining the reason for the appearance of such a region inside the fluid and, consequently, the reason for the generation of vorticity layers developing along the free streamlines. An alternative pointed out by van Dommelen and Shen (1983b) is to impose the requirement that $B_0 = 0$. In this case the solution is smooth and the corresponding flow is unseparated within the interaction region (Figure 5.10b).

Turning to a detailed description of the solution with $B_0 = 0$, we note that in reality (and this will be shown) separation occurs within a region that lies at some distance downstream from the interaction region.

Thus, taking $B_0 = 0$, after substitution of (5.3.31) into (5.3.30) and subsequent integration, we obtain

$$u_0^* = (a_0^2|\psi^*|^2 - 2p^*)^{1/2}, \quad y_0^* = g^*(x^*) + (2a_0)^{-1}\ln\left(\frac{u_0^* + a_0\psi^*}{u_0^* - a_0\psi^*}\right).$$
$$(5.3.32)$$

Using the expressions (5.3.17), we match the solutions (5.3.27) for region IV with the solution (5.3.29), (5.3.32) as $\Psi \to \Psi_S - 0$ and $\psi^* \to -\infty$. As a result we find that

$$g^* = -(2a_0)^{-1}\ln(-p^*/2a_0^2) + b_0. \qquad (5.3.33)$$

As $\psi^* \to \infty$ the solution for region III matches with the solution (5.3.25), (5.3.26), and so

$$G^* = -(2a_0)^{-1}\ln(-p^*/2a_0^2) + g^*(x^*).\qquad(5.3.34)$$

The expressions (5.3.33) and (5.3.34) determine the required function $G^*(x^*)$:

$$G^* = -a_0^{-1}\ln(-p^*) + a_0^{-1}\ln(2a_0^2) + b_0.\qquad(5.3.35)$$

In the solution obtained, as follows directly from (5.3.32) and (5.3.35), the displacement effect (or streamline slope, which is the same thing)

$$G^{*'} = \lim_{\psi^* \to \infty}\left(\frac{\partial y_0^*}{\partial x^*}\right) = -\frac{p^{*'}}{a_0 p^*}\qquad(5.3.36)$$

arises from the slow flow in the nonlinear locally inviscid region III and is transmitted through the main part of the boundary layer to the outer flow.

The flow in regions II–IV (Figure 5.9) is locally inviscid, and therefore to satisfy the no-slip condition at the body surface it is necessary to introduce a wall layer (region V) where the effect of viscous stresses is essential. From the expansion (5.3.19) it follows that as $x \to -0$ the stream function $\mathrm{Re}^{1/2}\psi$ in the viscous sublayer varies according to the law $|x|^{1/2}$, and therefore ψ becomes of order $\mathrm{Re}^{-2/3}$ within the interaction region, that is, for $|x| = O(\mathrm{Re}^{-1/3})$. However, we do not dwell upon a detailed analysis of the solution, since, just as in the boundary layer as $x \to -0$, the flow in this region plays a secondary role in the interaction process. We mention only that this solution (which guarantees satisfying the no-slip condition at $\psi = 0$ and matching with the solutions in regions 5 and IV), due to linearity of the equations obtained, may be written in explicit form (Sychev, 1979, 1983).

The system of relations describing the flow in the boundary layer of the interaction region is not closed. To close it and to find the function $p^*(x^*)$, as has been done many times in previous chapters, we examine region I of the outer potential flow, whose transverse dimension is of the order of the longitudinal dimension of the interaction region, that is, the region where the expansion (5.3.24) becomes invalid.

According to (5.3.25) and (5.3.36), the variable part of the pressure and the vertical velocity component at the outer edge of the boundary layer are quantities of order $\mathrm{Re}^{-1/6}$. Therefore, introducing for the region under consideration the variables

$$x^* = \mathrm{Re}^{1/3}x, \quad y^* = \mathrm{Re}^{1/3}y,\qquad(5.3.37)$$

we represent the solution in the form

$$u = U_{00} + \mathrm{Re}^{-1/6} U_0^*(x^*, y^*) + O(\mathrm{Re}^{-1/3}\ln \mathrm{Re}),$$
$$v = \mathrm{Re}^{-1/6} V_0^*(x^*, y^*) + O(\mathrm{Re}^{-1/3}\ln \mathrm{Re}), \qquad (5.3.38)$$
$$p = p_{00} + \mathrm{Re}^{-1/6} P_0^*(x^*, y^*) + O(\mathrm{Re}^{-1/3}\ln \mathrm{Re}).$$

As a result of substituting these expansions into the Navier–Stokes equations, we find that $U_0^* = -U_{00}^{-1} P_0^*$, and the functions $P_0^*(x^*, y^*)$ and $U_{00} V_0^*(x^*, y^*)$ are related to each other by the Cauchy–Riemann equations. Consequently,

$$W^*(z^*) = P_0^* + iU_{00}V_0^*, \quad z^* = x^* + iy^* \qquad (5.3.39)$$

is an analytic function. From matching with the solution in the main part of the boundary layer, and using the expression (5.3.36), we obtain

$$\mathrm{Im}\, W^*|_{y^*=0} = U_{00}^2 G^{*'}(x^*) = -\frac{U_{00}^2 p^{*'}}{a_0 p^*}. \qquad (5.3.40)$$

As $|z^*| \to \infty$, the solution in region I must match with the solution describing the flow on the scale of the body as $|z| \to 0$ (region 1). Therefore, making use of the expansion (5.3.24), we write

$$W^* = i2kz^{*1/2} + i\lambda^* z^{*-1/2} + i\lambda_1 z^{*-1} + O(z^{*-3/2}),$$
$$|z^*| \to \infty, \quad \lambda_1 = -U_{00}^2/2a_0. \qquad (5.3.41)$$

It is easy to see that as $\arg z^* \to \pi$, this expansion allows matching with the solution in the boundary layer ahead of the interaction region. Taking $\arg z^* \to 0$ and using (5.3.40), we find that as $x^* \to \infty$,

$$G^{*'} = 2kU_{00}^{-2} x^{*1/2} + \lambda^* U_{00}^{-2} x^{*-1/2}$$
$$- (2a_0)^{-1} x^{*-1} + O(x^{*-3/2}), \quad p^* \to 0. \qquad (5.3.42)$$

In the expansion (5.3.41), in addition to the terms that enter into (5.3.24), there are others containing the constant λ^*. Their appearance is due to indeterminacy in the choice of origin for the longitudinal coordinate in the interaction region. The value of λ^* cannot be found from the solution of the local problem. For definiteness we will take $\lambda^* = 0$ in the following, keeping in mind that the transition to the solution with $\lambda^* \neq 0$ is accomplished by replacing the variable x by $x - \lambda^* k^{-1} \mathrm{Re}^{-1/3}$.

From the solution of the boundary-value problem (5.3.40), (5.3.41) for the analytic function (5.3.39), it follows (see Chapter 1, Section 5) that at $y^* = 0$ its real part, which evidently determines the pressure

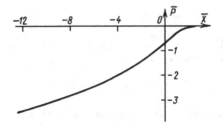

Fig. 5.11 Pressure distribution in the interaction region.

distribution in the boundary layer, is related to its imaginary part by the following integral:

$$U_{00}^2 G^{*\prime} = 2kx^{*1/2}\theta(x^*) - \frac{1}{\pi} \int_{-\infty}^{\infty} \left[\frac{p^*(\tau) + 2k(-\tau)^{1/2}\theta(-\tau)}{\tau - x^*} \right] d\tau$$

(5.3.43)

in the sense of its principal value. Here $\theta(x^*)$ is the Heaviside function. This expression completes the system of relations (5.3.25)–(5.3.29), (5.3.32)–(5.3.35) that describes the flow in the interaction region.

Equations (5.3.35) and (5.3.43), used for determining $G^*(x^*)$ and $p^*(x^*)$, contain positive constants a_0, k, U_{00}. However, with the help of the transformation

$$x^* = (ka_0)^{-2/3}U_{00}^{4/3}\overline{X}, \quad p^* = 2a_0^{-1/3}(kU_{00})^{2/3}\overline{P},$$
$$G^* = 2a_0^{-1}\overline{G} + a_0^{-1}\ln[a_0^{7/3}(kU_{00})^{-2/3}] + b_0,$$

(5.3.44)

these constants can be eliminated, and then the equations for $\overline{G}(\overline{X})$ and $\overline{P}(\overline{X})$ assume the form

$$\overline{G} = -\frac{1}{2}\ln(-\overline{P}), \quad \overline{G}^\prime = \overline{X}^{1/2}\theta(\overline{X}) - \frac{1}{\pi}\int_{-\infty}^{\infty}\left[\frac{\overline{P}(\tau) + (-\tau)^{1/2}\theta(-\tau)}{\tau - \overline{X}}\right]d\tau.$$

(5.3.45)

The solution of these equations is found numerically. Figure 5.11 shows a plot of the function $\overline{P}(\overline{X})$. We note that the solution of the problem (5.3.39)–(5.3.41) can be obtained not only by reducing it to (5.3.45) but also with the help of a series expansion of the function $W^*(z^*)$ (van Dommelen and Shen, 1983b).

From equations (5.3.45) and the expansion (5.3.42), taking into account the transformation (5.3.44), it follows immediately that as $x^* \to \infty$,

$$p^* = -2kx^{*1/2}\exp[-\gamma_0 x^{*3/2} + o(1)],$$

$$G^* = \gamma_0 a_0^{-1} x^{*3/2} - (2a_0)^{-1}\ln x^* - a_0^{-1}\ln(k/a_0^2) + b_0 + o(1), \quad (5.3.46)$$

$$\gamma_0 = \frac{4ka_0}{3U_{00}^2}.$$

Let us turn now to an analysis of the solution obtained. In the region of slow flow in the boundary layer (region III), where $u = O(\mathrm{Re}^{-1/12})$, the velocity profile has a minimum at $\Psi = \Psi_S$, and according to (5.3.28), (5.3.29), and (5.3.32) we have $u_0^*(x^*, 0) = (-2p^*)^{1/2}$. This means, in turn, that $u_0^*(x^*, 0)$ takes positive values throughout the interaction region, and as $x^* \to \infty$, in accordance with (5.3.46), this function tends to zero exponentially. Hence, with the additive constant B_0 in (5.3.31) equal to zero, a smooth solution is obtained that contains no point where the longitudinal velocity component vanishes.

Thus, contrary to the usual situation that we encountered in previous chapters, when the separation point is located in the interaction region, there is no separation inside the interaction region in the flow under consideration.

This distinctive property of the separation mechanism on a moving surface is explained by the fact that the slow flow inside the interaction region, where one would expect the appearance of reverse flow, is inviscid. The absence of the effect of internal friction forces (in the leading terms of the expansion) is related to the fact that the region is located not on the body surface but at a distance of the order of the boundary-layer thickness.

Thus, despite the effect of a large self-induced pressure gradient $(\partial p/\partial x = O(\mathrm{Re}^{1/6}))$, the flow in the interaction region remains unseparated. The interaction results only in a decrease of the longitudinal velocity component and a pronounced thickening of the boundary layer. Such flow behavior near the separation point agrees with the experimental measurements of velocity profiles obtained by Ludwig (1964) (see Figure 5.4).

The flow beyond the interaction region. Separation of the boundary layer. The fluid deceleration in the central part of the boundary layer must, obviously, lead to the fact that the longitudinal component of the velocity vector will finally become zero at some point located downstream from the interaction region, and a region of reverse flow will lie beyond that point. It is clear that this must take place in the part of the boundary layer that is a continuation of region III. (In Figure 5.12 it is shown as region B.)

As we have seen for the analysis of the solution in region III, the

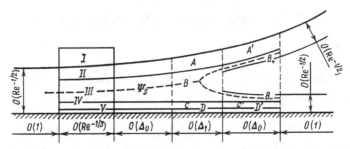

Fig. 5.12 Asymptotic flow structure in the vicinity of a separation point on a moving surface.

effect of viscous stresses is manifested here only in later terms of the expansions. Since the pressure gradient, according to (5.3.46), tends to zero exponentially as $x^* \to \infty$, it is expected that in region B the effect of internal friction forces will now become important in the leading approximation. (One can show this formally by considering subsequent terms of the asymptotic expansions in region III.)

Making use of these ideas, we obtain some preliminary estimates of the flow parameters, which allow us to turn to construction of the solution in region B and other regions lying beyond the interaction region.

However, we first make one small comment concerning the method of matched asymptotic expansions.

It is well known that for the matching of uniformly valid asymptotic expansions, the existence of overlapping domains of validity is essential (Kaplun, 1967). As a rule, in constructing asymptotic expansions, for the independent variables one chooses coordinates related to the flow geometry. The use of such variables in this and previous chapters did not lead to any difficulties in matching, because of the possibility of introducing overlap domains (although these were not specified). In some cases the characteristic regions do not have overlap domains in the original variables, but there is an overlap in certain other specially chosen variables. This is due to the fact that the characteristic regions prove to be wider in these new variables than in the original variables. We will see that such a situation does arise in exactly the flow we are considering.

Instead of x^*, let us take the function p^* as an independent variable in region III. Then the asymptotic expansion (5.3.46) will have the form

$$x^*(p^*) = X^*(\chi) = \gamma_0^{-2/3} \left(\chi^{2/3} + \tfrac{2}{9}\chi^{-1/3}\ln\chi + \tfrac{2}{3}\beta_0\chi^{-1/3} \right)$$
$$+ O(\chi^{-2/3}), \quad \chi = -\ln(-p^*), \tag{5.3.47}$$

$$\beta_0 = \ln(2k\gamma_0^{-1/3})$$

as $p^* \to -0$ ($\chi \to \infty$). Since the function $p^*(x^*)$ tends to zero exponentially as $x^* \to \infty$, then in the region B introduced above, the pressure will be smaller in order of magnitude than its value in the interaction region; that is,

$$p^* = \sigma \tilde{p}_0 = \text{Re}^{1/6}(p - p_{00}). \tag{5.3.48}$$

Here, $\sigma = \sigma(\text{Re})$ is a small parameter whose value tends to zero as $\text{Re} \to \infty$, and $\tilde{p}_0$ denotes the pressure distribution in region B. The parameter $\sigma(\text{Re})$ characterizes the difference in the orders of magnitude for the pressure in regions III and B.

Now substituting (5.3.48) into (5.3.47), we find that in region B

$$x = \text{Re}^{-1/3}x^* = \Delta_0 s_0 + \Delta_1 X(\tilde{p}_0) + o(\Delta_1),$$

$$\Delta_0 = \text{Re}^{-1/3}\left[(\ln \sigma^{-1})^{2/3} + \tfrac{2}{9}(\ln \ln \sigma^{-1})(\ln \sigma^{-1})^{-1/3}\right], \tag{5.3.49}$$

$$\Delta_1 = \text{Re}^{-1/3}(\ln \sigma^{-1})^{-1/3}, \quad s_0 = \gamma_0^{-2/3} = \left(\frac{3U_{00}^2}{4ka_0}\right)^{2/3}$$

From matching the functions $x^*(p^*)$ and $X(\tilde{p}_0)$ for regions III and B as $p^* \to -0$ and $\tilde{p}_0 \to -\infty$, it follows that

$$X = -\frac{2}{3}s_0\ln(-\beta_1\tilde{p}_0) + o(1) \text{ as } \tilde{p}_0 \to -\infty, \quad \beta_1 = (2k)^{-1}\gamma_0^{1/3}. \tag{5.3.50}$$

Thus, the region B, in which $p - p_{00}$ is of order $\sigma\text{Re}^{-1/6}$, is located at a distance $\Delta_0 s_0 > O(\text{Re}^{-1/3})$ from the origin of coordinates (Figure 5.12), and therefore regions III and B do not have overlap domains in the original variables.

Now considering the solution in region III as $p^* \to -0$ as the inner limit of an outer asymptotic expansion, and substituting the expression (5.3.48) into (5.3.32), we find that in region B

$$u_0^* = O(\sigma^{1/2}), \quad \psi^* = O(\sigma^{1/2}). \tag{5.3.51}$$

These estimates obviously ensure the balance of inertia terms in the equations of motion with the pressure gradient, and also, as we can see, the matching conditions with the main and lower parts of the boundary layer.

Using the expressions (5.3.28) and (5.3.29), and the estimates (5.3.48), (5.3.49), and (5.3.51) obtained for the flow parameters, we can represent the solution in region B in the original variables as follows:

$$x = \Delta_0 s_0 + \Delta_1 X, \quad \psi = \mathrm{Re}^{-1/2}\Psi_S + \sigma^{1/2}\mathrm{Re}^{-7/12}\widetilde{\Psi},$$

$$u = \sigma^{1/2}\mathrm{Re}^{-1/12}\widetilde{u}_0(X, \widetilde{\Psi}) + o(\sigma^{1/2}\mathrm{Re}^{-1/12}),$$

$$y = \widetilde{G}(\mathrm{Re}) + \mathrm{Re}^{-1/2}\widetilde{Y}_0(X, \widetilde{\Psi}) + o(\mathrm{Re}^{-1/2}),$$

$$p = p_{00} + \sigma\mathrm{Re}^{-1/6}\widetilde{p}_0(X, \widetilde{\Psi}) + o(\sigma\mathrm{Re}^{-1/6}).$$

$$(5.3.52)$$

Substituting these expansions into the Navier–Stokes equations, and keeping in mind the balance mentioned above between inertia and viscous terms, we arrive at the Prandtl boundary-layer equations

$$\widetilde{u}_0\frac{\partial\widetilde{u}_0}{\partial X} + \frac{\partial\widetilde{p}_0}{\partial X} = \widetilde{u}_0\frac{\partial}{\partial\widetilde{\Psi}}\left(\widetilde{u}_0\frac{\partial\widetilde{u}_0}{\partial\widetilde{\Psi}}\right), \quad \frac{\partial\widetilde{p}_0}{\partial\widetilde{\Psi}} = 0, \quad \frac{\partial\widetilde{Y}_0}{\partial\widetilde{\Psi}} = \frac{1}{\widetilde{u}_0}. \quad (5.3.53)$$

For the small parameter $\sigma(\mathrm{Re})$, the transcendental equation

$$\sigma(\ln\sigma^{-1})^{2/3} = \mathrm{Re}^{-1/2}. \quad (5.3.54)$$

is obtained.

The function $\widetilde{G}(\mathrm{Re})$ from (5.3.52) still remains arbitrary, and for its determination it is necessary to consider the flow in region C, which is a continuation of the lower part of the boundary layer (Figure 5.12).

The pressure changes that occur in the interaction region do not introduce changes in the velocity profile in region IV. Therefore, the functions $\overline{U}_0(\Psi)$ and $\overline{Y}_0(\Psi)$ from (5.3.27) will be retained as the leading terms of the expansions in region C. Using the expansions (5.3.17) for these functions as $\Psi \to \Psi_S - 0$, and matching with the solution in region B, we find that as $\widetilde{\Psi} \to -\infty$,

$$\widetilde{u}_0 \to -a_0\widetilde{\Psi}, \quad \widetilde{Y}_0 \to -a_0^{-1}\ln(-\widetilde{\Psi}) + o(1), \quad (5.3.55)$$

and also that

$$\widetilde{G} = \mathrm{Re}^{-1/2}[(12a_0)^{-1}\ln(\sigma^{-6}\mathrm{Re}) + b_0]. \quad (5.3.56)$$

We note that to satisfy the no-slip condition at the body surface it is necessary to introduce the wall layer (region D), in which the effect of viscous stresses appears, but which as before does not influence the central part of the boundary layer. Moreover, in the interaction region the pressure gradient tends to zero as $x^* \to \infty$, and therefore changes in pressure in the flow downstream will not influence the flow (in the leading terms) in the wall region. Here and everywhere downstream, up to the region $x = O(1)$, the solution for region V as $x^* \to \infty$, written in appropriate variables, will remain valid.

To determine the asymptotic behavior of the solution at the outer edge of region B, we consider the flow in the upper part of the boundary layer (region A).

The mutual dependence of the streamline slopes and pressure changes at the outer edge of the boundary layer within the interaction region is not present outside this region. Moreover, the exponential decrease of the pressure gradient toward zero, accompanied by the increasing streamline slopes in the upper part of the boundary layer as $x^* \to \infty$ (see the corresponding expressions (5.3.46) and (5.3.42)), leads to the fact that downstream of the interaction region they have different orders of magnitude. Thus, in region A the streamline slopes are completely determined by the flow within the interaction region and, consequently, are defined by the expressions (5.3.25) and (5.3.26) as $x^* \to \infty$; the pressure changes that occur here are found to have no influence on the streamline slopes. This follows formally from the validity of the expansion (5.3.24) that determines the streamline slopes at the outer edge of the boundary layer for $|z| > O(\mathrm{Re}^{-1/3})$. Therefore, rewriting the expressions (5.3.25), (5.3.26), and (5.3.42) in the variable X from (5.3.52), we have:

$$Y = (6a_0)^{-1}\ln(\sigma^{-6}\mathrm{Re}) + \frac{3}{2}\frac{\gamma_0^{2/3}}{a_0}X + Y_0(\Psi)$$
$$- a_0^{-1}\ln(a_0^2\gamma_0^{1/3}/k) + b_0 + o(1). \qquad (5.3.57)$$

As in region C, $U_0(\Psi)$ is retained as the leading term in the expansion for the velocity profile. Now using the expressions (5.3.15), we carry out the matching of the solutions in regions A and B. As a result, taking into account (5.3.56), we have

$$\tilde{u}_0 \to a_0\tilde{\Psi}, \quad \tilde{Y}_0 \to a_0^{-1}\ln\tilde{\Psi} + \frac{3}{2}\frac{\gamma_0^{2/3}}{a_0}X + a_0^{-1}\ln(a_0^2\gamma_0^{1/3}/k) + o(1)$$
$$(5.3.58)$$

as $\tilde{\Psi} \to \infty$.

To obtain the initial condition for region B, we analyze the behavior of the solution to equation (5.3.53) as $X \to -\infty$.

From the expression (5.3.50), it follows that

$$\tilde{p}_0 \to -2k\gamma_0^{-1/3}\exp(-2\gamma_1 X) \text{ as } X \to -\infty, \quad \gamma_1 = \frac{3\gamma_0^{2/3}}{4}. \qquad (5.3.59)$$

Then for the longitudinal component of the velocity vector, taking into account its behavior (5.3.55), (5.3.58) as $|\tilde{\Psi}| \to \infty$, or on the basis of the

Bernoulli integral that holds up to region B, the solution can be written in the form

$$\tilde{u}_0 \to \exp(-\gamma_1 X)\tilde{F}_0(\tilde{\tau}), \quad \tilde{\tau} = \tilde{\Psi}\exp(\gamma_1 X). \qquad (5.3.60)$$

Substituting this expression together with (5.3.59) into (5.3.53), and integrating the equation obtained, we obtain

$$\tilde{F}_0 = (a_0^2|\tilde{\tau}|^2 + 4k\gamma_0^{-1/3})^{1/2}. \qquad (5.3.61)$$

From the third of equations (5.3.53), it follows that, as $X \to -\infty$, the solution for $\tilde{Y}_0(X, \tilde{\Psi})$ that satisfies the matching condition as $|\tilde{\Psi}| \to \infty$ has the form

$$\tilde{Y}_0 \to \gamma_1 a_0^{-1}X + \tilde{H}_0(\tilde{\tau}) + o(1),$$

$$\tilde{H}_0 = \frac{1}{2a_0}\ln\left(\frac{\tilde{F}_0 + a_0\tilde{\tau}}{\tilde{F}_0 - a_0\tilde{\tau}}\right) + \frac{1}{2a_0}\ln\left(\frac{a_0^2\gamma_0^{1/3}}{k}\right). \qquad (5.3.62)$$

We emphasize once again that matching the asymptotic expansions in regions III and B is carried out after the change to the new variables indicated above.

According to the original ideas about the character of the flow in region B, at a certain value of X the longitudinal velocity component becomes zero, and flow separation occurs. Therefore, as for the problems of separation from a fixed surface studied above, the system of boundary conditions for region B has to be supplemented by conditions imposed downstream, that is, as $X \to \infty$.

The asymptotic expansions as $X \to \infty$ can be obtained on the basis of the same considerations as used earlier (see Chapters 1 and 2).

In the upper part of the region considered, as $X \to \infty$, a mixing sublayer is developed that separates the main part of the boundary layer from the region of reverse flow.

The solution for the mixing sublayer is sought in the form

$$\tilde{u}_0 = X^p f_+(\eta_+) + o(X^p), \quad \eta_+ = \tilde{\Psi}/X^r, \qquad (5.3.63)$$

where p and r are certain constants. Substituting this expansion into (5.3.53) and keeping in mind the balance of viscous and inertia terms, we find that $p = 2r - 1$. From the condition at the outer edge of the mixing sublayer, it is known that $\tilde{u}_0 = O(\tilde{\Psi})$ as $\tilde{\Psi} \to \infty$, and therefore $p = r$. Thus,

$$p = r = 1. \qquad (5.3.64)$$

The equation for $f_+(\eta_+)$, obtained by substituting (5.3.63) into (5.3.53), takes the form

$$(f_+ f_+')' + \eta_+ f_+' - f_+ = 0. \qquad (5.3.65)$$

The pressure gradient does not enter the equation for the mixing sublayer since its value, as we will see, approaches zero as $X \to \infty$.

From the equation for $\widetilde{Y}_0(X, \widetilde{\Psi})$ in (5.3.53), as a result of substituting the expansion (5.3.63), (5.3.64) there, it follows that

$$\widetilde{Y}_0 = h_+(\eta_+) + g_+(X) + o(1), \quad h_+ = \int f_+^{-1} d\eta_+, \qquad (5.3.66)$$

where $g_+(X)$ is an arbitrary function.

The solution of the equation for $f_+(\eta_+)$ must satisfy the condition (5.3.58), which ensures matching with the main part of the boundary layer; written in the variables X and η_+, it has the form

$$f_+'(\infty) = a_0.$$

Moreover, we must satisfy the usual condition for the mixing layer: $\widetilde{U}_0(X, \widetilde{Y}_0) \to 0$ as $\widetilde{Y}_0 - g_+(X) \to -\infty$. In the variables X and η_+, this condition is written as

$$f_+(d_0) = 0, \quad d_0 = \text{const}.$$

The solution of equation (5.3.65) satisfying these conditions can be obtained in an explicit form:

$$f_+ = a_0(\eta_+ + a_0), \quad d_0 = -a_0. \qquad (5.3.67)$$

Diesperov (1986) proved the uniqueness of this solution.

By substituting the expression (5.3.67) into (5.3.66) and carrying out the matching with the solution in the main part of the boundary layer, we will have

$$\widetilde{Y}_0 = \frac{3}{2}\gamma_0^{2/3} a_0^{-1} X + a_0^{-1}\ln X + h_+(\eta_+)$$
$$+ a_0^{-1}\ln(a_0^2 \gamma_0^{1/3}/k) + o(1), \qquad (5.3.68)$$
$$h_+ = a_0^{-1}\ln(\eta_+ + a_0) \quad (X \to \infty).$$

In a completely analogous way the solution is constructed for the mixing sublayer that develops along the dividing streamline $\widetilde{\Psi} = 0$ in the lower part of region B as $X \to \infty$. As a result we obtain

$$\tilde{u}_0 = X f_-(\eta_-) + o(X), \quad \tilde{Y}_0 = -a_0^{-1}\ln X + h_-(\eta_-) + o(1);$$
$$f_- = a_0(\eta_- + a_0), \quad h_- = -a_0^{-1}\ln(\eta_- + a_0), \quad \eta_- = -\tilde{\Psi}/X.$$
$$(5.3.69)$$

According to the expression (5.3.58) for $\tilde{Y}_0(X, \tilde{\Psi})$, the region of reverse flow, located between the mixing sublayers, expands as $X \to \infty$ according to a linear law. Repeating the ideas from Chapter 1, the motion in this zone is due to the entraining effect of the mixing sublayers and, keeping in mind that the quantity of fluid entrained also grows linearly as $X \to \infty$, it is not difficult to obtain

$$\tilde{u}_0 = -a_0^2/\gamma_1 + o(1), \quad \tilde{p}_0 = -a_0^4/(2\gamma_1^2) + o(1),$$
$$\tilde{Y}_0 = \gamma_1 X/a_0 - \gamma_1\tilde{\Psi}/a_0^2 + (2a_0)^{-1}\ln(a_0^2\gamma_0^{1/3}/k) + o(1).$$
$$(5.3.70)$$

Thus, equations (5.3.53), together with the expressions (5.3.55), (5.3.58)–(5.3.64), and (5.3.67)–(5.3.70), represent a boundary-value problem whose solution describes the flow in region B.

These relations contain three arbitrary (from the viewpoint of local considerations) constants a_0, k, and U_{00}. However, with the help of the transformations

$$X = \gamma_1^{-1}X', \quad \tilde{\Psi} = a_0\gamma_1^{-1}\Psi', \quad \tilde{u}_0 = a_0^2\gamma_1^{-1}U',$$
$$\tilde{p}_0 = a_0^4\gamma_1^{-2}P', \quad \tilde{Y}_0 = a_0^{-1}Y' + (2a_0)^{-1}\ln(a_0^2\gamma_0^{1/3}/k), \quad (5.3.71)$$
$$\gamma_1 = \frac{3}{4}\gamma_0^{2/3}, \quad \gamma_0 = \frac{4ka_0}{3U_{00}^2},$$

the boundary-value problem is reduced to a form containing just one arbitrary constant

$$\alpha^* = \frac{k}{2U_{00}}\left(\frac{3}{a_0^3}\right)^{1/2} \tag{5.3.72}$$

If one also carries out the Prandtl (1938) transposition transformation

$$Y' = Z' + X', \tag{5.3.73}$$

then the problem becomes symmetric relative to the plane $Z' = 0$, and consequently the conditions applied as $Z' \to -\infty$ can be replaced by the usual conditions at a plane of symmetry: $\Psi' = \Psi'_{Z'Z'} = 0$ at $Z' = 0$. And, finally, a displacement of the origin of the coordinate system

$$X' = X'' + \ln \alpha^* \tag{5.3.74}$$

allows elimination of the constant α^*.

Thus, after the transformations (5.3.71)–(5.3.74), the boundary-value problem for region B, written in the variables X'' and Z', assumes the form

$$\frac{\partial \Psi}{\partial Z} \frac{\partial^2 \Psi}{\partial X \partial Z} - \frac{\partial \Psi}{\partial X} \frac{\partial^2 \Psi}{\partial Z^2} + \frac{dP}{dX} = \frac{\partial^3 \Psi}{\partial Z^3};$$

at $Z = 0$: $\Psi = \dfrac{\partial^2 \Psi}{\partial Z^2} = 0$;

as $Z \to \infty$: $\Psi \to \exp(Z - X)$; (5.3.75)

as $X \to -\infty$: $\Psi \to 2 \exp(-X) \sinh Z$, $P \to -2 \exp(-2X)$;

as $X \to \infty$: $\Psi \to X \Phi(N)$, $\Phi = \exp N - 1$,

$$N = Z - X - \ln X, \quad N = O(1);$$
$$\Psi \to -XS, \quad S = Z/X, \quad S = O(1),$$
$$P \to -1/2.$$

(Here and below the primes are omitted.)

The conditions at the plane of symmetry $(Z = 0)$ point to the fact that, for fixed values of X, the longitudinal velocity component Ψ_Z assumes its minimum value there. This implies that the separation point lies on the plane of symmetry, and therefore the Moore–Rott–Sears condition is satisfied:

$$\frac{\partial \Psi}{\partial Z} = \frac{\partial^2 \Psi}{\partial Z^2} = 0 \text{ at } X = X_S, \quad Z = 0.$$

In contrast to the classical formulation of a problem for the Prandtl boundary-layer equations, the pressure distribution $P(X)$ in (5.3.75) has to be found as part of the solution. The displacement effect $\vartheta(X)$ here is given at the outer edge of region B; according to the expansion of (5.3.75) as $Z \to \infty$,

$$\vartheta = \lim_{Z \to \infty} \left(-\frac{\partial \Psi}{\partial X} \Big/ \frac{\partial \Psi}{\partial Z} \right) = 1.$$

As first shown by Catherall and Mangler (1966), the solutions of boundary-value problems for the Prandtl equations with a given displacement thickness and the usual no-slip condition at the body surface behave in a regular way at a point of zero skin friction, and consequently can describe real fluid motions with flow reversal (Bogolepov and Neiland, 1971; Smith 1976).

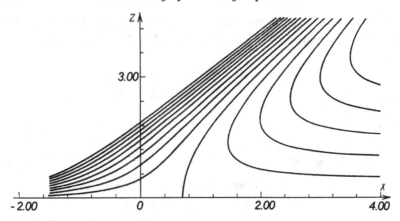

Fig. 5.13 Sketch of streamlines near the separation point.

The solution of the problem (5.3.75) turns out to be surprisingly simple and is written in explicit form:

$$\Psi = 2\exp(-X)\sinh Z - Z, \quad P = -1/2 - 2\exp(-2X).$$

From the expression for the stream function one finds the longitudinal coordinate of the separation point $X_S = \ln 2$ and the streamline field shown in Figure 5.13. (Neighboring streamlines in this figure are calculated at intervals in values of Ψ equal to 0.5.)

Downstream from the region where separation occurs, the detached boundary layer consists of the main part and the mixing sublayer (regions A' and B_+ in Figure 5.12), which are continuations of region A and the upper part of region B, respectively. After passing through the intermediate region where $x = O(\Delta_0)$, this separated part becomes an ordinary mixing layer having, on the length scale of the body, that is, for $x = O(1)$ and $y = O(1)$, the thickness of the original boundary layer. The shape of the streamline along which it develops is determined as $x \to +0$ by the expression (5.3.2). The lower part of the boundary layer, adjacent to the moving body surface (regions B_-, C', D'), also becomes a boundary layer with thickness of order $\mathrm{Re}^{-1/2}$. The motion in the region of reverse flow results from the entraining effect of the mixing layer and the wall layer.

The analysis described above, concerned with the flow beyond the interaction region, was given in the work of Sychev (1984, 1987a).

The asymptotic solution of the Navier–Stokes equations that describes the flow in the vicinity of a separation point on a moving surface contains a number of constants. Their values can be found from the solution for the entire separated-flow problem on the length scale of the body, that is, from a combined solution of the following two problems: the problem of external flow over the body and the problem of calculating the boundary layer on the entire body surface up to the separation point of the free streamline. The location of the latter is determined by the condition that the approaching boundary-layer profile is of a preseparation type and has the form (5.3.4), (5.3.5), (5.3.13)–(5.3.23); that is, the Moore–Rott–Sears condition is satisfied here. It is evident that the values are thereby obtained both for the constants a_0, b_0, c^*, Ψ_S characterizing this profile and for the constants k, p_{00} appearing in the expressions (5.3.1) and (5.3.3), since the latter are determined by assigning the position of the detachment point for the free streamline.

We note also that when the value of the body surface speed is decreased, as should be expected, we arrive at a flow in the vicinity of a the point of separation from a fixed surface (Sychev, 1972) (see Chapter 1). If we denote the body surface speed by ϵu_w^0 and take u_w^0 as some positive constant, while $\epsilon(\mathrm{Re})$ is a small parameter whose value tends to zero as $\mathrm{Re} \to \infty$, then the value of the constant k in (5.3.1) and (5.3.2) tends to zero as $\epsilon^{1/2}$. This means that the local solution for the limiting state of the flow field as $\mathrm{Re} \to \infty$, as in the case of separation from a fixed surface, satisfies the Brillouin–Villat condition of finiteness of the free-streamline curvature at a point of detachment.

For $\epsilon > O(\mathrm{Re}^{-1/8})$, the solution has almost the same structure as for finite values of the body surface speed. The difference consists in the fact that the characteristic regions shown in Figure 5.12 are found to be incorporated into the region where $|x| = O(\epsilon^3)$. The longitudinal scale of the interaction region here will be $|x| = O(\epsilon^{1/3}\mathrm{Re}^{-1/3})$, and the variable parts of the pressure and the longitudinal velocity component in the nonlinear locally inviscid part of the interaction region will be $O(\epsilon^{2/3}\mathrm{Re}^{-1/6})$ and $O(\epsilon^{1/3}\mathrm{Re}^{-1/12})$, respectively. For $\epsilon = \mathrm{Re}^{-1/8}$, the characteristic values both for the flow parameters and the scales of the regions are easily seen to approach the corresponding values for the triple-deck structure of laminar separation from a fixed surface. A more detailed description of this limit process is given in the work of Sychev (1979, 1983) mentioned above.

Wake breakdown. To conclude this section, we turn to consideration of the breakdown of a symmetric wake, which we will see is not

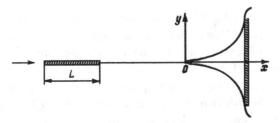

Fig. 5.14 The pattern of separated flow ahead of an obstacle, according to Chaplygin (1899).

essentially different from the separation at a moving surface considered above.

This phenomenon is observed any time that the wake behind a thin body, for example a plate aligned parallel to the oncoming flow, is influenced by an adverse pressure gradient, which can be established by the presence of another body (Figure 5.14). The process of breakdown is always associated with the fact that the effect of a positive pressure gradient leads to a strong deceleration of the fluid motion in the inner part of the wake, so that the flow speed finally becomes zero at some point located on the plane of symmetry.

The shear stress at the wake plane of symmetry is equal to zero, and therefore the Moore–Rott–Sears condition is satisfied at the breakdown point, where the velocity also becomes zero. Consequently, after some minor simplifications, the solution for the flow in the vicinity of the separation point on a moving surface is transformed into a solution that describes the flow near the wake breakdown point. The essence of these simplifications consists in the fact that the line of zero shear stress lying within the boundary layer ahead of the separation point becomes the plane of symmetry. Ahead of the breakdown point ($x \to -0$), one can identify two characteristic regions: the main part and the nonlinear locally inviscid region adjacent to the plane of symmetry. The solution in these regions will have the form (5.3.4), (5.3.5), (5.3.13)–(5.3.15) if one takes $\Psi_S = g(x) = 0$ in these expressions. (Because of the symmetry we consider only the flow in the upper half plane, while on the wake axis $\psi = y = 0$ as usual.) Thus, the breakdown of the wake is preceded by a growth of its transverse dimension according to a logarithmic law. The limiting state of the flow field as Re $\to \infty$ is described by Chaplygin's solution (1899) for the flow past a body with a cavity that begins with a cusp (Figure 5.14).

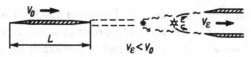

Fig. 5.15 Sketch of wake breakdown according to experimental observations of Werlé (1971) for Re $\approx 10^4$.

It is interesting to note that the effect of wake expansion during its breakdown was first discovered experimentally by Werlé (1971) with the help of flow visualization. The observed flow picture, taken from this work, is shown in Figure 5.15.

The solution of the wake breakdown problem is described in the work of Sychev (1978b, 1983, 1984).

5.4 Unsteady Boundary-Layer Separation

If the motion of a smooth body in a space filled with fluid begins from a state of rest, then, as already noted in Section 1 of this chapter, the flow field remains unseparated and potential during an initial period of time. Flow separation does not occur at once, but only after a short finite time when a boundary layer has already been established close to the surface. Therefore, before turning to an analysis of the flow in the vicinity of the separation point, we should consider the flow stage that precedes the appearance of separation.

Preseparation flow of an unsteady boundary layer. After the beginning of the motion of a body in a viscous fluid, the pressure distribution is given by the solution for a problem of steady unseparated flow (Blasius, 1908). A classical example is the flow around a circular cylinder which instantaneously acquires a certain finite constant speed. The equations for the flow in the unsteady boundary layer close to the surface together with appropriate initial and boundary conditions have the form

$$\frac{\partial u}{\partial t} + u\frac{\partial u}{\partial x} + V\frac{\partial u}{\partial Y} + \frac{dp_e}{dx} = \frac{\partial^2 u}{\partial Y^2}, \quad u = \frac{\partial \Psi}{\partial Y}, \quad V = -\frac{\partial \Psi}{\partial x};$$

$$u \to U_e(x) \text{ as } Y \to \infty, \quad t > 0;$$

$$u(t,x,0) = V(t,x,0) = u(t,0,Y) = u(t,\pi,Y) = 0 \text{ for } t \geq 0;$$

$$u = U_e(x) \quad \text{for } t = 0, \quad Y > 0;$$

$$U_e = 2\sin x, \quad dp_e/dx = -4\sin x \cos x, \quad 0 \leq x \leq \pi.$$

$$(5.4.1)$$

Here, $x = 0$ and $x = \pi$ are the coordinates of the front and rear stagnation points; $LU_\infty^{-1}t$ is the time; the previous notation is retained for the other variables, and the cylinder radius is taken as the characteristic body length L. The solution of problem (5.4.1), which describes the initial flow stage ($t \to +0$), has the form

$$\Psi = 2t^{1/2}U_e \sum_{n=0}^{\infty} t^n \varphi_n(x, \eta), \quad \eta = Y/2t^{1/2}. \qquad (5.4.2)$$

As a result of substituting (5.4.2) into (5.4.1), for the functions $\varphi_n(x, \eta)$ one obtains ordinary differential equations in the variable η, while x plays the role of a parameter; thus, for the leading terms of the expansion we have

$$\varphi_0 = \varphi_{00}(\eta), \quad \varphi_1 = U_e' \varphi_{10}(\eta). \qquad (5.4.3)$$

It should be noted that the function $\varphi_0(x, \eta)$ satisfies the heat conduction equation, and this means that the leading term in the expansion (5.4.2) describes a process of viscous diffusion.

The expressions for $\varphi_{00}(\eta)$ and $\varphi_{10}(\eta)$ were found by Blasius, and it turns out that at the instant of time $t = t_0 = [2(1 + 4/3\pi)]^{-1} = 0.3510$, the point of zero skin friction begins to move upstream along the cylinder surface from the rear stagnation point. Cowley (1983) calculated 51 terms of the expansion (5.4.2) and obtained the value $t_0 = 0.32191989$.[3]

One might expect the solution of the problem (5.4.1) to reach a steady state as $t \to \infty$, but this does not occur since a steady-state solution of this problem does not exist. The fact is that in the solution of the steady-flow equations for a boundary layer with pressure gradient, from (5.4.1), the skin friction becomes zero at $x = x_0 = 1.82298$ ($= 104.45°$) (Terrill, 1960), and as usual a nonremovable singularity appears here (see Chapter 4). The question arises: is there a finite instant of time t_S at which the singularity appears in the solution to the problem (5.4.1)? The answer to this question was given by van Dommelen and Shen (1980, 1982). On the basis of a numerical solution with the use of Lagrangian variables, they were the first to establish that when $t = t_S = 1.50225$ and $x = x_S = 1.9368(= 110.97°)$, a point appears in the boundary layer at which the Moore–Rott–Sears criterion is satisfied and a singularity appears in the solution. Shown in Figure 5.16 are graphs of the velocity profile and the vorticity $\omega = \partial u/\partial Y$, referred to its value on the body

---

[3] See the survey by Riley (1975) and the book by Telionis (1981) for information about other work in this area.

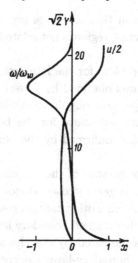

Fig. 5.16 Velocity and vorticity profiles in the boundary layer near the singular point (van Dommelen and Shen, 1982).

surface, at the instant of time $t = 1.4969$ and at $x = 111.04°$, that is, immediately before the singularity appears.

We should note that the application of Lagrangian variables has been found very useful in general for the study of unsteady boundary-layer flows (see Cowley et al., 1990).

When $t \to t_S - 0$, as in the case of a steady flow on a moving surface (Section 2 of this chapter), one can distinguish three characteristic regions within the boundary layer (van Dommelen and Shen, 1982). First of all, there is a "locally inviscid" zone near the Moore–Rott–Sears point. Here, the solution for $T = t_S - t \to +0$ (Elliott et al., 1983) has the form

$$x - x_S = KT + T^{3/2}\widetilde{x}, \quad Y = T^{-1/4}\widetilde{Y},$$
$$u = -K + T^{1/2}\widetilde{U}(\widetilde{x},\widetilde{Y}) + o(T^{1/2}), \quad V = T^{-5/4}\widetilde{V}(\widetilde{x},\widetilde{Y}) + o(T^{-5/4}),$$

where K is a positive constant that is found in the process of solving the problem (5.4.1), with the numerical value $K = 0.52$ (van Dommelen and Shen, 1982). At the outer edge of this region and close to the body surface are zones with thickness of the same order as the thickness of the original boundary layer, for which $u = O(1)$, so that as $\widetilde{Y} \to 0$ the function $\widetilde{U} = O(\widetilde{Y}^{-2})$.

The unbounded growth of the boundary-layer thickness as $T \to +0$ leads to an interaction process between the locally inviscid flow in the boundary layer and the external flow for $(t - t_S)\mathrm{Re}^{2/11} = \widehat{T} = O(1)$.

It is important to mention that the large pressure gradient ($\partial p/\partial x = O(\mathrm{Re}^{1/11})$) in the interaction region is not related to the original pressure gradient (5.4.1).

The solution of the problem for the interaction region as formulated by Elliott et al. (1983) was obtained by Cassel and Walker (1993). At some finite value $\widehat{T} = \widehat{T}_S$ a new singularity again occurs in the solution.

The dependence mentioned above for the boundary-layer thickness as $t \to t_S - 0$ is very well confirmed by the numerical solution of Ingham (1984).

The reason for the appearance of the singularity consists in the fact that the vorticity which is generated continuously at the body surface from the initial instant of the motion cannot, generally speaking, remain within the region of a fully attached boundary layer for an infinitely long period of time. This is indicated by the character of the singularity: unlimited growth of the boundary-layer thickness at $x = x_S$. After its appearance, that is, for $t > t_S$, the unsteady effects that previously were concentrated in the boundary layer will start to penetrate into the external potential flow.

Thus, the solution of van Dommelen and Shen provides the description of the flow up to the moment of appearance of the singularity. This fully corresponds to the ideas of Telionis and Tsahalis (1974) about the transition to a regime of separated flow accompanied by the appearance of a singularity at some finite instant of time. The results of the solution of this problem obtained with various numerical methods by Cowley (1983), Ingham (1984), and other authors (see Cowley et al., 1990) gave values of t_S and x_S close to those obtained by van Dommelen and Shen.

Experimental data indicating the character of the flow for impulsive motion of a cylinder are in good agreement with the results described above. The investigations of Prandtl (1927), Honji and Taneda (1969), and also other authors, based upon visual observations, show that immediately after the beginning of the motion an unseparated potential flow is established, and a region of reverse flow appears only in a thin wall layer near the rear stagnation point. Later in time the transverse scale of this region increases and then a vortex layer leaves the surface and rolls up into two symmetric concentrated vortices. Pressure measurements at the cylinder surface carried out by Schwabe (1935) indicate that during the initial stage the pressures are quite close to the distribution in (5.4.1).

Experimental data (Nagata, Minami, and Murata, 1979), and also the

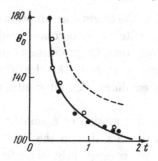

Fig. 5.17 Location of the point of zero skin friction on the surface of a cylinder: filled circles – van Dommelen and Shen (1980); open circles – Nagata et al. (1979) at Re = 600; solid line – Thoman and Szewczyk (1969) at Re = $2 \cdot 10^4$; dashed line – Collins and Dennis (1973) at Re = 50.

numerical solutions of Thoman and Szewczyk (1969) and of Collins and Dennis (1973), are in quite good quantitative agreement with the results of van Dommelen and Shen (1980). Shown in Figure 5.17 are values of the coordinate of the point of zero skin friction at various instants of time; with increasing Reynolds number, the solution of the Navier–Stokes equations is seen to approach the solution of the problem (5.4.1).

And, finally, the time when the singularity appears agrees very well with the time of the change in structure of the reverse flow which, according to experimental data of Bouard and Coutanceau (1980) for Re = 4750, occurs at $t = 1.50$ and is accompanied by an abrupt increase in the boundary-layer thickness with the formation of a second recirculation zone.

It should be mentioned that the results of other work in which problems of impulsive motion were studied also point to the appearance of a singularity. Thus, Ece et al. (1984) studied the flow around a circular cylinder which starts to rotate simultaneously with its translation, and Kravtsova and Ruban (1985) considered the case when it begins to oscillate in rotation. Cebeci et al. (1988) investigated the flow around the leading edge of a thin airfoil undergoing harmonic oscillations. The work of Chen (1983) considered the flow around an elliptic cylinder at angle of attack and, finally, Walker (1978), Doligalski and Walker (1984), and Peridier et al. (1991a) studied the boundary layer that is induced by a point vortex located at some distance from the body surface.

The flow in the vicinity of a separation point of an unsteady boundary layer. The appearance of a singularity in a solution of the boundary-layer equations for a given regular pressure distribution first of all points to the fact that the original assumption regarding the limiting state of the flow field as Re $\to \infty$ becomes invalid. Therefore, as we have seen in previous chapters, for problems of steady flow over bodies one has to seek the limiting state in the class of separated flows of an ideal fluid with free streamlines. A similar situation arises also in the case of unsteady flow.

In problems of impulsive motion of a smooth body the original assumption consisted in the fact that in the limit as Re $\to \infty$ the flow remains unseparated and, consequently, according to the classical approach of Prandtl, one can distinguish two characteristic regions: an external potential flow and a boundary layer at the wall. The condition of asymptotic matching of the solutions in these two regions gives the velocity and pressure distributions at the outer edge of the boundary layer. However, the appearance of a singularity in the solution of the boundary-value problem for the Prandtl equations at some instant of time $t = t_S$ gives evidence of the incompatibility of this solution with a prescribed pressure distribution at subsequent times.

For a description of the flow when $t > t_S$, it is necessary to reconsider the solution for the limiting flow state or, in other words, to abandon the assumption of unseparated flow. At the same time, since the flow starts from a condition of rest, then according to the Lagrange theorem for Re $= \infty$ it has to remain irrotational. In order to combine, within the framework of ideal-fluid theory, the irrotationality condition with violation of the condition of attached flow, Helmholtz (1858, 1868) introduced the idea of a vortex sheet or surface of tangential discontinuity, the appearance of which corresponds to flow separation. For real flows, according to Prandtl (1904, 1927), the vortex sheet has a viscous structure; that is, it is a boundary layer detached from the body surface, with its thickness decreasing in proportion to Re$^{-1/2}$ as Re $\to \infty$. As already noted, numerous experimental results show that the vortex sheet rolls up in the form of a spiral.

The construction of self-similar solutions which describe flows with a vortex sheet was initially proposed in the well-known work of Prandtl (1924) and Kaden (1931). The general equation describing the evolution of a vortex sheet in an ideal fluid was introduced and studied by Nikolskii (1957a,b). (Somewhat later Birkhoff (1962) gave this equation for the particular case when the vortex sheet has a periodic structure in some direction.)

One should note that all the analysis below is not formally related to the flow considered above with an attached boundary layer, since the

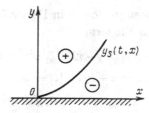

Fig. 5.18 Flow pattern near the point of detachment of a vortex sheet.

process of transition from unseparated to separated flow has not been understood up to the present time. This also concerns steady flows where the structure of the external potential flow must change discontinuously when a characteristic parameter of the problem reaches a critical value (see Chapter 4).

Thus, as we turn to the analysis of flow in the vicinity of a separation point of an unsteady boundary layer, the following assumption should be adopted from the beginning: we will seek a limiting state as Re $\to \infty$ in the class of separated potential flows with vortex sheets. We note that in previous chapters for consideration of steady motions the limiting state was sought in the class of separated flows with free streamlines, which according to Helmholtz represent the most simplified view of vortex sheets.

We introduce the following symbols for the dimensionless variables: t is the time, referred to LU_∞^{-1}; $O'x'$ and $O'y'$ are the axes of a Cartesian coordinate system, directed along the body surface and normal to it; u' and v' are the projections of the velocity vector onto these axes; ψ' is the stream function; and p' is the pressure. For the other dimensional and dimensionless parameters the previous notation is retained.

Let us consider the limiting solution as Re $\to \infty$. We introduce a noninertial coordinate system (Figure 5.18) with axes Ox and Oy fixed to the separation point moving along the surface. If $x_0(t)$ is the position of this point in the original coordinate system for Re $= \infty$, then evidently one can write:

$$x' = x_0(t) + x, \quad y' = y,$$
$$u' = \dot{x}_0(t) + u_0(t, x, y), \quad v' = v_0(t, x, y). \tag{5.4.4}$$

Here and below, the dot indicates differentiation with respect to t. The solution describing an ideal fluid flow near the point of detachment of a vortex sheet from a smooth surface was derived by A. A. Nikolskii and S. K. Betyaev (Betyaev, 1980). In the upper part of the flow plane, that is, above the vortex sheet (region $+$ in Figure 5.18), the expansion of the complex velocity has the form

$$u_0 - iv_0 = U_{00} - i2kU_{00}^{-1}z^{1/2} + O(z),$$
$$z = x + iy, \quad |z| \to 0, \tag{5.4.5}$$

where the functions $U_{00} = U_{00}(t) > 0$ and $k = k(t) > 0$ remain arbitrary, and their values are found from the solution of the global problem. As $x \to +0$, the form of the vortex sheet is defined by the expression

$$y = y_S(t, x) = \frac{4}{3}\frac{k}{U_{00}^2}x^{\mu} + O(x^{5/2}), \quad \mu = 3/2. \tag{5.4.6}$$

Between the body surface and the vortex sheet, that is, for $x > 0$ and $0 < y < y_S$ (region $-$), the correct representation $(|z| \to 0)$ is

$$u_0 - iv_0 = -\frac{1}{\mu+1}\frac{\dot{m}_0}{m_0}z + o(z), \quad m_0(t) = \frac{4k}{3U_{00}^2}; \tag{5.4.7}$$

therefore, as $x \to +0$,

$$u_0 = M_0 x + o(x), \quad M_0 = -\frac{1}{\mu+1}\frac{\dot{m}_0}{m_0}. \tag{5.4.8}$$

From the expressions (5.4.6) and (5.4.8) it follows directly that the fluid motion in this region is determined by a deformation of the tangential discontinuity. If the functions of time entering (5.4.5)–(5.4.8) are replaced by constants, then these expansions transform into (5.3.24) and (5.3.2) for steady flow near the detachment point of the free streamline, and the region where $0 < y < y_S$ for $x > 0$ becomes a stagnation zone.

The analysis of the preceding section showed that boundary-layer separation from a surface moving downstream at a constant speed occurs in a region which lies at a certain small distance (see the expressions (5.3.49) from the interaction region. Taking this circumstance into consideration, for the following it will be convenient for us to represent in advance the solution of the Navier–Stokes equations in the noninertial coordinate system Oxy in the following form:

$$x' = x_S(t, \text{Re}) + x, \quad y' = y,$$
$$u' = \dot{x}_S(t, \text{Re}) + u(t, x, y, \text{Re}),$$
$$v' = v(t, x, y, \text{Re}), \tag{5.4.9}$$
$$p' = -\ddot{x}_S(t, \text{Re})x + p(t, x, y, \text{Re}),$$
$$x_S = x_0(t) + \Delta_0(\text{Re})s_0(t), \quad s_0(t) > 0.$$

Here, $s_0(t)$ and $\Delta_0(\text{Re})$ are certain arbitrary functions; concerning Δ_0, we will assume that $\Delta_0 \to 0$ as $\mathrm{Re} \to \infty$, and therefore the representation (5.4.9) reduces to (5.4.4).

The functions u, v, p and u', v', p' in (5.4.9) satisfy the complete system of Navier–Stokes equations in the absence of a body force; p is the so-called modified pressure, which includes, along with the pressure, the effect of a body force $-\ddot{x}_S$ due to the noninertial coordinate system. (Henceforth, we will refer to it simply as the pressure.)

The no-slip conditions at the body surface

$$u' = v' = 0 \text{ at } y' = 0$$

in the new coordinate system, according to (5.4.9), have the form

$$u = -\dot{x}_S(t, \mathrm{Re}) = -\dot{x}_0(t) - \Delta_0(\mathrm{Re})\dot{s}_0(t) = u_w(t, \mathrm{Re}),$$
$$v = 0 \text{ at } y = 0; \quad \dot{x}_0(t) < 0.$$

Consequently, we come to consideration of the flow over a moving surface.

The solution of the problem for the external potential flow defines the velocity and pressure distributions at the outer edge of the unsteady boundary layer which develops along the body surface for $x' < x_S(t, \mathrm{Re})$ (or $x < 0$). Using the expression (5.4.5) and the Cauchy–Lagrange integral, we find that as $x \to -0$,

$$u_0|_{y=0} = U_e(t, x) = U_{00} + 2kU_{00}^{-1}(-x)^{1/2} + O[(-x)],$$
$$p_0|_{y=0} = p_e(t, x) = p_{00} - 2k(-x)^{1/2} + O[(-x)], \qquad (5.4.10)$$
$$\left.\frac{\partial p_0}{\partial x}\right|_{y=0} = k(-x)^{1/2} + O(1),$$

where $p_{00} = p_{00}(t)$ is determined from the solution of the global problem.

Turning to the expressions (5.4.10) and (5.3.1), (5.3.3) we see that all the differences in the outer boundary conditions, as well as in the conditions at the body surface, for an unsteady boundary layer, in comparison with the steady flow already considered, consist in the fact that U_{00}, k, and u_w are now functions of time. This, however, does not affect the leading terms in the asymptotic expansion of the solution as $x \to -0$, since the unsteady term in the equations of motion has a value $O(1)$, whereas the inertia terms and the pressure gradient have values of order $(-x)^{-1/2}$. Hence, within the interaction region also, where the pressure gradient is $O(\mathrm{Re}^{1/6})$ and the unsteady term as before is $O(1)$, the solution for the leading terms of the expansion will not undergo any changes. In other words, against the background of a large self-induced pressure gradient, the flow within the interaction region is quasi-stationary: all constants a_0, b_0, c^*, Ψ_S entering the solution now depend on time as a parameter.

The solution (5.3.28), (5.3.29), (5.3.32), (5.3.33) for the flow in the nonlinear locally inviscid region III , which gives the main contribution to the displacement effect of the boundary layer, in the variables x', y' takes the form

$$x' - x_0(t) = \mathrm{Re}^{-1/3}x^*, \quad y' = (12a_0)^{-1}\mathrm{Re}^{-1/2}\ln\mathrm{Re} + \mathrm{Re}^{-1/2}y_0^*,$$

$$\psi' = \mathrm{Re}^{-1/2}\left[\Psi_S + \dot{x}_0(t)y_0^* + \mathrm{Re}^{-1/12}\psi^*(t, x^*, y_0^*)\right] + o(\mathrm{Re}^{-7/12}),$$

$$p' = p_{00}(t) + \mathrm{Re}^{-1/6}p^*(t, x^*) + o(\mathrm{Re}^{-1/3}\ln\mathrm{Re}); \tag{5.4.11}$$

$$\psi^* = (-2p^*)^{1/2}a_0^{-1}\sinh[a_0 y_0^* - a_0 g^*(t, x^*)],$$

$$g^* = -(2a_0)^{-1}\ln(-p^*/2a_0^2) + b_0,$$

where $p^*(t,x^*)$ is found from (5.3.44), (5.3.45).

Downstream from the interaction region, according to (5.3.46), the pressure gradient tends to zero exponentially. Consequently, beyond this region the condition of quasi-steady flow will be violated. It is not difficult to show that this happens in a region B (Figure 5.12), where the effects of internal friction forces become important and where the separation point is located. Indeed, in the region B, which is an extension of the locally inviscid region III, let

$$p - p_{00} = \sigma(\mathrm{Re})\mathrm{Re}^{-1/6}\widetilde{P}_0.$$

Repeating the reasoning of the preceding section, we find that in this region the representations (5.3.49)–(5.3.51) are valid; that is, using the notation of (5.4.9) introduced earlier,

$$x = \Delta_1 X, \quad u = O(\sigma^{1/2}\mathrm{Re}^{-1/12}), \quad s_0(t) = \left(\frac{3U_{00}^2}{4ka_0}\right)^{2/3}, \tag{5.4.12}$$

and the functions $\Delta_0(\mathrm{Re})$, $\Delta_1(\mathrm{Re})$, and $\sigma(\mathrm{Re})$ are related through equations (5.3.49). We require now that in this region the unsteady term in the momentum equation, which according to (5.4.12) is $O(\sigma^{1/2}\mathrm{Re}^{-1/12})$, be of the same order as the inertia terms and the pressure gradient; that is,

$$\sigma^{1/2}\mathrm{Re}^{-1/12} = \sigma\mathrm{Re}^{-1/6}/\Delta_1.$$

Then $\sigma = \Delta_1^2\mathrm{Re}^{1/6}$ and, recalling the expression (5.3.49) for Δ_1, we arrive at the transcendental equation (5.3.54) for $\sigma(\mathrm{Re})$, which has been obtained from the condition of a balance between the pressure gradient and the viscous term. This implies that flow unsteadiness in the leading terms of the expansion appears at the same time as the effect of viscosity.

Using the estimates derived, and bearing in mind that region B, according to (5.3.52) and (5.3.56), lies at a distance

$$\mathrm{Re}^{-1/2}[(12a_0)^{-1}\ln(\sigma^{-6}\mathrm{Re}) + b_0]$$

from the lower part of the boundary layer, and that the coefficients in the expansion are now functions of t, the solution can be represented in the following form:

$$x = \Delta_1 X, \quad Y = a_0^*(t)\ln(\sigma^{-6}\mathrm{Re}) + b_0(t) + \widetilde{Y},$$

$$\Psi = \Psi_S(t) - \Delta_1\left[\dot{a}_0^*(t)\ln(\sigma^{-6}\mathrm{Re}) + \dot{b}_0(t)\right]X$$

$$+\Delta_1\widetilde{\Psi}_0(t, X, \widetilde{Y}) + o(\Delta_1), \quad (5.4.13)$$

$$u = \frac{\partial\psi}{\partial y} = \Delta_1\widetilde{U}_0(t, X, \widetilde{Y}) + o(\Delta_1),$$

$$p = p_{00}(t) + \Delta_1^2\widetilde{P}_0(t, X, \widetilde{Y}) + o(\Delta_1^2),$$

where $Y = \mathrm{Re}^{1/2}y$, $\Psi = \mathrm{Re}^{1/2}\psi$, and $a_0^* = (12a_0)^{-1}$. Substituting these expansions, together with the expressions (5.3.49) and (5.3.54) for Δ_0, Δ_1, and σ into the Navier–Stokes equations (1.1.2) of Chapter 1, we arrive at the Prandtl equations for an unsteady boundary layer:

$$\frac{\partial^2\widetilde{\Psi}_0}{\partial t\partial\widetilde{Y}} + \frac{\partial\widetilde{\Psi}_0}{\partial\widetilde{Y}}\frac{\partial^2\widetilde{\Psi}_0}{\partial X\partial\widetilde{Y}} - \frac{\partial\widetilde{\Psi}_0}{\partial X}\frac{\partial^2\widetilde{\Psi}_0}{\partial\widetilde{Y}^2} + \frac{\partial\widetilde{P}_0}{\partial X} = \frac{\partial^3\widetilde{\Psi}_0}{\partial\widetilde{Y}^3}, \quad \frac{\partial\widetilde{P}_0}{\partial\widetilde{Y}} = 0. \quad (5.4.14)$$

The fact that the distance between this region and the interaction region also changes with time was taken into consideration beforehand in (5.4.9) for the introduction of the coordinate system.

In regions A and C (Figure 5.12) adjacent to region B, as in a steady flow (see the preceding chapter), the velocity profiles that had developed in the boundary layer as a result of its interaction with the external flow will be retained as the leading terms of the expansions; the same can be said about the wall sublayer (region D). Therefore, the matching conditions for region B as $|\widetilde{Y}| \to \infty$ will remain the same as in the case of steady flow. Using the expressions (5.3.58) and (5.3.55), we obtain

$$\widetilde{\Psi}_0 \to ks_0^{1/2}a_0^{-2}\exp\left(a_0\widetilde{Y} - \frac{3X}{2s_0}\right) \quad (\widetilde{Y} \to \infty),$$

$$\widetilde{\Psi}_0 \to -\exp(-a_0\widetilde{Y}) \quad (\widetilde{Y} \to -\infty). \quad (5.4.15)$$

The Bernoulli integral remains valid upstream from region B, and therefore, from matching with the solution (5.4.11), it follows that as $X \to -\infty$ (see the expressions (5.3.59)–(5.3.62)),

$$\widetilde{\Psi}_0 \to 2k^{1/2}s_0^{1/4}a_0^{-1}\exp\left(-\frac{3}{4s_0}X\right)$$

$$\times \sinh\left[a_0\widetilde{Y} - \frac{3}{4s_0}X - \ln(a_0k^{-1/2}s_0^{-1/4})\right],$$

$$\widetilde{P}_0 \to -2ks_0^{1/2}\exp\left(-\frac{3}{2s_0}X\right). \tag{5.4.16}$$

As for the conditions downstream ($X \to \infty$), they undergo considerable changes in comparison with the steady case.

In steady motions, the slow reverse flow beyond the stagnation point is caused by the entraining action of the mixing layers. As a result, the pressure gradient (see Chapters 1, 2) downstream from the separation point tends to zero and does not enter the equation that describes the fluid motion in the viscous mixing layers; the variable part of the pressure on the scale of the body, as we have seen, is of order Re^{-1}. If the flow is unsteady, then beyond the separation point the motion is caused by the deformation of the vortex sheet. The appropriate solution has the form (5.4.7), (5.4.8); that is, the variable part of the pressure on the body scale now is of order unity and as $x \to +0$ ($0 < y < y_S$),

$$p_0 = p_{00} + p_{10}x^2 + o(x^2), \quad p_{10}(t) = -\tfrac{1}{2}(\dot{M}_0 + M_0^2). \tag{5.4.17}$$

Here, regardless of the exponent μ in the expression (5.4.6) that determines the displacement effect at the outer edge of the boundary layer beyond the separation point, the pressure gradient varies linearly with the longitudinal coordinate. As a result, the value of $\partial\widetilde{P}_0/\partial X$ as $X \to \infty$ within region B will change in proportion to X, and the effect of this pressure gradient will now appear in the leading terms of the expansion as $X \to \infty$.

Indeed, using the expressions (5.3.63), (5.3.64), (5.3.66), we represent the solution for the upper sublayer as $X \to \infty$ in the form

$$\widetilde{\Psi}_0 = \Phi_+(t, X) + Xf_+(t, \eta_+) + o(X),$$

$$\eta_+ = \widetilde{Y} + g_+(t, X), \quad \widetilde{P}_0 = \frac{\pi_0(t)}{2}X^2 + o(X^2). \tag{5.4.18}$$

Then, after substituting these expressions into (5.4.14), we obtain

$$\frac{\partial^2 f_+}{\partial t\,\partial\eta_+} + \left(\frac{\partial f_+}{\partial\eta_+}\right)^2 - f_+\frac{\partial^2 f_+}{\partial\eta_+^2} + \pi_0 = \frac{\partial^3 f_+}{\partial\eta_+^3}, \quad \frac{\partial g_+}{\partial t} = \frac{\partial\Phi_+}{\partial X}. \tag{5.4.19}$$

From matching with the solution in region A as $\eta_+ \to \infty$ (according to (5.4.15)), it follows that

$$f_+ = \exp(a_0\eta_+) + O(\eta_+^2), \quad \eta_+ \to \infty;$$
$$g_+ = -\alpha_0 X + \alpha_1 \ln X + \alpha_2, \tag{5.4.20}$$
$$\alpha_0 = \frac{3}{2a_0 s_0}, \quad \alpha_1 = -a_0^{-1}, \quad \alpha_2 = a_0^{-1}\ln(ks_0^{1/2}a_0^{-2}).$$

Now, using the second equation in (5.4.19), we find

$$\Phi_+ = -\dot{\alpha}_0 X^2/2 + \dot{\alpha}_1 X \ln X + (\dot{\alpha}_2 - \dot{\alpha}_1)X. \tag{5.4.21}$$

The solution in the lower sublayer is constructed in a similar way. It has the form

$$\tilde{\Psi}_0 = \Phi_-(t, X) - Xf_-(t, \eta_-) + o(X),$$
$$\eta_- = -\tilde{Y} + g_-(t, X); \quad \Phi_- = -\dot{\alpha}_1 X \ln X + \dot{\alpha}_1 X, \tag{5.4.22}$$
$$g_- = \alpha_1 \ln X; \quad \frac{\partial^2 f_-}{\partial t \partial \eta_-} + \left(\frac{\partial f_-}{\partial \eta_-}\right)^2 - f_-\frac{\partial^2 f_-}{\partial \eta_-^2} + \pi_0 = \frac{\partial^3 f_-}{\partial \eta_-^3},$$
$$f_- = \exp(a_0\eta_-) + O(\eta_-^2) \quad \text{as } \eta_- \to \infty.$$

At the inner boundaries of these layers, that is for $\eta_+ \to -\infty$ and $\eta_- \to -\infty$, as usually occurs on leaving the boundary layer, the vorticity must tend to zero, and therefore

$$\frac{\partial f_+}{\partial \eta_+} \to h_+(t), \quad \frac{\partial f_-}{\partial \eta_-} \to h_-(t). \tag{5.4.23}$$

Then from (5.4.19), (5.4.21), and (5.4.22) it follows that

$$h_+ = h_- = h_0(t), \quad \dot{h}_0 + h_0^2 + \pi_0 = 0. \tag{5.4.24}$$

Region B, as is immediately seen from the expression (5.4.15), expands as $X \to \infty$ according to a linear law. The coefficient of proportionality in this expansion $\alpha_0 = 3/(2a_0 s_0)$ is obviously a function of time, and changes in it lead to the fluid motion between the upper and lower sublayers as $X \to \infty$. From the expression for the pressure in (5.4.18), we can represent the solution here as

$$\tilde{\Psi}_0 = X^2 H_0(t)\zeta + o(X^2), \quad \tilde{\zeta} = \tilde{Y}/X = O(1),$$
$$\tilde{P}_0 = \pi_0(t)X^2/2 + o(X^2). \tag{5.4.25}$$

Substituting these expansions into (5.4.14), we find that

$$\dot{H}_0 + H_0^2 + \pi_0 = 0, \tag{5.4.26}$$

and from matching with the solutions for the sublayer it follows that

$$H_0 = h_0 = -\frac{\dot{\alpha}_0}{2\alpha_0}. \tag{5.4.27}$$

Thus, (5.4.14)–(5.4.16) and (5.4.18)–(5.4.27) represent a boundary-value problem whose solution describes the flow in the vicinity of a separation point moving upstream. If in these expressions the functions of time a_0, b_0, k, U_{00} are taken to be constant, then after the transformations (5.3.71)–(5.3.74) we arrive at the problem (5.3.75).

Downstream from region B the upper and lower sublayers, after passing through the intermediate region where $x' - x_0(t) = O(\Delta_0)$ (Figure 5.12), merge with the main and lower parts of the boundary layer, respectively. Here, they transform into the separated boundary layer and the boundary layer adjacent to the surface, with thicknesses of order $\mathrm{Re}^{-1/2}$, and where the unsteadiness and the pressure gradient are important. The solution for the potential-flow region located between the sublayers transforms into the solution (5.4.7), (5.4.8), (5.4.17).

If the longitudinal coordinate of the separation point, located in region B, is denoted by $X = s_1(t)$, then, according to the Moore–Rott–Sears criterion, at this point

$$\tilde{U}_0 = \frac{\partial \tilde{\Psi}_0}{\partial \tilde{Y}} = \dot{s}_1(t) , \qquad \frac{\partial \tilde{U}_0}{\partial \tilde{Y}} = 0 . \tag{5.4.28}$$

Thus, on the basis of the expressions (5.4.29) and (5.4.28), we find that the separation point is moving relative to the body surface with the velocity

$$u_S(t, \mathrm{Re}) = \dot{x}_0(t) + \Delta_0 \dot{s}_0(t) + \Delta_1 \dot{s}_1(t) + o(\Delta_1) , \tag{5.4.29}$$

where $s_0(t)$ is determined from the expression (5.4.12).

The local solution obtained is seen to contain arbitrary functions of time. These functions can be determined from a combined solution of the equation of Nikolskii (1957a) for the external potential flow and Prandtl's equations for the unsteady boundary layer that develops along the body surface. The position of the separation point on the scale of the body ($x' = x_0(t)$) is found from the condition that the Moore–Rott–Sears criterion is satisfied. In this way we will determine the functions $a_0(t)$, $b_0(t)$, $c^*(t)$, $\Psi_S(t)$ characterizing the boundary-layer profile at $x' = x_0(t) - 0$, as well as the values $U_{00}(t)$, $p_{00}(t)$, and $k(t)$ in the expansion of the solution for the outer region in the vicinity of the separation point. And finally, the function $s_1(t)$ in (5.4.28), (5.4.29) can be determined

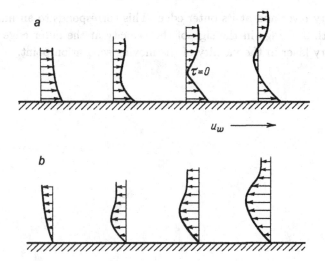

Fig. 5.19 Velocity profiles in the boundary layer near a separation point on a surface moving downstream and the corresponding velocity profiles in an unsteady boundary layer.

from the solution of the boundary-value problem for region B. The point of zero skin friction is located at a finite distance upstream from the separation point. The value of this distance depends upon t and will be determined from the solution of the boundary-layer equations for $x' < x_0(t)$.

To conclude, we comment on the qualitative behavior of the flow in the vicinity of a separation point moving upstream.

The basis of the analysis was the idea of an analogy between the occurrence of unsteady separation and separation from a moving surface. One of the initial conditions consisted in the fact that in a system of coordinates moving with the separation point, the longitudinal velocity component at the outer edge of the boundary layer and at the body surface assumes positive values: $U_{00}(t) > 0$ and $u_w(t, \mathrm{Re}) = -\dot{x}_0(t) + O(\Delta_0)$, $\dot{x}_0(t) < 0$. At the same time, no limitations whatever were imposed upon the relation between $U_{00}(t)$ and $\dot{x}_0(t)$. The velocity profiles in the boundary layer in the coordinate system attached to the moving separation point for the cases when $U_{00}(t) > -\dot{x}_0(t)$ and $U_{00}(t) < -\dot{x}_0(t)$ are shown in Figures 5.2a and 5.19a. The corresponding profiles for an unsteady boundary layer in the original coordinate system are shown in Figures 5.2b and 5.19b. It is clear that when $U_{00}(t) < -\dot{x}_0(t)$, the longitudinal velocity component assumes negative values everywhere in the boundary layer and at its outer edge. This corresponds to an unsteady flow with a change in the sign of the velocity at the outer edge of the boundary layer in the vicinity of the moving separation point.

6

The Asymptotic Theory of Flow
Past Blunt Bodies

6.1 The Background of the Problem

This chapter will be devoted to the analysis of one of the fundamental problems of hydrodynamics: ascertaining the limiting state for the steady flow behind a body of finite size as the Reynolds number Re → ∞ in cases where unseparated flow over the body is impossible, for example, behind a blunt body such as a circular cylinder or a plate placed normal to the oncoming flow. Although in reality such flows already become unsteady at Reynolds numbers of the order of 10^1–10^2, and undergo transition to a turbulent state with further increase in Reynolds number, the solution of this problem is of great interest in principle. Moreover, one might anticipate that such a solution would allow the study of fluid flows at moderate Reynolds numbers when a steady flow regime is still maintained but the methods of the theory of slow motion (Re < 1) are no longer applicable.

There are several points of view about possible ways of solving this problem. According to the first of these, already stated by Prandtl (1931) and then developed by Squire (1934) and Imai (1953, 1957b), the limiting flow configuration as Re → ∞ is the classical Kirchhoff (1869) flow with free streamlines and a stagnation zone that extends to infinity and expands asymptotically according to a parabolic law (Figure 6.1). This picture has been severely criticized both because of poor quantitative agreement with experimental results (for measurements of body drag) and also for some fundamental reasons. According to Batchelor (1956b), the most serious difficulties preventing realization of the Kirchhoff model are connected with the "open" configuration of the stagnation zone, since this makes it impossible for the flow to change continuously with a continuous change in Reynolds number from small values, when the wake

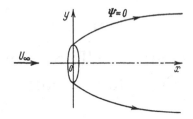

Fig. 6.1 Flow around a body according to Kirchhoff's model (1869).

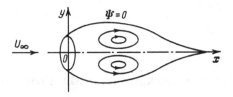

Fig. 6.2 Batchelor's model (1956b).

is closed, to large values. Besides, as noted in Batchelor's work, it is unlikely that there would occur such great flow nonuniformity at infinity as would follow from the Kirchhoff model.

To replace this model, Batchelor (1956b) proposed a flow picture with a closed region of recirculating fluid motion behind the body (Figure 6.2). Assuming the dimensions of this region and the characteristic speed of the motion in it to be finite, he concluded that the body drag is zero in the limit as the Reynolds number Re $\rightarrow \infty$. Here, the motion in the recirculation zone in the limit is an inviscid rotational flow with closed streamlines and (in the two-dimensional case) constant vorticity (Prandtl, 1904; Batchelor, 1956a). In general, there is a jump in the tangential velocity component, and hence in the Bernoulli constant, across the dividing streamline. The result is that the recirculation zone has a shape with a cusp at the rear.

The flow picture proposed by Lavrent'ev (1962) in a certain sense can be considered as a generalization of Batchelor's model. The main difference consists in allowing for the existence of stagnation zones within the closed wake region (L_1 and L_2 in Figure 6.3). Here, the flow between the streamlines bounding these zones and the outer boundary of the recirculation zone can be either rotational or irrotational. Due to the fact that the dimensions of the closed wake behind the body remain finite as Re $\rightarrow \infty$, as in Batchelor's model, the drag of the body in the limit will again be equal to zero.

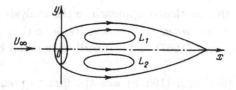

Fig. 6.3 Lavrent'ev's model (1962).

In considering models proposed for the description of flow fields behind blunt bodies, we might also mention earlier attempts at constructing flows with closed streamlines to simulate separated flows. The best
known are models by Zhukovskii (1907) and Föppl (1913). These, however, cannot be considered as limiting states of viscous flows as Re → ∞,
because they contain vorticity concentrated in the form of singularities
(point vortices).

We also exclude various models of ideal fluid flows with free streamlines and stagnation zones where the pressure is smaller than that at
infinity, because all of them are unavoidably related to these or other
methods for the artificial "closing" of cavities (e.g., see Gurevich, 1979;
Gogish, Neiland, and Stepanov, 1975; Wu, 1972).

Thus, the only candidate for the configuration of the limiting state
of the viscous fluid flow behind a blunt body as Re → ∞ that is free
from internal inconsistencies and leads to a nonzero value of the drag
coefficient is the Kirchhoff model. Therefore, it is necessary to come
back and study it in more detail. As already mentioned, the main arguments against this model were related to the idea of the unlikelihood of
the existence of an "open wake" extending behind the body to infinity.
However, such an interpretation of the Kirchhoff solution, in the light
of modern ideas about the nature of asymptotic solutions, is unjustified.
Indeed, if we assume that the zone of recirculating flow behind the body
is closed but that its length and width grow monotonically with increasing Reynolds number, then as Re → ∞ they can become infinitely large
in comparison with the body dimensions, so that on the scale of the
latter the flow field will approach the flow in the Kirchhoff model. Thus,
in considering the solution of the Navier–Stokes equations for the global
flow behind a body of finite size as Re → ∞, we should take into account
the existence of various characteristic scales, one of which is a characteristic body dimension[1] while two others characterize the dimensionless
length $\mathcal{L}$ and width $\mathcal{D}$ of the closed zone of recirculating flow in the wake

---

[1] This is a value of order unity in the adopted dimensionless variables.

behind it. It is this particular approach to the analysis of the problem considered in the case when $\mathcal{D}/\mathcal{L} \to 0$ as Re $\to \infty$ that was carried out in the work of Sychev (1967) and is described in detail below (Sections 2–5).

Let us note that Imai (1957b) was the first to point out the possibility of a flow for which the closed separation zone becomes longer in proportion to the Reynolds number, although he did not give the corresponding solution. His assumption about the linear growth of the closed wake behind the body was based on an analogy with the flow in a channel with moving walls and a fixed obstacle, which becomes a plane Poiseuille flow at distances of order Re from the obstacle. Roshko (1967) also assumed nonzero drag, in an argument similar to that of Sychev (1967) described below, in reaching the conclusion that the wake length should grow linearly with Reynolds number.

Finally, we will consider theoretical models for flows in which the condition $\mathcal{D}/\mathcal{L} \to 0$ is not met and it is assumed that both the length and the width of the recirculation zone of separated flow behind the body grow with increasing Reynolds number according to the same (linear) law. Among them are the model of Taganov (1970) and the models of Smith (1985) and Peregrine (1985). The most detailed analysis of this flow pattern is given in the work of Chernyshenko (1988). Taganov's work postulates (and Chernyshenko's work proves) the absence of a mixing layer between flows at different speeds (in the limit a tangential discontinuity) at the boundary of the recirculation zone; in the studies of Smith and Peregrine, it is assumed that a mixing layer (in the limit a tangential discontinuity) remains as Re $\to \infty$. Accordingly, in the first case the body drag as Re $\to \infty$ must tend to zero as Re^{-1} and in the second case to a finite value. Concerning the latter models, it should be noted that they are found to be in contradiction with certain well-known results of investigations of the asymptotic behavior of the solutions of the Navier–Stokes equations at large distances from a body. In fact, according to the well-known principle of Lagerstrom and Kaplun (e.g., see Lagerstrom, 1964), if we consider a flow field around a body in scaled variables $(x/\mathrm{Re}^\sigma,\ y/\mathrm{Re}^\sigma) = O(1)$, then for any $\sigma < 0$ and Re $\to 0$ or $\sigma > 0$ and Re $\to \infty$, the body will shrink into a point that is incapable of disturbing the flow.[2] This means that the relative thickness of the

---

[2] We note that for this reasoning the important principle is the choice of any identical scales for the independent variables x and y, that is, the maintaining of spatial isotropy. In the opposite case the situation loses physical reality.

recirculation zone of separated flow $\mathcal{D}/\mathcal{L}$ must tend to zero as Re $\rightarrow \infty$ and cannot remain of order unity. In essence, the Lagerstrom–Kaplun principle is the basis for the Oseen solution that represents the asymptotic state of the flow field at large distances from the body at any value of the Reynolds number (see Van Dyke, 1964; Lagerstrom, 1964).

If we turn to experimental studies related to the problem considered, it is necessary first of all to point out the difficulties associated with the realization of a regime of steady flow behind nonstreamlined bodies. From this point of view the most successful are the experimental investigations by Grove, Shair, Petersen, and Acrivos (1964) and Acrivos, Leal, Snowden, and Pan (1968). They were undertaken with the special purpose of establishing the basic patterns and trends in the changes of certain important parameters with increasing Reynolds number, for separated flows around blunt bodies. The major part of the experiments was carried out with circular cylinders and it was possible to preserve steady flow behind them up to Reynolds numbers of the order of 100 (and up to 300 when splitter plates were used). The main result of these tests consisted in the demonstration of a clear linear dependence of the length of the closed wake behind the body upon the Reynolds number. At the same time, the width of the wake increased much more slowly and remained, according to the observations of the authors, of the order of the body dimensions within the range of Reynolds numbers considered. The pressure coefficient in the cylinder base region became practically constant quite early (for Re ≈ 30) and approximately equal to -0.45. Finally, we should note that the experiments revealed a strong dependence of the flow parameters on the flow blockage coefficient, that is, the ratio of the cylinder diameter to the transverse dimension of the test section of the experimental facility. (See Section 4 of this chapter concerning the reasons for this effect.)

On the basis of an extrapolation of the experimental observations described, Acrivos, Snowden, Grove, and Petersen (1965) proposed a theoretical flow picture as Re $\rightarrow \infty$ with a closed wake having finite width and growing in length in proportion to the Reynolds number. With such assumptions the wake flow remains viscous even as Re $\rightarrow \infty$, and the overall drop in pressure is finite and uniformly distributed along the wake. (The latter follows from the condition of balance between the pressure gradient and viscous and inertia terms in the equations of motion.) However, as pointed out by Batchelor (his remark was included in the authors' paper), such a flow picture cannot be consistent with the inviscid flow external to the wake, since in the inviscid flow (which

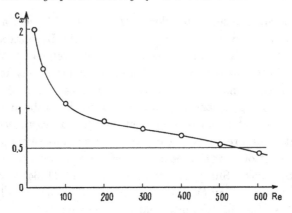

Fig. 6.4 The drag coefficient of a circular cylinder according to Fornberg (1980, 1985).

sees the wake as a thin body with relative thickness of order Re^{-1}) the pressure must return to a value corresponding to the undisturbed flow at distances of the order of the wake width. Thus, the theoretical model of Acrivos et al. is not self-consistent. At the same time the experimental results obtained by these authors are important for comparison with any theory claiming to be asymptotic.

The use of numerical methods for integrating the full Navier–Stokes equations is a relatively new direction for investigation of the separated flows behind nonstreamlined bodies. Although the problem of improving such methods has been given much attention recently, reliable quantitative results (for the case of an incompressible fluid) have so far been obtained only within a quite limited range of Reynolds numbers. Thus, in the work of Dennis and Chang (1970), the flow around a circular cylinder was calculated for a range of Reynolds numbers up to 100. For the same problem the volume of data and accuracy of results obtained by Fornberg (1980, 1985) for Reynolds numbers up to 300 and 600 have not yet been surpassed.

The studies mentioned have confirmed the linear law of growth of the length of the closed wake with increasing Reynolds number, and the work of Fornberg moreover shows that the drag coefficient of a circular cylinder, at least in the range of Reynolds numbers up to 300, approaches asymptotically the value 0.50, in agreement with Brodetsky's (1923) solution obtained according to the Kirchhoff model (see Figure 6.4, taken from his work).

We will see below that the results of both experimental and numerical investigations in this range of Reynolds numbers are in good agreement with the results of calculations based upon the asymptotic theory of separated flow proposed in the work of Sychev (1967). However, for $\mathrm{Re} > 300$ there occurs a marked deviation from the dependence $c_x(\mathrm{Re})$ mentioned, toward a decrease in the value of c_x. Simultaneously, a more intensive growth in the width of the closed separation zone begins, so that the flow field starts to evolve in the direction of the picture that supposes linear growth with respect to Re of both the longitudinal and the transverse dimensions of the separation region.

Below we will first consider in detail the flow scheme corresponding to $\mathcal{D}/\mathcal{L} \to 0$ and later the possibility of the existence of a flow for $\mathcal{D}/\mathcal{L} = O(1)$.

6.2 Formulation of the Problem. Initial Assumptions and Estimates

To approach the analysis of the problem of the limiting form for the separated flow behind a blunt body as the Reynolds number $\mathrm{Re} \to \infty$, we start by considering the initial assumptions that will provide the basis for the entire further investigation.

We will consider the two-dimensional steady flow around a body placed in a uniform incompressible fluid flow. For convenience we take the body to be symmetric with respect to an axis parallel to the velocity vector of the undisturbed flow. We will use dimensionless variables for all relations. For this purpose, as reference values for the independent variables, the velocity components, and the pressure perturbation, we choose a characteristic body dimension L, the value of the undisturbed flow speed U_∞, and twice the dynamic pressure ρU_∞^2, respectively. The Reynolds number of the problem will be defined as

$$\mathrm{Re} = \frac{U_\infty L}{\nu},$$

where ν is the kinematic viscosity of the fluid.

As the main initial assumption (and in the long run the only one) concerning the limiting flow state, we will take the body drag coefficient as $\mathrm{Re} \to \infty$ to approach a nonzero value. This hypothesis is in the best agreement with numerous experimental investigations and results of numerical calculations. It was the basis of the asymptotic theory proposed by Sychev (1967), and referred to earlier, for flow around blunt

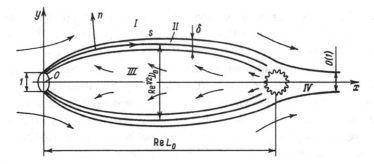

Fig. 6.5 Flow pattern behind a blunt body at high Reynolds numbers (Sychev, 1967).

bodies at high Reynolds numbers. Our presentation below will follow the derivation given in that work.

We begin with a general analysis of the problem. First of all we will show that from the assumption of finite body drag as Re → ∞ it follows that the separation zone is closed, and also that its length referred to a characteristic body dimension grows with increasing Reynolds number according to a linear law. To this end we consider the flow pattern shown in Figure 6.5. The region III of recirculating flow behind the body is separated from the external potential flow I by the viscous mixing layer II, across which the velocity changes from the small values characteristic of the fluid motion in the separation zone up to values of the order of the undisturbed flow speed.

To analyze the flow in the mixing layer II, we introduce the orthogonal curvilinear coordinate system Osn with the axis Os directed along the dividing (zero) streamline $n = 0$. Taking the origin of coordinates to be located at some point on the body surface, we can show (using standard estimates of boundary-layer theory) that the dimensionless thickness δ of the viscous mixing layer when $s \gg 1$ is $\delta = O[(s/\mathrm{Re})^{1/2}]$. It follows that the dimensionless thickness of the mixing layer becomes of order unity when the length $\mathcal{L}$ of the boundary of the separation zone becomes of order Re. Here, the necessity for the closing of the separation zone at such distances behind the body is a simple consequence of the absence (in this zone) of sources or sinks of fluid. Now if for $\mathcal{L} = O(\mathrm{Re})$ the mixing-layer thickness is $\delta = O(1)$, then the same thickness will characterize the far wake (IV, Figure 6.5) which is formed by the merging of the mixing layers and which consists of fluid elements moving outside the dividing streamlines (for $n > 0$). Since it follows from the momentum theorem that the body drag is directly proportional to the momentum deficit in the wake, the drag coefficient will tend to a finite limit. If the length of the closed wake behind the body grew with increasing Reynolds number either more rapidly or more slowly than predicted by the linear law, then the body drag would, respectively, tend to infinity or to zero. Thus, the condition of finite body drag as Re → ∞ determines (in order of magnitude) the length of the boundary of the closed separation zone. Roshko's (1967) argument showing that $\mathcal{L} = O(\text{Re})$ was likewise based on momentum-flux considerations.

Especially important for further analysis of the problem is the determination of the shape of the closed wake and, prior to that, of an estimate for the transverse dimension $\mathcal{D}$ of this region. As already mentioned, the ratio $\mathcal{D}/\mathcal{L}$ should tend to zero as Re → ∞. Then the limit process Re → ∞ for fixed values of the variables x/Re, y/Re (i.e., on the length scale of the closed wake) transforms the actual body into a point, the closed wake into a finite segment of the axis Ox, and the potential flow outside it into the undisturbed flow. We will use the rectangular Cartesian coordinate system Oxy, with $y = 0$ along the flow axis of symmetry and the origin located inside or on the surface of the body (Figure 6.5), to investigate the fluid motion in the wake. We denote the dimensionless velocity components in this coordinate system by u and v, and the pressure coefficient by $2p$.

Let

$$\frac{\mathcal{D}}{\mathcal{L}} = \epsilon(\text{Re}) \to 0, \quad \text{Re} \to \infty. \tag{6.2.1}$$

Then, on the basis of the small-perturbation theory for the external potential flow, $p = O(\epsilon)$ at the surface of the closed separation zone throughout its entire length, that is, for $X = \text{Re}^{-1}x = O(1)$. For the flow inside this zone, where $Y = (\epsilon\text{Re})^{-1}y = O(1)$, on the basis of the continuity equation we take $v/u = O(\epsilon)$. Now, if we apply the component of the momentum equation[3] along the axis Oy, after introducing the estimates obtained (and using the obvious limitation on the maximum of the longitudinal velocity component u), it is easy to establish that $\partial p/\partial y \leq O(\epsilon\text{Re}^{-1})$, that is, that the relative transverse pressure change is $\Delta p/p \leq O(\epsilon)$ and can be neglected. Thus, throughout the region of recirculating flow the pressure p is (in the first approximation) a function only of X, and its value is of order ϵ. It follows that the charac-

---

[3] As the original equations we should use the Navier–Stokes equations, for example the system (1.1.2) of Chapter 1 (for $\kappa = 0$).

teristic velocity of the recirculating fluid motion in the closed wake must also tend to zero. Then this will be a stagnation zone in the limit, and immediately behind the body, in a region with characteristic dimensions of the same order as the body dimensions, the limiting flow will be a Kirchhoff flow.

It is important to emphasize that such a limiting flow state (with free streamlines and an open wake) should be considered here as a local limit valid only in the region $(x, y) = O(1)$ and not as the limiting state for the entire flow field up to infinity. As already mentioned, such an interpretation of the Kirchhoff scheme does not lead to any violation of the natural condition that perturbations decay at infinity.

As is well known, the asymptotic form of the free streamlines according to the Kirchhoff scheme is parabolic: $y \sim \sqrt{x}$ as $x \to \infty$; then, from the condition of matching the solution for the region $(x, y) = O(1)$ as $x \to \infty$ with the solution for the region $(X, Y) = O(1)$ as $X \to 0$, it is not difficult to verify that the small parameter (6.2.1) introduced earlier will be $\epsilon(\mathrm{Re}) = \mathrm{Re}^{-1/2}$. From this it follows that for a closed wake the pressure p as well as the ratio of transverse and longitudinal components of the velocity vector will be of order $\mathrm{Re}^{-1/2}$. It remains to determine the order of magnitude of either of these components. This is easy to do if one makes use of the following idea. The normal component of the velocity induced by the mixing layer II (as follows from the usual estimates of boundary-layer theory) is proportional to $(s\mathrm{Re})^{-1/2}$, so that it will become of order Re^{-1} at distances $x = O(\mathrm{Re})$ behind the body. It is this estimate that one should use for the transverse velocity component v in the closed region of the wake III, because it is only in this case that the flow here will be determined by the entraining action of the mixing layer, and that we will obtain the necessary agreement with results from investigation of the flow in the separation region immediately behind the separation point (Chapter 1). Thus, for the velocity components in the closed wake region we obtain: $u = O(\mathrm{Re}^{-1/2})$, $v = O(\mathrm{Re}^{-1})$. If now we substitute the derived estimates for all the variables into the first momentum equation (6.2.1) (projected on the axis Ox), then it is easy to establish that for a balance of inertia terms and pressure gradient in the closed wake it is necessary that the variable part of the pressure coefficient be of order Re^{-1}; at the same time its complete value, as we have seen, is of order $\mathrm{Re}^{-1/2}$. This result immediately allows us to reach a conclusion about the form of the closed region of recirculating flow. For the pressure coefficient $2p$ to have a constant value in the first approximation when an external inviscid flow passes over this region (as a thin body with relative thickness $\mathrm{Re}^{-1/2}$), it is necessary for the boundary of the region to have an elliptical shape in the first approximation. Indeed, it is known from the theory of a thin airfoil in an incompressible flow that only for flow around a thin ellipse will the variable part of the pressure coefficient at the surface be a second-order quantity (Van Dyke, 1964).

Thus, denoting by $L_0 = \mathrm{Re}^{-1}\mathcal{L}$ and $D_0 = \mathrm{Re}^{-1/2}\mathcal{D}$, respectively, the length and width of the closed wake region in the plane of the variables X and Y, we can write the equation of its boundary in the form

$$Y = F_0(X) + o(1), \quad F_0 = D_0\left[(X/L_0)(1 - X/L_0)\right]^{1/2}. \qquad (6.2.2)$$

The pressure coefficient in the first approximation will be equal to $-2\mathrm{Re}^{-1/2}D_0/L_0$.

Here we conclude the general analysis of the problem and pass to a more detailed study by constructing appropriate asymptotic expansions for the various flow regions.

6.3 The Flow in the Mixing Layer and the Separation Zone

We begin the analysis of the flow by considering the mixing layer that divides the closed separation region (the closed wake) from the external potential flow. At distances of order Re downstream from the body, the thickness of the mixing layer will be finite, so that dimensionless independent variables of order unity can be defined here by the relations

$$S = \mathrm{Re}^{-1}s, \quad N = n, \qquad (6.3.1)$$

where s, n are the dimensionless orthogonal curvilinear coordinates introduced earlier, linked to the dividing streamline ($n = 0$). As the dimensionless unknown functions we will take the stream function ψ and the pressure p, which, according to the estimates of the preceding section, will be represented by the following asymptotic expansions:

$$\psi = \psi_0(S, N) + \mathrm{Re}^{-1/2}\psi_1(S, N) + \cdots,$$
$$p = \mathrm{Re}^{-1/2}p_0 + \mathrm{Re}^{-1}p_1(S, N) + \cdots. \qquad (6.3.2)$$

Substituting (6.3.1) and (6.3.2) into the original system of Navier–Stokes equations,[4] we obtain for the function $\psi_0(S, N)$ the boundary-layer equa-

---

[4] Equations (1.1.2), (1.1.3) of Chapter 1, where instead of x, y one should use the notation s, n.

tion for a mixing layer with zero pressure gradient:

$$\frac{\partial \psi_0}{\partial N}\frac{\partial^2 \psi_0}{\partial S \partial N} - \frac{\partial \psi_0}{\partial S}\frac{\partial^2 \psi_0}{\partial N^2} = \frac{\partial^3 \psi_0}{\partial N^3}, \quad p_0 = \text{const.} \tag{6.3.3}$$

As Re $\to \infty$ the external potential flow approaches the undisturbed flow, whereas the flow speed in the separation zone tends to zero. Therefore, the boundary conditions for this equation have the form

$$\begin{aligned}
\frac{\partial \psi_0}{\partial N} &\to 1, \quad N \to \infty, \\
\psi_0 &= 0, \quad N = 0, \\
\frac{\partial \psi_0}{\partial N} &\to 0, \quad N \to -\infty.
\end{aligned} \tag{6.3.4}$$

The initial thickness of the mixing layer (as $S \to 0$) should be taken equal to zero in the first approximation, since for $\text{Re}S = O(1)$, that is, immediately behind the body (in a region of the order of its dimensions) it is of order $\text{Re}^{-1/2}$.

Then the solution of the problem (6.3.3), (6.3.4) can be described by a well-known self-similar solution:

$$\psi_0 = S^{1/2} f_0(\eta), \quad \eta = N/S^{1/2} \tag{6.3.5}$$

and substituting (6.3.5) into (6.3.3), (6.3.4) reduces the flow problem to the following:[5]

$$\begin{aligned}
2f_0''' + f_0 f_0'' &= 0, \\
f_0'(\infty) = 1, \quad f_0(0) = 0, \quad f_0'(-\infty) &= 0.
\end{aligned} \tag{6.3.6}$$

The numerical solution of (6.3.6) allows us to determine the asymptotic values of the stream function (Lock, 1951):

$$\psi_0 \sim -K_0 S^{1/2}, \quad K_0 = 1.2386 \text{ as } N \to -\infty, \tag{6.3.7}$$

as well as the dimensionless shear stress at the dividing streamline $\psi_0 = 0$:

$$\tau_0 = \left.\frac{\partial^2 \psi_0}{\partial N^2}\right|_{N=0} = S^{-1/2} f_0''(0), \quad f_0''(0) = 0.1996. \tag{6.3.8}$$

These results will be needed later.

We now consider the flow within the closed separation zone. According

---

[5] Diesperov (1985) was the first to prove the existence and uniqueness of the solution of this problem.

to the results of the analysis in Section 2, as independent variables of order unity for this region we should take

$$X = \mathrm{Re}^{-1}x, \quad Y = \mathrm{Re}^{-1/2}y, \tag{6.3.9}$$

and the dependent flow variables may be represented by the following asymptotic expansions:

$$
\begin{aligned}
\psi &= \Psi_0(X, Y) + \mathrm{Re}^{-1/2}\Psi_1(X, Y) + \cdots, \\
p &= \mathrm{Re}^{-1/2}P_0 + \mathrm{Re}^{-1}P_1(X, Y) + \cdots,
\end{aligned}
\tag{6.3.10}
$$

while, as has been established, P_0 should be taken equal to the value

$$P_0 = p_0 = -D_0/L_0. \tag{6.3.11}$$

Substituting (6.3.9) and (6.3.10) into the original system of Navier–Stokes equations, we obtain equations for the first approximation in the form

$$\frac{\partial \Psi_0}{\partial Y}\frac{\partial^2 \Psi_0}{\partial X \partial Y} - \frac{\partial \Psi_0}{\partial X}\frac{\partial^2 \Psi_0}{\partial Y^2} + \frac{\partial P_1}{\partial X} = 0, \quad \frac{\partial P_1}{\partial Y} = 0. \tag{6.3.12}$$

Thus, in the first approximation the flow in the closed wake is an inviscid flow with longitudinal pressure gradient.

The integral of (6.3.12) will be

$$\frac{1}{2}\left(\frac{\partial \Psi_0}{\partial Y}\right)^2 + P_1(X) = H_0(\Psi_0), \tag{6.3.13}$$

with the function $H_0(\Psi_0)$ arbitrary and to be determined by conditions in the closure region of the separation zone.

One of the boundary conditions for $\Psi_0(X, Y)$ is obtained by matching with the solution for the mixing layer. Let us establish the relation between S, N and X, Y. From geometrical considerations with regard to (6.2.2) it is clear that

$$
\begin{aligned}
S &= X + O(\mathrm{Re}^{-1}), \\
\mathrm{Re}^{-1/2}N &= Y - F_0(X) + O(\mathrm{Re}^{-1}).
\end{aligned}
\tag{6.3.14}
$$

Then, equating the limiting expressions for the stream function (6.3.7) in the viscous mixing layer as $\mathrm{Re} \to \infty$ and $N \to -\infty$ and in the recirculation region as $\mathrm{Re} \to \infty$ and $Y \to F_0(X)$, we obtain

$$\Psi_0[X, F_0(X)] = -K_0 X^{1/2}. \tag{6.3.15}$$

Another (obvious) boundary condition in the flow plane of symmetry will be

$$\Psi_0(X, 0) = 0. \tag{6.3.16}$$

In order to determine the function $P_1(X)$ we would need the solution for the external potential flow around the closed wake region in the second approximation, which takes into account the displacement effect of the mixing layers and the laminar wake behind this region (for $X > L_0$). The mathematical difficulties in obtaining such a solution would be great. We are therefore limited to application of the above results.

In summary, we note first of all that the recirculating flow within the closed wake is found to be a flow for which all the streamlines are in part located within the viscous mixing layer. This is an essential difference from a Prandtl–Batchelor flow, and therefore this is not a flow with constant vorticity. With increasing distance downstream, a greater number of streamlines will enter the viscous mixing layer, so that at $X = L_0$ the fluid entrained in this layer (for $N < 0$) will include all the mass that takes part in the recirculating motion, and the separation zone is closed.

The flow singularities in the region where the mixing layers join together will be studied separately in Section 5, and now we turn to an analysis of the flow in the forward part of the wake and consider the possibility of matching the local asymptotic expansions for this region with the global solution. For this purpose it is necessary first of all to study the behavior of the latter as $X \to 0$. The boundary condition (6.3.15) indicates its singular character in the vicinity of the point $X = Y = 0$. Bearing in mind that here the shape of the boundary (6.2.2) has the form

$$Y = D_0(X/L_0)^{1/2} + \cdots, \quad X \to 0, \tag{6.3.17}$$

and also that the only combination of independent variables that does not change its order in approaching the region $(x, y) = O(1)$ of the order of the body dimensions is

$$\lambda = Y/X^{1/2} = y/x^{1/2}, \tag{6.3.18}$$

it is natural to represent the first approximations to the functions in (6.3.10) as $X \to 0$ in the form

$$\Psi_0 = X^{1/2}\Psi_{00}(\lambda) + \cdots, \quad P_1 = P_{10} + \cdots. \tag{6.3.19}$$

Substituting these expressions into the equations (6.3.12), since the pressure gradient $P_1'(X)$ is bounded as $X \to 0$, leads to the equation $\Psi_{00}'' = 0$,

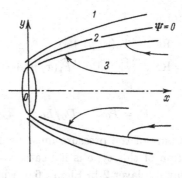

Fig. 6.6 The flow in the near wake behind a blunt body.

which, taking into account (6.3.16), has the solution

$$\Psi_{00} = A_0\lambda. \tag{6.3.20}$$

Thus, as the body is approached, the flow in the closed wake tends toward a uniform flow with constant speed $\mathrm{Re}^{-1/2}A_0$, which obviously must be negative since the fluid in the region considered cannot move in the positive direction of the axis Ox. Indeed, according to the solution (6.3.19), (6.3.20) the stream function on the wake boundary (6.3.17) will assume the value $\Psi_0 = X^{1/2}A_0D_0/L_0^{1/2}$ and, comparing it with the condition (6.3.15), we find

$$A_0 = -K_0L_0^{1/2}/D_0. \tag{6.3.21}$$

The results obtained determine the inner limit $(X \to 0)$ of the outer asymptotic expansion for the solution in the wake behind the body and thus (on the basis of the matching principle) the outer limit $(x \to \infty)$ of the inner asymptotic expansion for the solution in the region $(x, y) = O(1)$ of the order of the body dimensions. In the limit as $\mathrm{Re} \to \infty$ the flow in this region must tend to a Kirchhoff flow with constant pressure on the boundary of the stagnation zone, whose asymptotic form is parabolic (Figure 6.6):

$$y = y_S(x) = a_0x^{1/2} + \cdots, \quad x \to \infty. \tag{6.3.22}$$

Taking into account this circumstance, as well as the form of the asymptotic expansions (6.3.10) and (6.3.19), we can assume that for the near-wake region the correct asymptotic representations of the functions ψ

and p will have the form

$$\psi = \mathrm{Re}^{-1/2}\widehat{\Psi}_0(x, y) + \cdots,$$
$$p = \mathrm{Re}^{-1/2}\widehat{P}_0 + \mathrm{Re}^{-1}\widehat{P}_1(x, y) + \cdots. \tag{6.3.23}$$

Here we should take

$$\widehat{P}_0 = P_0 = -D_0/L_0; \tag{6.3.24}$$

that is, the variable part of the pressure must be of order Re^{-1}. The possibility of a solution of this type is indicated by the following considerations. In the mixing layer 2 in Figure 6.6, which divides the flow region 3 in the near wake from the external potential flow 1, the longitudinal pressure gradient will be absent in the first approximation, and therefore the flow here will be described by a system of relations analogous to (6.3.3) and (6.3.4),[6] obtained by taking

$$s, \widehat{N} = \mathrm{Re}^{1/2}n, \tag{6.3.25}$$

as independent variables of order unity, while the unknown functions are represented in the form

$$\psi = \mathrm{Re}^{-1/2}\widehat{\psi}_0(s, \widehat{N}) + \cdots, \quad p = \mathrm{Re}^{-1/2}\widehat{p}_0(s, \widehat{N}) + \cdots. \tag{6.3.26}$$

However, a significant difference between this flow and the flow in the mixing layer of the closed wake considered above will be the transverse pressure drop due to the effect of centrifugal forces. Substituting (6.3.25), (6.3.26) into the second equation of the system (1.1.2) of Chapter 1, we obtain

$$\frac{\partial \widehat{p}_0}{\partial \widehat{N}} = \kappa \widehat{u}_0^2 = \kappa \left(\frac{\partial \widehat{\psi}_0}{\partial \widehat{N}} \right)^2,$$

where $\kappa(s) = O(1)$ is the dimensionless curvature of the free streamline in the Kirchhoff flow. Thus, to achieve constant pressure (with accuracy up to terms of order $\mathrm{Re}^{-1/2}$) in the inner flow region (as $\widehat{N} \to -\infty$), it is necessary for the position of the dividing streamline ($\widehat{N} = 0$) in the mixing layer (in the first approximation) to be such as to ensure that changes in the pressure coefficient at the outer boundary of the mixing layer (as $\widehat{N} \to \infty$) are exactly balanced by the effect of centrifugal forces. The existence of a solution of this type was demonstrated in the work of Imai (1957b).

---

[6] This system should now be supplemented by a condition defining the initial velocity profile in the mixing layer.

As $s \to \infty$ ($x \to \infty$), the curvature $\kappa(s)$ will tend to zero as $s^{-3/2}$ and the transverse pressure drop will vanish. The thickness of the mixing layer increases here according to the law $x^{1/2}$, as does the width of the near wake. Hence, it follows that the external potential flow will tend toward the flow past a parabola, while the pressure in the mixing layer will approach a constant value. This value will be (6.3.24), since the speed of the undisturbed flow in which the parabola is located should be taken equal to $U_\infty(1 + \mathrm{Re}^{-1/2}D_0/L_0)$ rather than U_∞, in accordance with the well-known result of the asymptotic thin-airfoil theory (Van Dyke, 1964). This result clearly shows that only the consideration of Kirchhoff flow as a local limit leads to the appearance of a term of order $\mathrm{Re}^{-1/2}$ in the asymptotic representation of the pressure coefficient in the wake behind the body. Taking this limit to be "global" would lead to the conclusion that the value of the pressure coefficient in the wake is of order Re^{-1}, which is exactly the same as in the work of Imai (1957b).

Here, we will not consider higher approximations in the expansion (6.3.19) of the solution for the closed wake as $X \to 0$ or in the behavior of the functions (6.3.23) as $x \to \infty$. We note only that, since the leading terms of the latter can be represented in the form

$$\widehat{\Psi}_0 = x^{1/2}\widehat{\Psi}_{00}(\lambda) + \cdots, \quad \widehat{P}_1 = \widehat{P}_{10} + \cdots$$

as $x \to \infty$, matching of the asymptotic expansions (6.3.10) and (6.3.23) will be ensured if

$$\widehat{\Psi}_{00} = \Psi_{00}, \quad \widehat{P}_{10} = P_{10}.$$

6.4 Drag of the Body and Parameters of the Separation Zone

It is well known from the theory of ideal fluid flows with free streamlines that the asymptotic form of the free streamlines as $x \to \infty$ is related to the body drag. This relationship is determined by the dependence of the coefficient a_0 in the expression (6.3.22) upon the drag coefficient c_{x0} calculated according to the Kirchhoff theory, and has the form (e.g., see Gurevich, 1979)

$$a_0 = \sqrt{(2/\pi)c_{x0}}. \tag{6.4.1}$$

Comparing the asymptotic representation (6.3.22) with the expression (6.3.17) for the boundary of the closed wake as $X \to 0$, and using the matching principle, we find that $a_0 = D_0/\sqrt{L_0}$ and, therefore,

$$D_0/\sqrt{L_0} = \sqrt{(2/\pi)c_{x0}}. \tag{6.4.2}$$

Thus, we obtain one of the relations linking the limiting value of the body drag coefficient with the geometric parameters of the separation zone. Another relation may be found from the following considerations. From the momentum theorem, the body drag is equal to the momentum deficit in the far wake for $X > L_0$. On the other hand, the value of this deficit in the limit as Re $\to \infty$ must be equal to the sum of the integrals of the friction forces along the dividing streamlines $Y = \pm F_0(X)$, since the contribution of the surface pressure forces, which are due to the displacement effect of the mixing layers and the far wake, will be vanishingly small (of order $\text{Re}^{-1/2}$).

Then, using the solution (6.3.8) derived earlier for the mixing layer, we can write the following expression for the limiting value of the drag coefficient[7]:

$$c_{x0} = 4 \int_0^{L_0} \tau_0 dS = 8\sqrt{L_0} f_0''(0) = 1.60\sqrt{L_0}. \qquad (6.4.3)$$

As a result, the relations (6.4.3), (6.4.2), and (6.3.11) allow us to write the final formulas for determining the geometric parameters and the pressure coefficient for the recirculation zone in the form

$$\mathcal{L}\text{Re}^{-1} = L_0 = 0.39 c_{x0}^2, \quad \mathcal{D}\text{Re}^{-1/2} = D_0 = 0.50 c_{x0}^{3/2},$$
$$c_{p0} = 2\text{Re}^{-1/2} P_0 = -2.55 c_{x0}^{-1/2} \text{Re}^{-1/2}, \qquad (6.4.4)$$

that is, to express them, in a first approximation, in terms of the body drag coefficient according to the Kirchhoff theory.

The results of the asymptotic theory (Sychev, 1967) are found, as first shown in the work of Smith (1979a), to be in good agreement with the existing experimental data of Grove, Shair, Petersen, and Acrivos (1964) and of Acrivos, Leal, Snowden, and Pan (1968), as well as with the numerical calculations based on the full Navier–Stokes equations carried out in the work of Dennis and Chang (1970) for the case of a circular cylinder. These comparisons may be supplemented by the results of calculations by Fornberg (1980). By way of illustration, Figures 6.7–6.9 show the appropriate dependence for the length and width of the closed separation zone and the pressure coefficient behind a circular cylinder as functions of the Reynolds number $\text{Re} = U_\infty d/\nu$ (d is the cylinder diameter). The solid curves correspond to the asymptotic theory and are drawn in accordance with the formulas (6.4.4) for a value of the drag coefficient $c_{x0} = 0.50$, corresponding to the flow around a cylinder

---

[7] For calculation of the drag coefficient, it should be recalled that the value of the shear stress has been referred to twice the dynamic pressure.

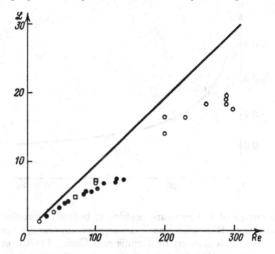

Fig. 6.7 Dependence of the length of the closed separation zone on the Reynolds number Re: solid line – the asymptotic theory; open squares – results of numerical calculations by Dennis and Chang (1970); open circles – calculations of Fornberg (1980); filled circles – experimental data of Acrivos et al. (1968).

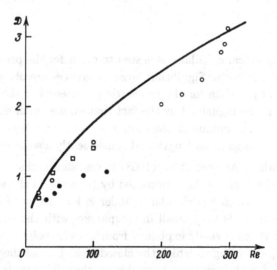

Fig. 6.8 Dependence of the width of the closed separation zone on the Reynolds number Re: solid line – the asymptotic theory; open squares – results of numerical calculations by Dennis and Chang (1970); open circles – calculations of Fornberg (1980); filled circles – experimental data of Grove et al. (1964).

according to the Kirchhoff scheme with smooth separation (Brodetsky, 1923). The best agreement between the asymptotic theory, experimental

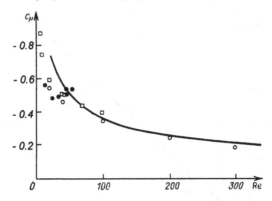

Fig. 6.9 Dependence of the pressure coefficient behind a circular cylinder on the Reynolds number Re: solid line – the asymptotic theory; open squares – results of numerical calculations by Dennis and Chang (1970); open circles – calculations of Fornberg (1980); filled circles – experimental data of Grove et al. (1964).

data, and numerical calculations is seen to occur for the pressure coefficient behind the body. Significant large differences are observed in the comparison of the data for the geometric parameters of the separation zone. This can be explained by the fact that even a small displacement of the dividing streamline in the closure region of the wake leads to a considerable change in its length and consequently also in its width.

In the work of Acrivos et al. (1968) it was shown that a significant effect upon the entire wake is imparted by the walls of the wind-tunnel test section in which the circular cylinder is located, even in the case when its diameter is very small in comparison with the tunnel width. This circumstance is easily explained from the viewpoint of our asymptotic theory, according to which the closed wake behind the body must grow as $Re^{1/2}$ with increasing Reynolds number. This leads to the fact that already at Re = 100 the closed wake will be thicker than the body by one order; correspondingly, the influence of the wind-tunnel walls will also increase. This effect, constraining the development of the separation zone, seems to have led the authors to the conclusion mentioned in Section 1 that the width of the separation zone behind the body remains of the order of the body dimensions with increasing Reynolds number.

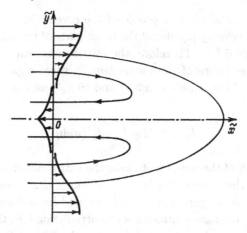

Fig. 6.10 Initial profile for reverse jet.

6.5 Separation-Zone Closure and Unsteadiness

Up to now nothing has been said about the flow in the closure region of the separation zone for $X = \mathrm{Re}^{-1}x \to L_0$. Here, the mixing layers acquire a thickness of the order of the body dimensions and merge to form a complex rotational flow. Since the external potential flow around the closed separation zone, which is like a thin airfoil of elliptical shape, has a stagnation point at $X = L_0$, the dimensionless pressure p here is recovered up to values of order unity. This means that the pressure change in the region of wake closure will be of order unity. But then the changes in the dimensionless velocity components (u, v) here will be of the same order. Hence, it follows that the flow in the region where the mixing layers merge and where

$$|\tilde{x}| = |x - \mathcal{L}| = O(1), \quad |\tilde{y}| = |y| = O(1), \qquad (6.5.1)$$

can be considered as locally inviscid.

It appears that a reverse flow in the form of a jet has to develop from the fluid carried by the inner parts of the mixing layers (where $\psi_0 < 0$). This follows from the Bernoulli equation for a streamline which belongs to the region indicated. Indeed, since as $\tilde{x} \to -\infty$ the transverse pressure gradient in the separation zone tends to zero, the value of the speed of each fluid element after reversal of its motion will approach the value of its speed before the reversal. Then from the condition of matching the local solution for the closure region as $\tilde{x} \to -\infty$ with the global solution for the closed wake region as $X \to L_0 - 0$, we find that on the scale of the latter the initial velocity profile of the reverse jet is established by reversal of the velocity profiles of the inner parts of the mixing layers for $S = L_0$ (Figure 6.10). Therefore, the initial momentum of the jet will be equal to the values of the momentum in these regions at $X = L_0$; with the help of the solution (6.3.5) and (6.3.6) we can express this in the form

$$I_0 = 2\sqrt{L_0} \int_{-\infty}^{0} f_0'^{2}(\eta)d\eta. \qquad (6.5.2)$$

Since the width of the reverse jet along the entire closed separation zone will remain of the order of the body dimensions, this momentum will be transferred to the region $(x, y) = O(1)$ located immediately behind the body. This circumstance introduces a contradiction into the asymptotic flow theory, since it does not correspond to the Kirchhoff scheme with a stagnation zone behind the body. The contradiction was, in particular, pointed out in the work of Messiter (1975). However, a more detailed study shows that the formation of a reverse jet is impossible (Sychev, 1982).

Let us consider the balance of external forces applied to the boundaries of the recirculating fluid volume within the closed separation zone, projected on the direction of the undisturbed flow as $\mathrm{Re} \to \infty$. Such forces are the integrals of shear stresses along the dividing streamlines and the integral of the pressure forces (projected on the axis Ox) along the closed contour bounding the separation region (but not enclosing the body!). The sum of these oppositely directed forces must equal zero, that is,

$$2 \int_{0}^{L_0} \tau_0 dS = \oint p \, dy. \qquad (6.5.3)$$

We note that the main contribution (as $\mathrm{Re} \to \infty$) to the longitudinal component of the pressure forces $\oint p \, dy$ is introduced in the closure region of the separation zone $(\widetilde{x}, \widetilde{y}) = O(1)$ where the pressure is recovered up to a value of order unity. The sum $2 \int_{0}^{L_0} \tau_0 dS$ of the integrals of the shear stresses along the dividing streamlines, on the one hand, is equal to the dimensionless body drag $(c_{x0}/2)$, as already noted (see (6.4.3)). On the other hand, the integrals of the friction forces along the dividing streamlines determine the fluid momentum within the inner regions of the mixing layers at $X = L_0$; that is, the following equality must hold:

$$I_0 = 2 \int_{0}^{L_0} \tau_0 dS, \qquad (6.5.4)$$

where I_0 is determined by the integral (6.5.2). From (6.5.3) and (6.5.4) we find that

$$\oint p \, dy = I_0. \tag{6.5.5}$$

This implies that the integral of the pressure forces has the value that is necessary to cancel the momentum in the inner regions of the mixing layers but half as large as required for turning back the flow to form a reverse jet.

Thus, the analysis of equilibrium conditions for the slowly recirculating fluid volume in the separation zone (the closed wake) shows that the formation of a reverse jet is impossible. The fluid carried by the inner parts of the mixing layers will be brought to rest by the forces associated with recovery of the pressure in the closure region of the separation zone. But this in turn implies that steady motion in the region $(\tilde{x}, \tilde{y}) = O(1)$ is impossible. In this case the argument based on use of the Bernoulli integral also becomes invalid. Unsteadiness of the local flow in the region where the mixing layers merge seems to require an interpretation as an irregular motion associated with intense fluid mixing accompanied by strong dissipation of kinetic energy. This local flow unsteadiness may to a certain extent be likened to phenomena that occur in the region of cavity closure when a body is moving in a liquid at small cavitation numbers. Just as in this latter case, the results obtained above from an asymptotic analysis of a global steady flow picture are not invalidated, since the details of the flow in the region of wake closure do not have appreciable influence upon the flow field as a whole. This circumstance provides some justification for the use of such models of separated flows as that of Riabouchinsky (1920), proposed by Roshko (1967) for these purposes.

In conclusion we note that the analysis carried out for the separated flow behind a blunt body characterized by a finite drag as Re $\to \infty$ points to the possibility that flow unsteadiness may appear for steady boundary conditions not as a result of hydrodynamic instability but because of the absence of steady solutions of the Navier–Stokes equations as Re $\to \infty$.

6.6 On the Zero-Drag Theory

The analysis carried out in Sections 2–5 is concerned with solutions satisfying the Lagerstrom–Kaplun principle. However, it does not exclude the possibility of existence of steady solutions that do not satisfy this principle. Let us discuss such solutions in some detail.

As already mentioned, the flow model which assumes that both the longitudinal and transverse dimensions of the closed separation zone as Re → ∞ increase according to a linear law was first proposed by Taganov (1970). This assumption means that the relative thickness $\mathcal{D}/\mathcal{L}$ of the separation zone tends to a finite limit as Re → ∞, whereas on the scale of this zone the body shrinks to a point located at the forward end of the zone. Then the pressure gradient along the dividing streamline and the speed of the recirculating flow remain finite. According to the Prandtl–Batchelor theorem, the vorticity of the flow inside the separation zone must be constant. Such flows, usually referred to as potential-vortex flows, were studied and calculated in detail in the work of Sadovskii (1970, 1971, 1973). The family of potential-vortex flows depends on one parameter: the strength of the jump in tangential velocity (the jump in the Bernoulli constant) at the boundary of the separation zone. In Taganov's work (1970), based on qualitative considerations, preference is shown for a limiting state in which the velocity jump is absent and only a jump in the vorticity occurs. Then, by introducing a viscous layer at the boundary of the separation zone, it is easy to establish that the integral of friction forces along the dividing streamline will be of order Re^{-1}, and consequently that the body drag coefficient c_x will be of the same small order. To determine a numerical value for this coefficient, Sadovskii (1973) applied a method based on determination of the dissipation of mechanical energy in the separation zone and in the external potential flow. As a result, the value $c_x = 46.5\pi\mathrm{Re}^{-1}$ was obtained. It is important to note that the value of the drag coefficient does not depend on the shape of the body. The relative thickness $(\mathcal{D}/\mathcal{L})$ of the separation zone is then equal to 0.60.

In a series of later studies this flow scheme was corroborated and further developed. To begin with, we should discuss the results of Fornberg's (1985) numerical calculations. As mentioned earlier, the evolution of the flow behind a circular cylinder at Reynolds numbers Re > 300 is characterized by the beginning of linear growth not only of the length but also of the width of the separation zone, with simultaneous changes in the character of the dependence of the drag coefficient upon Reynolds number (Figure 6.4). Thus, at Re = 600 the value of c_x becomes equal to 0.43, smaller than in the Kirchhoff scheme. The shape of the separation zone and the picture of the streamlines at a Reynolds number Re = 600, shown in Figure 6.11, look very much like the potential-vortex flow of Sadovskii.

Fornberg's results prompted the studies of Smith (1985) and Pere-

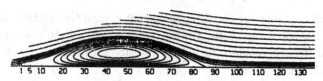

Fig. 6.11 Streamlines in the flow behind a circular cylinder for Re = 600 (Fornberg, 1985).

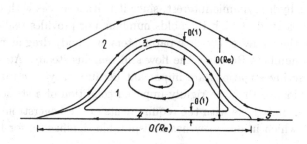

Fig. 6.12 Flow pattern on the scale of the separation zone (Smith, 1985).

grine (1985). Their work made no assumption about the absence of a jump in the tangential velocity for the limiting state of the flow, and considered Sadovskii's solution (1971, 1973) with a jump in the Bernoulli constant. As a result, the body drag obtained in the limit is finite if the length and the width of the separation zone increase according to a linear law. In Smith's work (1985) the structure of such a flow was studied in detail. On a global scale of order Re (Figure 6.12), this includes the inviscid vortex core (1), the external potential flow (2), and also viscous regions with thickness $O(1)$ at the boundary (3) of the separation zone, in a reverse flow (4) near the plane of symmetry of the core, and in the wake (5) beyond the separation zone. Thus, the inner parts of the mixing layer (3), together with the reverse jet (4), are zones of a cyclic boundary layer, and then the realization of the proposed scheme depends upon the possibility of the existence of such a cyclic flow.

This question was studied in detail by Chernyshenko (1988). His work contains a demonstration (by contradiction) of the impossibility of existence of a cyclic boundary layer with a finite change in the longitudinal velocity component, so that in the limit the potential-vortex flow must have a zero jump in the Bernoulli constant. This result lends support to the scheme of Taganov (1970).

Without dwelling upon details of the investigation carried out in the work of Chernyshenko (1988), we note that as a result, in addition to the values given above for the drag coefficient c_x and the relative thickness $\mathcal{D}/\mathcal{L}$ of the separation zone, the author determines both the length and width of the separation zone behind the circular cylinder: $\mathcal{L} = 0.098\text{Re}$, $\mathcal{D} = 0.059\text{Re}$.

Thus, the theory based upon the assumption that as Re $\to \infty$, the relative thickness $\mathcal{D}/\mathcal{L}$ of the separation zone tends to a finite limit turns out in the end to be a zero-drag theory. It can hardly be of any significant hydrodynamic interest, since it neither agrees with the flows actually observed at high Reynolds numbers nor provides realistic estimates for the values of the flow parameters and body drag at moderate Reynolds numbers Re when the flow still remains steady. At the same time, these investigations are important because they give rather convincing evidence that the rigidly imposed condition of a stationary solution is incompatible with the requirements of the Lagerstrom–Kaplun principle, which in our view has indisputable significance for hydrodynamics.

6.7 The Transition to a Thin Body

The final section of this chapter will be devoted to considering regimes of separated flow that represent a transition between the flow past a thick body considered just above (with relative thickness remaining finite as Re $\to \infty$) and the flow past a thin body with a local separation zone in the vicinity of the trailing edge, considered in Chapter 3. The presentation here[8] borrows extensively from the work of Cheng and Smith (1982), which was concerned with the study of these intermediate regimes.

Let a symmetric obstacle (airfoil) be placed at zero angle of attack in an unbounded uniform fluid flow. We choose a rectangular Cartesian coordinate system Oxy with the origin coinciding with the airfoil leading edge and the longitudinal axis directed downstream (Figure 6.13). The airfoil shape (its upper surface) will be given by the relation $y = hf(x)$, $0 \leq x \leq 1$, and we will take $h \to 0$ as Re $\to \infty$. The longitudinal and transverse velocity components and the pressure will be denoted by u, v, and p, respectively. (As before, the variables introduced are dimensionless; the speed of the uniform oncoming flow is taken as the reference velocity and the airfoil chord as the reference length.)

If the relative thickness of the airfoil remains finite at arbitrarily large Reynolds numbers, then, as shown in the previous sections, the poten-

---

[8] This section was written by S. N. Timoshin.

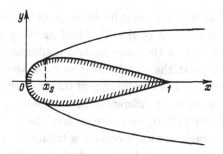

Fig. 6.13 Flow around an airfoil according to the Kirchhoff model.

tial flow around the airfoil is determined by the solution of the Kirchhoff problem. We can assume that the Kirchhoff scheme is appropriate not only for $h = O(1)$ but also for smaller (but not too small) airfoil thicknesses. As we will see, the range of variation in the parameter h where this assumption is applicable depends upon the shape of the body.

For the condition $h \ll 1$, the Kirchhoff problem may be solved within the framework of the linear theory. For this purpose we represent the flow variables in the form

$$u = 1 + hu_1(x, y) + \cdots , \quad v = hv_1(x, y) + \cdots , \quad p = hp_1(x, y) + \cdots .$$
$$(6.7.1)$$

Substitution of (6.7.1) into the Navier–Stokes equations shows that the unknown functions satisfy the system of linearized Euler equations, from which it follows that $p_1 + iv_1$ is an analytic function in the upper half-plane of the complex variable $z = x + iy$. In accordance with the conditions of no flow through a solid boundary and constant pressure at a free streamline,

$$v_1(x, 0) = 0 \quad \text{for } x < 0 ,$$
$$v_1(x, 0) = f'(x) \quad \text{for } 0 < x \leq x_S , \qquad (6.7.2)$$
$$p_1(x, 0) = 0 \quad \text{for } x \geq x_S ,$$

where x_S is the position of the separation point of the free streamline.

Moreover, the function $p_1 + iv_1$ becomes zero at infinitely large distances and is bounded at the point $x = x_S$, $y = 0$ (Figure 6.13).

The solution of the mixed boundary-value problem (6.7.2) is given by the formula of Keldysh and Sedov (1937). In particular, the pressure distribution at the airfoil surface for $0 < x < x_S$ has the form

$$p_1(x, 0) = -\frac{\sqrt{x_S - x}}{\pi} \int_0^{x_S} \frac{f'(t)\,dt}{\sqrt{x_S - t}(x - t)} . \qquad (6.7.3)$$

The integral (6.7.3) is understood in the sense of the principal value.

We see that the solution of the Kirchhoff problem in the linear approximation, as well as in the more general nonlinear case, contains an undetermined constant: the coordinate $x = x_S$ of the separation point of the free streamline. The elimination of this indeterminacy is accomplished in different ways for different classes of profiles.

Following Messiter (1979), we consider first the simplest example of a profile having a surface with a corner: a triangular profile $f(x) = x$, $0 \leq x \leq 1$. Because boundary-layer separation is inevitable at a corner of a body, we should take $x_S = 1$ in the relation (6.7.3). Now we can establish the limits of applicability of the solution. For this purpose we note that according to (6.4.4) the recirculation zone has length of order $c_x^2 \mathrm{Re}$, where c_x is the drag coefficient of the body, while the Kirchhoff theory is valid only under the condition that the length of this zone is large in comparison with the characteristic profile dimension, that is, for $c_x \gg \mathrm{Re}^{-1/2}$. The drag of the airfoil may be determined by calculating the sum of the forces acting on its surface. The friction forces acting on the solid surface provide a contribution to the value of c_x that is, as usual, $O(\mathrm{Re}^{-1/2})$. The integral of the pressure forces is found from (6.7.1) and (6.7.3); as a result, it is seen that

$$c_x = h^2 c_{x1} + \cdots, \quad h \to 0,$$
$$c_{x1} = \frac{2}{\pi} \left(\int_0^{x_S} \frac{f'(t)dt}{\sqrt{x_S - t}} \right)^2. \tag{6.7.4}$$

For a profile of triangular form, $f'(x) = 1$ and $c_{x1} = 8/\pi$. From this and the inequality obtained above, it follows that the Kirchhoff scheme may be applied for a range of thicknesses $\mathrm{Re}^{-1/4} \ll h \lesssim 1$.

If $h = O(\mathrm{Re}^{-1/4})$, then the length of the separation zone is found to be of the same order as the profile chord length. Potential flows of this type with constant pressure in the separation region are described by the well-known scheme of Chaplygin (1899).

Based on qualitative considerations, Messiter (1979) determined that in the range of thicknesses $\mathrm{Re}^{-5/8} \ll h \ll \mathrm{Re}^{-1/4}$, the separation zone has length $O(h\mathrm{Re}^{1/4})$. Thus, for $h \sim \mathrm{Re}^{-5/8}$ the zone of separated flow becomes localized in the viscous sublayer of the interaction region in the vicinity of the airfoil trailing edge.

Let us return to the relation (6.7.3) and consider the flow over an airfoil with a smooth shape. In order to determine the position of the separation point and thereby to eliminate the arbitrariness from the solution, we use one of the main results of Chapter 1. The study of the local flow structure in the vicinity of the boundary-layer separation point showed that at the surface of a smooth body, ahead of the separation point, the pressure is represented in the form

$$p = -\text{Re}^{-1/16} \alpha a_0^{9/8} x_S^{-9/16} \sqrt{x_S - x} + \cdots, \quad \text{Re} \to \infty,$$

$$x_S - x \to +0, \quad \alpha = 0.42, \quad a_0 = 0.3321 \tag{6.7.5}$$

(taking into account the fact that the boundary layer on a thin airfoil is described by the Blasius solution in the leading approximation).

The solution (6.7.3) also has a singularity at the separation point. Calculating the limiting value of the integral as $x \to x_S - 0$, and recalling the original form of the expansion (6.7.1), we can show that

$$p = -\frac{h}{\pi} \sqrt{x_S - x} \left[\int_0^{x_S} \frac{f'(t) - f'(x_S)}{(x_S - t)^{3/2}} dt - 2 \cdot f'(x_S) x_S^{-1/2} \right] + \cdots,$$

$$h \to 0, \quad x_S - x \to +0. \tag{6.7.6}$$

Now let us compare the expressions (6.7.5) and (6.7.6), assuming that the separation point is located at a finite distance from the leading edge $(x_S \sim 1)$. If $h \gg \text{Re}^{-1/16}$, then the coefficient of the singular term in (6.7.6) should be set equal to zero. In essence this means that in the approximation considered, the Brillouin–Villat condition is met at the point of free-streamline separation. The condition of absence of a singularity in the solution (6.7.3) yields an equation for determining the abscissa of the separation point

$$\int_0^{x_S} \frac{f'(t) - f'(x_S)}{(x_S - t)^{3/2}} dt - 2f'(x_S) x_S^{-1/2} = 0. \tag{6.7.7}$$

If $h = \text{Re}^{-1/16} h_1$ and $h_1 = O(1)$, then the Brillouin–Villat condition is not satisfied; the coordinate x_S must be found from the equation

$$\int_0^{x_S} \frac{f'(t) - f'(x_S)}{(x_S - t)^{3/2}} dt - 2f'(x_S) x_S^{-1/2} = \pi \alpha a_0^{9/8} x_S^{-9/16} h_1^{-1}. \tag{6.7.8}$$

Since equation (6.7.7) represents a limiting form of (6.7.8) at large values of h_1, in the following we will analyze the latter more general equation.

Let us consider an airfoil with wedge-shaped leading and trailing edges: $f(x) = c(x - x^2)$, $0 \le x \le 1$. Equation (6.7.8) assumes the following form:

$$2h_1 c x_S^{1/16} (4x_S - 1) = \pi \alpha a_0^{9/8}.$$

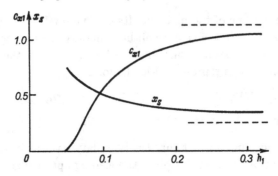

Fig. 6.14 Dependence of the coordinate x_S of the separation point and the drag coefficient c_{x1} upon the thickness parameter h_1 for a profile with wedge-shaped edges: the dashed lines show the limiting values as $h_1 \to \infty$.

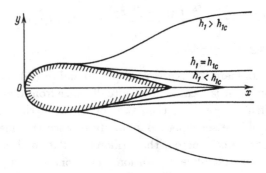

Fig. 6.15 The shape of the free streamlines for various values of the thickness parameter.

The dependence of the coordinate of the separation point and the value of c_{x1} upon the thickness parameter h_1 is shown in Figure 6.14 for $c = 2$. For large h_1, the position of the separation point is determined by the Brillouin–Villat condition (i.e., from equation (6.7.7)). A reduction in the airfoil thickness is accompanied by a displacement of the separation point in the direction of the trailing edge and by a simultaneous reduction in the drag of the airfoil. When a critical thickness $h_{1c} = 0.048$ is reached, the coefficient c_{x1} becomes zero.

According to Cheng and Smith (1982), the transformation of the flow with further reduction in profile thickness has the following form. Together with a decrease in c_{x1} there occurs an abrupt reduction in the size of the recirculation zone, and the pressure there increases. For $h_{1c} > h_1 = O(1)$, the dimensions of the separation zone become comparable with the corresponding airfoil dimensions, and for the drag co-

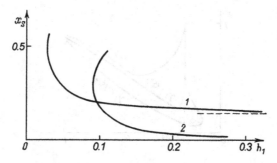

Fig. 6.16 Position of the separation point on airfoils with cusped trailing edge.

efficient of the airfoil we have the estimate $c_x = O(\text{Re}^{-1/2})$. The flow structure on the scale of the airfoil for h_1 close to h_{1c} is shown in Figure 6.15.

On a thinner profile $(h \ll \text{Re}^{-1/16})$, the separation region encloses the trailing edge, and has length of order $h^2 \text{Re}^{1/8}$, so that for $h = O(\text{Re}^{-1/4})$ the scales of the separation zone coincide with the corresponding scales of the interaction region in order of magnitude (see Chapter 3, Section 2).

The example considered shows that for a smooth profile with thickness $h = O(\text{Re}^{-1/16})$, the separation point is displaced downstream of the Brillouin–Villat point through a distance comparable with the airfoil chord. The result of this displacement of the separation point is a transformation of the global picture of the separated flow. We recall that for a profile with a sharp corner the Kirchhoff scheme is applicable for a much wider range of the parameter h.

In cases where the profile edges are different from wedge-shaped, the solution of equation (6.7.8) may have certain special properties.

Thus, it has been found that the location of the separation point is not single-valued for profiles having a cusped trailing edge. Two examples of such solutions are shown in Figure 6.16. Curve 1 corresponds to a profile with a sharp nose: $f(x) = c_1 x(1-x)^2$, $c_1 = 27/4$. The dashed line shows the limiting position of the separation point for $h_1 \gg 1$. Curve 2 is drawn for a profile with rounded leading edge: $f(x) = c_2 x^{1/2}(1 - x)^2$, $c_2 = 5\sqrt{5}/8$. Here, along with the range of nonuniqueness of the solution, another interesting feature is discovered: at large values of h_1 the separation point is found to be near the profile leading edge.

The reasons for displacement of the separation point to the vicinity of the profile nose were discussed by Timoshin (1985). It was found that this phenomenon is controlled by the details of the shape of the forward

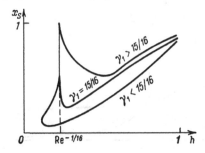

Fig. 6.17 Schematic representation of the dependence of the location of the separation point on the airfoil thickness.

part of the body, which would seem unimportant at first glance. We assume that, as $x \to +0$, $f(x) = b_0 x^{\gamma_0} + b_1 x^{\gamma_1} + \cdots$, where $b_0 > 0$, $\gamma_0 > 0$, $\gamma_1 > \gamma_0$, and b_1 are certain constants. If $\gamma_0 > 1/2$ (zero leading-edge radius of curvature) or $\gamma_0 = 1/2$ but $b_1 > 0$, then the separation point remains at a finite distance from the nose of the profile. The analysis of the flow in these cases fits the pattern described above. But if $\gamma_0 = 1/2$ and $b_1 < 0$, then one can no longer say that $x_S = O(1)$ for small h. Shown in Figure 6.17 is the qualitative form of the curves $x_S(h)$ for such profiles. These curves originate at the Brillouin–Villat point for an airfoil with thickness $h \sim 1$. With decreasing h, the separation point of the free streamline is displaced to the leading edge according to the law

$$x_S = \left[-\frac{\pi b_0^2}{4 b_1 \gamma_1 g(\gamma_1)} \right]^{1/\gamma_1} h^{1/\gamma_1} + \cdots, \quad h \to 0,$$

where

$$g(\gamma_1) = 2 - \int_0^1 \frac{t^{\gamma_1 - 1} - 1}{(1 - t)^{3/2}} \, dt.$$

It can be shown that $g(\gamma_1) > 0$ if $\gamma_1 > 1/2$.

The rule described above retains its form up to thicknesses $h = O(\mathrm{Re}^{-\beta \gamma_1})$, $\beta = (32\gamma_1 - 15)^{-1}$, where the separation point is located closest to the leading edge: $x_S = O(\mathrm{Re}^{-\beta})$. From the latter domain the curve $x_S(h)$ according to the law

$$x_S = \left[-\frac{\pi \alpha a_0^{9/8}}{b_1 \gamma_1 g(\gamma_1)} \right]^{\frac{16}{16\gamma_1 - 15}} (h \mathrm{Re}^{1/16})^{\frac{16}{15 - 16\gamma_1}} + \cdots,$$

reaches the domain $h \sim \mathrm{Re}^{-1/16}$, $x_S \sim 1$, where the solution is determined by equation (6.7.8). As seen in Figure 6.17, for $1/2 < \gamma_1 < 15/16$

the dependence of the location of the separation point upon profile thickness has nonmonotonic character, and consequently the solution is found to be nonunique for a certain range of thicknesses.

Several comments concerning the location of the free-streamline breakaway point on strongly blunted profiles $(0 < \gamma_0 < 1/2)$ were made in the studies mentioned, but a detailed examination of this question up to the present time is still lacking.

The results described above may be generalized to a nonsymmetric class of flows. The first steps in this direction were made independently by Cheng (1984) and Khrabrov (1985), who considered flows around profiles with thickness of order $\text{Re}^{-1/16}$ placed in the flow at an angle of attack of the same order of smallness. If the location of the separation point for a symmetric problem is defined by equation (6.7.8) (Kirchhoff's scheme), then in the case of an asymmetric flow one obtains a system of two equations for the coordinates of the separation points on the upper and lower profile surfaces. Numerical solution of this system has shown, in particular, that for a symmetric profile with a parabolic nose and a cusped trailing edge there exists an asymmetric solution even for zero angle of attack. This is apparently related to the nonuniqueness of the solution of the symmetric problem. Further development of the theory for such flows is contained in the work of Cheng and Lee (1986) and Lee and Cheng (1990).

7

Numerical Methods for Solving
the Equations of Interaction Theory

It follows from the previous chapters that one of the key elements in the analysis of flow separation from a solid body surface is the investigation of the flow behavior in the region of boundary-layer interaction with the external inviscid flow. Although the interaction region is normally very small, it plays a major role in the separation phenomenon because of the mutual influence of the near-wall viscous flow and external inviscid flow in this region, with a sharp pressure rise prior to the separation point leading to very rapid deceleration of fluid particles near the wall and ultimately to the appearance of the reverse flow downstream of the separation. The complexity of the physical processes in the interaction region is accompanied, as might be expected, by mathematical difficulties in solving the equations that describe the flow in this region. While everywhere outside the interaction region the solution may often be obtained in analytical or at least self-similar form, the analysis of the interaction region requires that special numerical methods be used.

Numerical solution of the interaction problem serves not only to provide meaningful physical information on the development of events in the interaction region, but in many cases also appears to provide the only way of being sure that the solution for this problem really exists, and hence that the entire asymptotic structure of the flow, anticipated in the course of the asymptotic analysis of the Navier–Stokes equations, is self-consistent. For some flows, the solution of the interaction problem might exist only within a certain range of the variation of the corresponding flow parameter and be nonunique in part of this range, as is the case, for example, for the flow over the leading edge of a thin airfoil (Chapter 4). It is not clear whether the nonuniqueness of the solution may always be interpreted as separated-flow hysteresis. Likewise the nonexistence of the solution beyond a critical value of the flow parameter may not necessarily imply some kind of a bursting process analogous to that at the leading edge of a thin airfoil. Nevertheless, the information on the existence of the solution and its nonuniqueness is absolutely vital for an understanding of the flow behavior. To provide this information the numerical methods used in the theory of the interaction of the boundary layer with the external inviscid flow must be quite sophisticated. During the first decade after the fundamental papers on the interaction theory were published by Neiland (1969), Stewartson and Williams (1969), Messiter (1970), and Sychev (1972), not only was there no numerical investigation of the nonuniqueness of boundary-layer separation, but even the simple case of incipient separation represented considerable difficulties.

This is why the numerical methods were first designed for boundary-layer interaction with a supersonic external flow, in which case the interaction law relating the pressure and the displacement thickness of the boundary layer is much simpler than that for incompressible flows. Taking into account the way the methods have been developed historically and the obvious similarity in the basic ideas used to calculate the interaction region with subsonic and supersonic flows, both these cases will be considered in what follows.

7.1 Statement of the Problem

Despite the variety of physical situations that lead to the interaction of the boundary layer with the external inviscid flow, the formulation of the mathematical problem for the interaction region appears to be universal to a certain degree. For the flow in the near-wall viscous layer the Prandtl boundary-layer equations

$$u\frac{\partial u}{\partial x} + v\frac{\partial u}{\partial y} = -\frac{dp}{dx} + \frac{\partial^2 u}{\partial y^2}, \quad \frac{\partial u}{\partial x} + \frac{\partial v}{\partial y} = 0 \qquad (7.1.1)$$

always apply.

The no-slip boundary conditions at the surface of the solid body may always be written in the form

$$u = v = 0 \quad \text{at } y = 0 \qquad (7.1.2)$$

if an orthogonal curvilinear coordinate system is used with x measured along the body surface and y normal to it.

The form of the other boundary conditions depends on the particular model of interaction considered. If it is the conventional triple-deck model we have been dealing with in this book, then (7.1.2) should be supplemented by the upstream condition

$$u = y \quad \text{as } x \to -\infty \tag{7.1.3}$$

and the condition of matching with the solution in the main part of the boundary layer

$$u = y - \delta(x) + o(1), \quad y \to \infty, \tag{7.1.4}$$

where $\delta(x)$ is the increment in the displacement thickness of the boundary layer relative to its value upstream of the interaction region.

If $\delta(x)$ were known, it would be possible to find the pressure distribution $p(x)$ along the interaction region by making use of the relevant interaction law. In a supersonic flow it is given by the Ackeret formula

$$p = d\delta/dx + dY_0/dx, \tag{7.1.5}$$

while in a subsonic flow the Hilbert integral from the linear potential-flow theory gives the relationship between p and δ,

$$p = \frac{1}{\pi} \int_{-\infty}^{\infty} \frac{d\delta/ds + dY_0/ds}{s - x} ds, \tag{7.1.6}$$

where the function $Y_0(x)$ represents the shape of the body surface.

However, in the interaction strategy, neither the displacement thickness $\delta(x)$ nor the pressure $p(x)$ is known in advance. Instead, both are to be found in the course of solving the Prandtl equations (7.1.1) combined with the interaction law either in the form of equation (7.1.5) or equation (7.1.6). This change in the mathematical formulation of the problem is fundamental. First, it makes it impossible for the Goldstein (1948) singularity to appear at the separation point and, second, it changes the character of the dependence of the solution of the Prandtl equations on the boundary conditions, requiring an additional condition which is "responsible" for the upstream propagation of disturbances through the boundary layer.

The phenomenon of upstream propagation of disturbances was first discovered experimentally. Among the most intriguing were the experiments on the interaction of an oblique shock-wave with a boundary-layer in a supersonic flow.[1] They revealed an obvious discrepancy between the experimental results and predictions of the classical boundary-layer theory. If the classical Prandtl theory were valid, then the flow inside

---

[1] See, for example, the papers by Ackeret, Feldmann, and Rott (1946), Liepmann (1946), and Chapman, Kuehn, and Larson (1956).

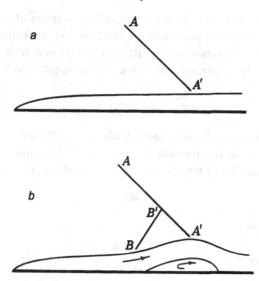

Fig. 7.1 Shock-wave interaction with a boundary-layer and upstream influence.

the boundary layer might be described by the equations (7.1.1) with prescribed pressure gradient. But in that case the boundary-layer equations (7.1.1) appear to be of parabolic type. It means that their solution at any point in the flow does not depend on variations in boundary conditions downstream of this point. Exactly the same assertion applies for the external inviscid flow. Since the inviscid flow outside the boundary layer is governed by the Euler equations, which are known to be hyperbolic for supersonic flow speed, no dependence of the solution on the downstream boundary conditions would be expected here. Thus, neither the flow inside the boundary layer nor in the external inviscid region should "feel" any downstream perturbations. In particular, this means that the inviscid flow everywhere upstream of the shock wave AA' (see Figure 7.1a) and the flow inside the boundary layer everywhere upstream of the point A' should be just the same as they would be if there were no shock wave at all. On the other hand, experimental data show that the shock impinging on the boundary layer at the point A' always causes a pressure rise well upstream of A'. If the shock wave is strong enough, it may even cause the boundary layer to separate and to produce the secondary shock BB' due to the thickening of the boundary layer prior to separation (Figure 7.1b). The system of primary AA' and secondary BB' shock waves is usually referred to as a "λ-structure."

This phenomenon may easily be described in terms of the triple-deck interaction theory. Suppose that the amplitude of perturbations is small and therefore the momentum equation in (7.1.1) may be linearized with respect to the undisturbed flow $u = y$, $v = 0$ to take the form

$$y\frac{\partial u}{\partial x} + v = -\frac{dp}{dx} + \frac{\partial^2 u}{\partial y^2}.$$

This equation is to be combined with the continuity equation from (7.1.1), subject to the boundary conditions (7.1.2) and the supersonic interaction law (7.1.5) for the case of a flat surface ($Y_0 \equiv 0$)

$$p = d\delta/dx.$$

The solution is

$$p = ce^{\lambda x}, \quad u = y + ce^{\lambda x}\frac{\lambda^{1/3}}{\text{Ai}'(0)}\int_0^{\lambda^{1/3}y}\text{Ai}(z)dz. \qquad (7.1.7)$$

Here Ai(z) is the Airy function, and λ is a universal positive constant related to the derivative of the Airy function at zero value of the argument:

$$\lambda = [-3\text{Ai}'(0)]^{3/4},$$

while c is the disturbance amplitude. The amplitude evidently cannot be found from the boundary conditions (7.1.2), (7.1.3), (7.1.4), but requires an additional condition which specifies, say, the location of the separation point or the value of the pressure at some point on the solid body surface. If neither of these is known in advance, as normally is the case for real flows, but the separation zone produced by the shock or wall irregularity is known to be closed, then the condition of disturbance attenuation downstream of the separation zone may be applied as $x \to +\infty$. In any case, the flow not only in the separation region but also upstream of separation appears to be influenced by the effect of upstream propagation of disturbances.

Remarkably the solution (7.1.7) had already been obtained by Lighthill in 1953, sixteen years before the triple-deck theory of the interaction between a laminar boundary layer and the external inviscid flow was formulated. To analyze the upstream disturbance propagation through the boundary layer, Lighthill (1953) subdivided the flow into two regions: (1) a viscous near-wall layer, which coincides with the corresponding layer in the triple-deck theory, and (2) an inviscid part of the flow that includes two inviscid layers from the triple-deck structure, the main part of the boundary layer and the external flow region. Despite the "non-rational" approach used by Lighthill (1953), his analysis resulted not only in the absolutely correct form of the solution (7.1.7), but also provided the estimate for the longitudinal extent of the interaction region: $\Delta x = O(\mathrm{Re}^{-3/8})$.

Two-tiered interaction models are still widely used in boundary-layer calculations. But in contrast to Lighthill's approach, more often the two regions that lie inside the boundary layer, the viscous near-wall layer and the main part of the boundary layer, are considered together as one composite region with the classical Prandtl equations used to describe the flow in this region. In contrast to equations (7.1.1), they may include variable density ρ. The boundary condition (7.1.4) is replaced by

$$u = u_e(x) \quad \text{as } y \to \infty, \tag{7.1.8}$$

where $u_e(x)$ is the flow velocity at the outer edge of the boundary layer. Note that the boundary condition (7.1.4) is used in the triple-deck interaction theory not only to specify the behavior of u at large values of y but also to determine $\delta(x)$. With (7.1.4) replaced by (7.1.8), one has to write the conventional integral for displacement thickness

$$\delta = \int_0^\infty \left(1 - \frac{\rho u}{\rho_e u_e} \right) dy. \tag{7.1.9}$$

This expression is to be combined either with the supersonic (7.1.5) or the subsonic (7.1.6) interaction law to complete the formulation of the interaction problem (7.1.1), (7.1.2), (7.1.5)–(7.1.6), (7.1.8), (7.1.9), which is usually referred to as a "model" interaction problem, in distinction to the asymptotic problem represented by (7.1.2)–(7.1.6).

No rational mathematical arguments (based, say, on asymptotic analysis of the Navier–Stokes equations) have been given to support the model approach. But numerous results of calculations for different flows prove its ability to provide good agreement with experimental data in the description of flow separation and of the influence of "viscosity effects" on the aerodynamic performance of airfoils, aircraft wings and other bodies. Therefore, it is not surprising that the model formulation of the interaction problem has attracted considerable attention from researchers, especially those interested in the engineering applications of the theory. Numerical methods for solving the model interaction problem started to be developed in the late 1960s, and many ideas put forward in this way appear to be well suited for both the model and the asymptotic formulations of the problem. In the present chapter we will use either of these forms, depending on which one was used in the original description of the corresponding method given by the author of the method.

7.2 Marching and Inverse Methods

When the pressure gradient is given, one usually applies marching methods to solve the boundary-layer equations, for example, the Crank–Nicolson (1947) method. This involves successive transition from one computation line $x = x_{j-1}$ orthogonal to the body surface to another line $x = x_j$ located farther downstream. The system of finite-difference equations for the values of the flow functions at the nodes (x_j, y_k) of the computational grid on each such line is solved after linearization using the Thomas technique, which is repeated a number of times in order to attain the necessary accuracy in the approximation of the nonlinear terms in the boundary-layer equations. Direct application of such an approach to solve the interaction problem is apparently impossible because of the dependence of a solution in the interaction region on the downstream boundary condition which cannot be taken into account in a marching technique. It follows from (7.1.7) that upstream propagation of disturbances takes place even for a boundary layer attached to the solid body surface and even with supersonic flow outside the boundary layer. The effect can only be enhanced with boundary-layer separation or when the flow outside the boundary layer is subsonic.

But if a flow without separation (or upstream of separation) is considered with supersonic external flow, when the interaction is characterized by local dependence of the induced pressure on the streamline slope at the outer edge of the boundary layer (7.1.5), it is easy to incorporate an additional iteration procedure into the marching method, which would make the pressure agree with the derivative of the boundary-layer displacement thickness at the section $x = x_j$ under consideration. Such an approach was used by Neiland (1969, 1971a) and by Stewartson and Williams (1969) in their first publications devoted to supersonic flow separation from a plane surface. To avoid the arbitrariness in the choice of the disturbance amplitude c in Lighthill's solution (7.1.7), Neiland used the pressure p as an independent variable instead of the longitudinal coordinate x. Stewartson and Williams performed their calculations in the usual coordinates (x, y). They specified an initial condition at a sufficiently large negative value of x, using equations (7.1.7) for this purpose. Since the formulation of the interaction problem (7.1.1)–(7.1.4) and (7.1.5) is invariant with respect to translation along the x-axis, the disturbance amplitude c in Lighthill's solution can be chosen arbitrarily. Indeed, if the disturbance amplitude c is represented in the form $c = \text{sign}(c)e^{\lambda x_0}$, it is easy to conclude that a change in c is equivalent to a displacement x_0 along the x-coordinate. Thus, the problem of the interaction of a boundary layer on a flat surface with a supersonic outer flow has two universal solutions. The first solution describes flows with a pressure increase that leads to separation, and the second describes flows which undergo an expansion in the interaction region (Neiland, 1974).

However, in the case of flow over a curved surface whose shape is specified by the equation $y = Y_0(x)$, the flow evolution in the interaction region no longer obeys a universal law. The obstacle to its application is the interaction condition, which is now written in the form

$$p = \delta' + Y_0'. \tag{7.2.1}$$

Nevertheless, the universal law can be used in many cases as a solution fragment. An example of such a flow is given by the problem of the interaction of a shock wave with a boundary layer and the mathematically equivalent problem of flow over a corner of a body contour for which

$$Y_0 = 0 \quad \text{for } x < 0; \quad Y_0 = \theta x \quad \text{for } x > 0.$$

Neiland (1971b), in analyzing these flows, pointed out that in both cases the solution obeys the universal law up to the point $x = 0$. In this case, the parameter c must be chosen so that in the continuation of the solution downstream of the point $x = 0$ all disturbances should decay on "exiting" from the interaction region. In other words, the condition

$$u = y \quad \text{as } x \to +\infty$$

must be used here as an additional boundary condition.

The same approach was used by Daniels (1974a,b) to calculate the symmetric flow near the trailing edge of a flat plate and the asymmetric supersonic flow over a flat plate at an angle of attack.

The problem becomes more complicated for subsonic speeds at the outer edge of the boundary layer, when the interaction condition fails to be local and the induced pressure depends on the distribution of displacement thickness along the entire interaction region. The dependence expressed by the integral (7.1.6) makes it necessary that the calculation procedure include global iterations over the whole flow field, in addition to local iterations at each section $x = x_j$.

The first numerical solution of the interaction problem for incompressible flows was constructed by Jobe and Burggraf (1974), who considered the flow over the trailing edge of a flat plate at zero incidence. The method of calculation that they proposed consists of two procedures carried out successively. The first is the calculation of the flow in the viscous wall layer, accomplished by a marching method but not with the conventional formulation. The calculation of the flow inside the boundary layer is carried out not for a prescribed pressure distribution but for a given boundary-layer displacement thickness. This method of solving the boundary-layer equations is called an inverse method. As for the pressure, it is determined in the process of solving the boundary-layer equations and then used to provide initial data for the second procedure, involving inversion of the integral (7.1.6). It is represented in the form:

$$\delta'(x) = -\frac{1}{\pi} \int_{-\infty}^{\infty} \frac{p(s)}{s - x} ds, \qquad (7.2.2)$$

which allows one to find a new distribution of displacement thickness from a known pressure distribution along the interaction region. The convergence of the iteration process constructed in this way is achieved using the underrelaxation

$$\delta_j^{n+1} = r\delta_j^* + (1 - r)\delta_j^n,$$

with the parameter r being about $r = 0.2$. Here, the superscript n denotes the iteration number, whereas the subscript j shows the location x_j of the point of interest on the solid body surface. Correspondingly, δ_j^n and δ_j^{n+1} are the "old" and the "new" values of δ at x_j, while δ_j^* is the result of the integration in (7.2.2).

The method of Jobe and Burggraf (1974) has been used to solve a number of problems in the theory of interaction between a boundary layer and the external flow. In particular, Chow and Melnik (1976) applied this method in considering the asymmetric flow over the trailing edge of a flat plate at an angle of attack. What is even more important, the method has allowed us to proceed with the calculation of boundary-layer separation in incompressible flows. The nonstandard approach to the solution of the boundary-layer equations at each iteration is of fundamental significance here. If the boundary-layer flow were calculated in the usual way, for a given pressure gradient, the Goldstein singularity would inevitably appear at the point of zero skin friction. The utilization of the boundary-layer displacement thickness as the known function eliminates this singularity from the solution and allows its continuation into the region of reverse flow.

The Goldstein singularity gives rise to nonanalyticity in the behavior not only of the boundary-layer displacement thickness but also of other flow variables, in particular the skin friction. Therefore, it is possible to find many ways of constructing a solution of the boundary-layer equations that do not contain the Goldstein singularity. Among them, as already noted, is the one offered by Nature herself, in establishing a definite relationship between induced pressure and boundary-layer displacement thickness, which was in essence realized in calculations of Neiland (1969, 1971a) and of Stewartson and Williams (1969), who required its fulfillment at each section $x = x_j$.

The construction of a numerical method for the solution of the interaction problem must correctly take into account another property of the boundary-layer equations. If separation does not occur and the longitudinal velocity component u in the viscous wall layer remains positive, then the boundary-layer equations possess the usual properties for equations of parabolic type: the values of the unknown flow variables at any point (x, y) of the region of integration are determined uniquely by boundary conditions imposed upstream of it.[2] Flow separation is accompanied by the development of a region of reverse flow with $u < 0$, where the direction of transmission of disturbances reverses and the solution of the boundary-layer equations becomes dependent upon the distributions of the flow variables downstream of the point (x, y) under consideration. It is clear that under these conditions any method in which the unknown flow functions at each section $x = x_j$ are determined by their distributions at the preceding section $x = x_{j-1}$ is at variance with the properties of the differential equations, which, as often happens, results in instability of the calculation.

The simplest method of suppressing the instability of the marching method in a region of reverse flow was proposed by Reyhner and Flügge-Lotz (1968). They noted that the flow velocity u in this region is small as a rule, and suggested using the approximate boundary-layer equations with the convection term $u\, \partial u/\partial x$ removed from the momentum equation to describe the flow inside the region.

An example of such an approach to the solution of the model problem of the interaction of a boundary layer with a supersonic external flow

---

[2] It is assumed here that one of the following functions is specified in advance: pressure gradient, displacement thickness, skin friction, etc.

is the investigation of Werle, Polak, and Bertke (1973). The first to apply the approximation of Reyhner and Flügge-Lotz to the solution of the asymptotic equations of interaction theory was Williams (1975). He turned again to the problem of self-induced boundary-layer separation in supersonic flow over a flat plate, and recalculated the flow downstream of the separation point, refining it successively in the following way. In the first approximation he used for the solution of the equations of interaction theory the same method as in the work of Stewartson and Williams (1969), supplementing it with the approximation of Reyhner and Flügge-Lotz. Since the solution found in this way in the region of reverse flow is not accurate, he recalculated it using an ordinary marching method carried out in this region in the direction opposite to the main flow. The resulting distribution of the longitudinal velocity component u was then used to calculate the convective term $u\,\partial u/\partial x$ in the reverse flow for the solution of the interaction problem in the second approximation, etc.

Smith (1977) combined the method of Reyhner and Flügge-Lotz with the method of Jobe and Burggraf (1974) to construct the solution of Sychev's problem (1972) of laminar separation of a stream of incompressible fluid from a smooth part of a body surface, although it was done without refining the solution in the region of reverse flow. Another result of fundamental importance was obtained by Gittler and Kluwick (1987) using this approach. They considered the problem of the hysteresis of the separated flow on the surface of an axisymmetric compression ramp in supersonic flow near the point where a cylindrical part of the body changes to a conical shape. The initial velocity profile was taken in a form similar to (7.1.7), with the parameter c found from the condition of decay of disturbances downstream of the corner. This condition turns out to be satisfied, within a certain range of surface corner angles, for three different values of the parameter c, so that the local separation zone in the vicinity of the corner can take three different forms for the same flow conditions.

The work of Gittler and Kluwick (1987), like their previous investigation on the same subject (Kluwick, Gittler, and Bodonyi, 1985), was devoted to flows in which the interaction condition has a form differing from that in (7.1.5)–(7.1.6). This circumstance was not, however, an obstacle to the application of the numerical methods described above. Moreover, the marching method, with shooting toward downstream conditions, proved to be useful also for the solution of other problems of interaction theory, for example, for the problem of self-induced separation of a viscous wall jet (Smith and Duck, 1977) with $p = \delta''$, for the calculation of jet flow over the corner of a body (Merkin and Smith, 1982), and for the solution of the problem of the flow at the edge of a circular disk rotating in its plane within fluid at rest (Smith, 1978).

Iteration methods. The idea of using a variable computational stencil in the iteration process to change the manner of approximating the equations depending on local flow properties gave a new impetus to the development of calculation methods for separated flows. This idea was taken from the work of Murman and Cole (1971), who proposed a calculation method for the solution of the Kármán equation of transonic small-perturbation theory, a method that later became widely known. They suggested recalculating the velocity potential φ in each iteration, proceeding successively from one calculation line $x = x_j$ to another line located further downstream. The finite-difference method of approximating the Kármán equation at grid nodes (x_j, y_k) was chosen based on the local value of the Mach number. If it exceeded unity, a "hyperbolic" stencil with backward differences for derivatives with respect to x was used; for subsonic velocity, it was replaced by an "elliptic" one and the x derivatives were represented by central differences.

The resulting algebraic equations for the values of the potential on the line $x = x_j$ can be solved by using the Thomas technique. To do this, it is necessary to know the values of the function φ at grid nodes not only upstream of the line $x = x_j$, where the solution has already been constructed, but also downstream of it at the section $x = x_{j+1}$. Here, the distribution of the potential is taken from the solution of the problem in the preceding approximation. The iterative calculation proceeds until the maximum variation of the potential in the flow field becomes smaller than some value $\epsilon = 10^{-5}$ to 10^{-6} assigned in advance.

Klineberg and Steger (1974) were the first to demonstrate the applicability of such an approach to the solution of the Prandtl equations. They chose the method of approximating the derivative $\partial u / \partial x$ in the convection term $u \, \partial u / \partial x$ of the momentum equation (7.1.1) to depend on the sign of the longitudinal velocity component u: for $u > 0$, it was represented by backward differences and for $u < 0$, by forward differences. Central differences were used for the first and second derivatives with respect to y. The corresponding computation stencils are shown in Figure 7.2. To suppress oscillations occurring in the calculations, Klineberg and Steger introduced another, intermediate, stencil allowing

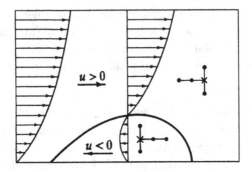

Fig. 7.2 Computation stencils used for finite-difference approximation of the boundary-layer equation.

the backward difference to be "switched over" smoothly to a forward one, as u varies from 0.01 to -0.01.

Similar work was published by Carter (1974). In contrast to Klineberg and Steger, he rewrote, from the very beginning, the Prandtl equations in terms of the vorticity $\omega = \partial u / \partial y$ and the stream function ψ, having differentiated the momentum equation (7.1.1) with respect to y:

$$u\frac{\partial \omega}{\partial x} + v\frac{\partial \omega}{\partial y} = \frac{\partial^2 \omega}{\partial y^2}, \quad \frac{\partial^2 \psi}{\partial y^2} = \omega. \tag{7.2.3}$$

This seemingly insignificant modification resulted in a considerable increase in the convergence rate of the iteration process and made unnecessary the use of a transition stencil at points with small values of u.

Carter, like his predecessors Klineberg and Steger, considered inverse problems of boundary-layer theory: the solution of the Prandtl equations was constructed by specifying either the skin friction $\tau(x) = \omega|_{y=0}$ or the displacement thickness $\delta(x)$. In order to proceed to the solution of the interaction problem, Ruban (1976a) supplemented Carter's method with recalculation of the skin friction in accordance with the condition of interaction of the boundary layer with the external flow. Considering the onset of boundary-layer separation in a supersonic flow over a compression ramp, he made use of the well-known consequence of the Prandtl equation

$$\frac{\partial^2 u}{\partial y^2}\bigg|_{y=0} = \frac{dp}{dx} \tag{7.2.4}$$

and, having expressed the pressure by equation (7.1.5), he represented the interaction condition in the form

$$\left.\frac{\partial \omega}{\partial y}\right|_{y=0} = \delta''. \tag{7.2.5}$$

The iteration process was organized in the following way. First, the values of the vorticity $\omega_{j,k}$ are determined at interior grid points, as is done in Carter's method, using the skin-friction distribution $\omega_{j,1}$ found in the preceding iteration (or given as a zeroth approximation). Then the stream function is found by integrating the second equation of the system (7.2.3), supplemented by the no-slip conditions at the wall

$$\psi = \frac{\partial \psi}{\partial y} = 0 \quad \text{at } y = 0.$$

Now it is possible to determine the distribution of displacement thickness from equation (7.1.4) rewritten in the form

$$\delta(x) = -\lim_{y \to \infty} \left(\frac{\partial \psi}{\partial y} - y\right).$$

It remains to use the interaction condition (7.2.5). If the derivative on the left-hand side is represented in a finite-difference form having second-order accuracy, we obtain

$$\omega_{j,1} = \frac{4}{3}\omega_{j,2} - \frac{1}{3}\omega_{j,3} - \frac{2}{3}\Delta y \delta_j''.$$

The form of the finite-difference representation for the derivative δ'' in the interaction condition is of primary importance. The point is that it is the interaction that is responsible for upstream propagation of disturbances through a boundary layer that has no reverse flow. The description of this effect is made possible by a central-difference representation of δ'', providing the dependence of the solution in the interaction region on the values of the displacement thickness at the right-hand grid boundary.

Elliott and Smith (1986) used a variable stencil in the marching method, selecting the parameter c in Lighthill's solution (7.1.7) from the condition of disturbance decay as $x \to \infty$. They considered the problem of boundary-layer separation near the trailing edge of a flat plate at incidence in a supersonic stream. As a result, the solution of Daniels (1974b) was extended to angles of attack corresponding to separated flow over the upper surface of the plate.

In order to calculate incompressible flows with separation in the interaction region, Carter and Wornom (1975) and Ruban (1976b) used

Carter's (1974) method, in which the solution of the boundary-layer equations is constructed for a given displacement thickness. This was combined with the recalculation of the displacement thickness at each iteration in accordance with the method of Jobe and Burggraf (1974). Carter and Wornom analyzed the separation occurring in a cavity on the body surface, and Ruban (1976b, 1977) calculated separation in the vicinity of a sharp bend in the body contour and near the trailing edge of a thin symmetric airfoil. Dijkstra (1979) described still another example of the application of this method. He investigated boundary-layer separation from a surface with a forward-facing step of small height.

Time-relaxation methods. The global iterations over the whole flow field involved in the solution of the interaction problem make the above iteration methods similar to time-relaxation methods, their similarity being manifested not only in the difficulty of the calculations but, what is especially important, also in the method of approximating the equations of motion. As a rule, the usual unsteady boundary-layer equations are the basis for time-relaxation methods, and it is therefore required that the finite-difference stencil be varied when the flow direction in the boundary layer changes, as follows directly from the Courant condition.

A time-relaxation method was first applied to the solution of the interaction problem in the work of Werle and Vatsa (1974). The authors added an artificial unsteady term $h\,\partial\delta/\partial t$ on the right-hand side of the momentum equation (7.1.1), with the coefficient h varying during the computation, and sought the solution of the steady problem as the limit of an unsteady problem, with t tending to infinity. They divided each time step into two half-steps, using one to determine the velocity vector field and the other to determine the boundary-layer displacement thickness. At the first half-step $t^* = t^n + \Delta t/2$ the momentum equation was represented in the form

$$\left(\frac{\partial^2 u}{\partial y^2} - u\frac{\partial u}{\partial x} - v\frac{\partial u}{\partial y}\right)^* = \left(\frac{dp}{dx}\right)^n - h\frac{\delta^* - \delta^n}{\Delta t/2} \tag{7.2.6}$$

where the pressure gradient is calculated according to the Ackeret formula $p = d\delta/dx$. If, as before, the derivatives with respect to y on the left-hand side of equation (7.2.6) are represented by central differences and, for the derivative with respect to x, backward differences are used when $u_{j,k}^n > 0$ and forward differences when $u_{j,k}^n < 0$, the result is a system of algebraic equations for $u_{j,k}^*$. The solution on each line $x = x_j$ can be constructed by the Thomas technique.

In order to obtain a second-order approximation at each step, Werle and Vatsa made use of the following relation at the second half-step:

$$\left(\frac{\partial^2 u}{\partial y^2} - u\frac{\partial u}{\partial x} - v\frac{\partial u}{\partial y}\right)^* = \left(\frac{dp}{dx}\right)^{n+1} - h\frac{\delta^{n+1} - \delta^*}{\Delta t/2}. \qquad (7.2.7)$$

Using the Ackeret formula for the pressure and setting $y = 0$ in (7.2.7), one obtains

$$\left(\frac{d^2\delta}{dx^2}\right)^{n+1} - \frac{2h}{\Delta t}\delta^{n+1} = \left(\frac{\partial^2 u}{\partial y^2}\right)^* - \frac{2h}{\Delta t}\delta^*.$$

It remains to represent $d^2\delta/dx^2$ using central differences and, applying the Thomas technique, to find the distribution of displacement thickness δ_j^{n+1} at time t^{n+1}.

Werle and Vatsa applied their method to the analysis of boundary-layer separation in the vicinity of a corner of a body contour. They proceeded from a model formulation of the interaction problem. Later, Werle and Verdon (1979) used this method for solving the asymptotic problem of supersonic flow separation near the trailing edge of an airfoil with thickness, and Napolitano, Werle, and Davis (1979) reformulated it in conformity with the solution of the asymptotic equations of interaction theory in the case of subsonic speed at the outer edge of the boundary layer. They left the first half-step unchanged, and at the second step obtained an integral equation for the function δ^{n+1} using the relation (7.1.6) instead of (7.1.5). This was done in the following way. If (7.2.6) is subtracted from (7.2.7), with h taken for simplicity to be $h = 1$, it is found that

$$\left(\frac{dp}{dx}\right)^{n+1} = \left(\frac{dp}{dx}\right)^n + \frac{\delta^{n+1} - 2\delta^* + \delta^n}{\Delta t/2}. \qquad (7.2.8)$$

Substituting the relation (7.2.8), differentiated with respect to x beforehand, into the integral (7.2.2) of thin-airfoil theory, we obtain

$$\frac{d^2\delta^{n+1}}{dx^2} = -\frac{1}{\pi}\int_{-\infty}^{\infty}\frac{dp^n/ds}{s-x}ds - \frac{2}{\pi\Delta t}\int_{\infty}\frac{\delta^{n+1} - 2\delta^* + \delta^n}{s-x}ds. \qquad (7.2.9)$$

The solution of this equation written in finite-difference form can be obtained by inversion of the corresponding matrix. However, the iterative approach proposed by Napolitano, Werle, and Davis (1979) is faster. After the integrals are approximated by finite sums, terms proportional to δ_{j-1}^{n+1}, δ_j^{n+1} and δ_{j+1}^{n+1} are extracted from the right-hand side of equation (7.2.9). Transposing them to the left-hand side of this equation and using central differences for $d^2\delta^{n+1}/dx^2$, we obtain the system of algebraic equations

$$\left(\frac{1}{\Delta x^2} + \frac{2\ln 3}{\pi\Delta t}\right)\delta_{j-1}^{n+1} - \frac{2}{\Delta x^2}\delta_j^{n+1} + \left(\frac{1}{\Delta x^2} - \frac{2\ln 3}{\pi\Delta t}\right)\delta_{j+1}^{n+1} = H_j .$$

$$(7.2.10)$$

The solution is sought by the method of successive approximations: the remaining part H_j of the integrals is calculated based on the values of δ_j^{n+1} in the preceding iteration, and a new approximation for δ_j^{n+1} is found from the solution (7.2.10), which is constructed using the ordinary Thomas technique. Underrelaxation with the parameter $r = 0.2$ to 0.3 is used to provide the convergence of this internal iteration process.

Napolitano, Werle, and Davis demonstrated the possibilities of their method by solving the problem of boundary-layer separation in a cavity on the surface of a flat plate. They succeeded, of course, in determining only the limiting steady state of the flow. Unsteady processes occurring in the region of interaction between the boundary layer and the external flow can be investigated by the method of Jenson et al. (1975) and Rizzetta et al. (1978). As the basic equations of motion they used the ordinary unsteady boundary-layer equations, which differ from the equations (7.1.1) by the term $\partial u/\partial t$ on the left-hand side of the momentum equation. Differentiating this equation with respect to y, one can write the interaction problem (7.1.1)–(7.1.5) in the form

$$\frac{\partial\omega}{\partial t} + u\frac{\partial\omega}{\partial x} + v\frac{\partial\omega}{\partial y} = \frac{\partial^2\omega}{\partial y^2},$$

$$\omega = 1 \quad \text{as } x \to -\infty \text{ and } y \to \infty, \qquad (7.2.11)$$

$$\left.\frac{\partial\omega}{\partial y}\right|_{y=0} = \frac{\partial Y_0}{\partial x} - \frac{\partial^2}{\partial x^2}\int_0^\infty (\omega - 1)dy .$$

The boundary condition for the derivative of the vorticity at the body surface $(y = 0)$ is the interaction law. It is formulated here for a supersonic external flow based on the Ackeret formula (7.2.1) and the relations (7.2.4), (7.1.4).

Jenson, Burggraf, and Rizzetta used the condition

$$\omega|_{y=0} = 1 \quad \text{as } x \to \infty$$

as an additional boundary condition determining the coefficient c of the eigenfunction in the solution (7.1.7) of Lighthill. They considered the local separation zone formed in the vicinity of a corner in the body contour, assuming that all disturbances decay far downstream and the velocity profile returns to its initial state $u = y$.

Each time step was calculated in two stages. First, the values of the vorticity $w_{j,k}^{n+1}$ were determined for a new instant of time t^{n+1} at interior grid points, and then its distribution $w_{j,1}^{n+1}$ was found at the body surface. This second problem was solved with the help of the interaction condition from (7.2.11). If the integral on the right-hand side is represented according to the trapezoidal rule and central differences are used for the second derivative with respect to x, then for all $j = 2, \ldots, N-1$ we have

$$\frac{w_{j,2} - w_{j,1}}{\Delta y} = -\frac{\Delta y}{\Delta x^2} \left[\frac{1}{2}(w_{j+1,1} - 2w_{j,1} + w_{j-1,1}) \right.$$
$$+ \left. \sum_{k=2}^{M-1} (w_{j+1,k} - 2w_{j,k} + w_{j-1,k}) \right] + \left(\frac{\partial Y_0}{\partial x} \right)_j .$$

With the known values of the vorticity at interior grid points, this system of equations, supplemented by the conditions

$$w_{1,1} = w_{N,1} = 1 ,$$

can be solved for $w_{j,1}$ by the Thomas technique.

Noting that the interaction law leads to serious restrictions on the time step in the time-relaxation method, Ruban (1978) suggested treating it in a fully implicit way. To this end, he developed a special technique allowing one to calculate the vorticity at interior points in the flow field and on the surface of the body, not successively, but simultaneously. For the momentum equation from (7.2.11), the following finite-difference scheme was used:

$$\frac{w_{j,k}^{n+1} - w_{j,k}^n}{\Delta t} + |u_{j,k}^n| \frac{w_{j,k}^n - w_{j-l,k}^n}{\Delta x} + |v_{j,k}^n| \frac{w_{j,k}^{n+1} - w_{j,k-m}^{n+1}}{\Delta y}$$
$$= \frac{w_{j,k+1}^{n+1} - 2w_{j,k}^{n+1} + w_{j,k-1}^{n+1}}{\Delta y^2}, \quad k = 2, \ldots, M-1; \quad j = 2, \ldots, N-1;$$

where $l = \text{sign}\left(u_{j,k}^n\right)$, $m = \text{sign}\left(v_{j,k}^n\right)$.

For a given value $w = 1$ of the vorticity at the outer edge of the viscous layer, this system of equations allows application of the Thomas technique at each section $x = x_j$:

$$w_{j,k}^{n+1} = R_k w_{j,k-1}^{n+1} + Q_k ,$$

so that the solution at interior grid points turns out to depend linearly on $w_{j,1}^{n+1}$:

$$w_{j,k}^{n+1} = A_{j,k} w_{j,1}^{n+1} + B_{j,k} . \qquad (7.2.12)$$

The flow speed at the outer edge of the viscous layer,

$$\int_0^{y_{\max}} \omega \, dy = F_j \omega_{j,1}^{n+1} + G_j \,, \tag{7.2.13}$$

possesses the same property.

Substituting (7.2.13) into the right-hand side of the interaction condition (7.2.11), and using the relation (7.2.12) to write the derivative $\left. \frac{\partial \omega}{\partial y} \right|_{y=0}$ in terms of $\omega_{j,1}^{n+1}$, we obtain

$$a_j \, \omega_{j-1,1}^{n+1} - b_j \, \omega_{j,1}^{n+1} + c_j \, \omega_{j+1,1}^{n+1} + d_j = 0 \,, \tag{7.2.14}$$

where

$$a_j = F_{j-1}, \quad b_j = 2F_j + \frac{\Delta x^2}{\Delta y}(1 - A_{j,2}), \quad c_j = F_{j+1},$$

$$d_j = G_{j+1} - 2G_j + G_{j-1} + \frac{\Delta x^2}{\Delta y} B_{j,2} - \left(\frac{\partial Y_0}{\partial x}\right)_j .$$

After solving this system of equations by the Thomas technique, one can return to equation (7.2.12) and recover ω throughout the flow field.

The above method enabled Ruban (1978) to investigate the process of formation of a local separation zone in the boundary layer in the presence of an incident shock wave. Zhuk and Ryzhov (1979) and Zhuk (1982) used this method to analyze the separation provoked by a shock that is moving along a surface. Later, Kravtsova and Ruban (1988) applied the method to calculate boundary-layer separation upstream of the base of a body, and Zhuk (1988) used it to investigate the flow structure in the interaction region on a permeable surface, when gas is injected or sucked through the wall.

In order to consider incompressible flows, Korolev (1980a,b) replaced the Ackeret formula by the integral of thin-airfoil theory in the interaction condition. As a result, an integral equation determining the distribution of vorticity along the wall was obtained instead of (7.2.14). The solution at each time step was constructed by using the method of successive approximations. This approach was used by Korolev to solve Sychev's problem (1972) of laminar separation from a smooth body surface and to investigate symmetric separation near the trailing edge of a thin elliptic airfoil.

7.3 Fast Iteration Methods

Semi-inverse methods. For subsonic speed at the outer edge of the boundary layer, the application of the time-relaxation method to calculate the flow in the interaction region requires that special measures be taken to suppress oscillations occurring in the solution. Their appearance is not due to one technique or another for the finite-difference approximation of the equations of interaction theory, but to the fundamental properties of these equations. As shown by Smith (1979b), Zhuk and Ryzhov (1980), and Mikhailov (1981), the "subsonic" interaction theory describes real flow fluctuations, namely Tollmien–Schlichting waves at the lower branch of the boundary-layer stability curve.

These waves represent an essentially unsteady form of fluid motion. They cannot appear if the flow is calculated using an iteration method based on the steady equations of interaction theory. The method of Carter and Wornom (1975) and Ruban (1976b) is of just this kind. It is referred to as an "inverse" iteration method, because at each iteration the calculation of the flow inside the boundary layer in this method is carried out not for a given pressure distribution but for a given boundary-layer displacement thickness δ^n. The pressure is found as part of the solution of this problem and then is used to provide initial data for the external flow. Here also an inverse problem is solved: a new distribution of displacement thickness δ^* is found using the formula (7.2.2). One should keep in mind that this iteration procedure becomes convergent only with the underrelaxation

$$\delta_j^{n+1} = r\delta_j^* + (1-r)\delta_j^n$$

where the parameter r depends on the particular problem being considered, but in all cases, when flows having a sufficiently extended separation zone are dealt with, one may not take r larger than 0.1.

The situation can be improved by changing the approach to the outer inviscid flow and using a "semi-inverse" method. Le Balleur (1978) and Carter (1978, 1979), the authors of this method, kept the calculation procedure unchanged for the flow inside the boundary layer, and for the outer flow they turned from the solution of an inverse problem to the solution of a direct problem. In each iteration, for a given distribution of displacement thickness $\delta^n(x)$, they determined two pressure functions simultaneously, one function $p_v(x)$ from the boundary-layer equations and the other function $p_{in}(x)$ from an analysis of the external inviscid flow. It is clear that they do not coincide with each other, and the problem consists in finding, from $\delta^n(x)$ and the pressure difference $p_v(x) - p_{in}(x)$, a new distribution of displacement thickness $\delta^{n+1}(x)$ for which this difference becomes smaller.

The recalculation formula suggested by Le Balleur proved not altogether successful. It does not eliminate the necessity of applying underrelaxation in the iteration procedure. However, if the formula of Carter is used,

$$\delta^{n+1} = \delta^n \frac{u_{e,v}}{u_{e,in}}$$

where $u_e(x)$ is the velocity at the outer edge of the boundary layer, the calculation can be carried out using the overrelaxation

$$\delta^{n+1} = r\,\delta^n \frac{u_{e,v}}{u_{e,in}} + (1-r)\delta^n,$$

which provides the most rapid convergence of the iteration process for $r = 1.2$ to $r = 1.8$.

Another advantage of the semi-inverse method is that it allows the inclusion of a subroutine in the calculation program designed to solve the usual problem of inviscid flow over a solid body. This possibility was already demonstrated in the first work of Carter (1978) devoted to transonic flow separation near the junction line of an ellipsoid and a circular cylinder. The semi-inverse method was used also for the solution of other problems in the theory of the interaction of a boundary layer with the outer inviscid flow. In particular, Kwon and Pletcher (1979) applied it to the analysis of separation in the middle part of an airfoil, and Davis and Carter (1986) to the investigation of a short separation zone at the leading edge of an airfoil. In both cases, the instability of the separated flow and the transition to turbulence upstream of the reattachment point were taken into account.

Quasisimultaneous methods. As far as the convergence rate of the iterative process is concerned, both the inverse and semi-inverse methods are in this respect significantly inferior to the "quasisimultaneous" method of Veldman (1981).[3] In its structure the latter is reminiscent of early methods of calculating supersonic flows in the interaction region, where the solution of the boundary-layer equations is constructed successively, step by step, for sections $x = x_j$. At each such section the pressure p_j is calculated simultaneously with the displacement thickness δ_j based on the interaction condition (7.1.5). For subsonic velocity in the external flow, this condition is expressed in terms of the integral of thin-airfoil theory (7.1.6). To represent it in a discrete form, Veldman (1981) used a simple rule of rectangles, according to which the integrand in

[3] See also the work of Veldman and Dijkstra (1981).

each interval (x_i, x_{i+1}) is assumed to be constant and is taken at the midpoint of the interval of integration:

$$\int_{x_1}^{x_N} \frac{\delta'(s)}{s - x_j} ds = \sum_{i=1}^{N-1} \frac{\delta'_{i+1/2}}{i - j + 1/2}.$$

If $\delta'_{i+1/2}$ is represented as

$$\delta'_{i+1/2} = \frac{\delta_{i+1} - \delta_i}{\Delta x},$$

then for the pressure at the point x_j we have

$$p_j = \sum_{i=2}^{N-1} \alpha_{ij}\delta_i + \beta_j, \tag{7.3.1}$$

where

$$\alpha_{ij} = \frac{\gamma}{(i-j)^2 - 1/4}, \quad \beta_j = \frac{\gamma\delta_N}{N - j - 1/2} - \frac{\gamma\delta_1}{3/2 - j}, \quad \gamma = \frac{1}{\pi\Delta x}.$$

Of primary importance in the method of Veldman is the way in which the interaction law (7.3.1) is used for the solution of the boundary-layer equations. As before, the solution is constructed successively at sections x_j. But now neither the pressure p_j nor the displacement thickness δ_j is given in advance. Instead, a linear relationship,

$$p_j^{n+1} - \alpha_{jj}\delta_j^{n+1} = \sum_{i=2}^{j-1} \alpha_{ij}\delta_i^{n+1} + \sum_{i=j+1}^{N-1} \alpha_{ij}\delta_i^n + \beta_j^n, \tag{7.3.2}$$

is prescribed between them. Because the equations for the outer invis-cid flow are elliptic, the right-hand side of the relation (7.3.2) includes summation not only over points with $i < j$, where the solution has al-ready been constructed, but also downstream of x_j. This leads to the necessity of carrying out global iterations. In Veldman's method they are accomplished through the use of overrelaxation: after the solution of the boundary-layer equations supplemented by the condition (7.3.2) has been obtained for all j, a new approximation for the displacement thickness is recalculated according to the formula

$$\delta_j^{n+1} = r\,\delta_j^* + (1 - r)\delta_j^n$$

with $r = 1.5$.

Having applied the quasisimultaneous method to boundary-layer separation in a cavity on the surface of a flat plate, Veldman and Dijkstra (1981) found that it provides convergence to an accuracy of

$$\max_j \left| \delta_j^{n+1} - \delta_j^n \right| < 10^{-4}$$

after 17 global iterations. To solve the same problem by an inverse method, Carter and Wornom (1975) needed 64 iterations, while Kwon and Pletcher (1979), using a semi-inverse method, obtained a solution after 16 iterations, but to an accuracy of

$$\max_j \left| \delta_j^{n+1} - \delta_j^n \right| < 5 \cdot 10^{-4}.$$

Cebeci, Stewartson, and Williams (1981) used the method of Veldman to analyze short separation zones arising at the leading edge of a thin airfoil. Smith and Merkin (1982) applied it to the solution of the asymptotic equations of interaction theory. They investigated separation occurring in an incompressible flow near a corner of a body, near the wedge-shaped trailing edge of a thin airfoil, and in certain other situations. In addition, Barnett and Verdon (1987) used this method to analyze subsonic turbulent flow in the vicinity of the trailing edge of an airfoil.

A detailed comparison of the quasisimultaneous and semi-inverse methods was carried out by Davis and Werle (1982) for the example of flow over the trailing edge of a flat plate. This work also contains the results of calculations of boundary-layer separation near the elliptic trailing edge of a thin airfoil and in a cavity on a flat surface. Davis and Werle introduced a number of refinements in the calculation procedure. Of particular interest is their method of representing the integral in the interaction law. Since a model problem for the interaction region was considered, they wrote the integral of thin-airfoil theory in the form

$$u_e(x) = 1 + \frac{1}{\pi\sqrt{\mathrm{Re}}} \fint_{-\infty}^{\infty} \frac{d(u_e\delta)/ds}{x - s} ds. \tag{7.3.3}$$

The convergence rate of the global iteration in the quasisimultaneous method is directly related to the form of the dependence of the pressure p or, equivalently, of the longitudinal velocity component u_e, on the displacement thickness δ of the boundary layer. In the particular case of supersonic external flow, when the interaction law (7.1.5) is local, there is no need at all to use global iterations to adjust the pressure for the displacement effect of the boundary layer. Therefore, it may be expected that the more "local" is this dependence between p and δ, the higher should be the convergence rate. In this connection it becomes desirable to choose the form for the finite-difference representation of the integral (7.3.3) in such a way that the contribution of the immediate vicinity of the singular point $s = x_j$ in the finite sum replacing the integral in (7.3.3) is as "pronounced" as possible. To attain this goal, Davis and Werle (1982) first performed an integration by parts in (7.3.3) and represented the result in the following symbolic form:

$$u_e(x) = 1 - \frac{1}{\pi\sqrt{\mathrm{Re}}} \int_{-\infty}^{\infty} \frac{u_e\delta}{(x-s)^2} ds\,.$$

Assuming further that $u_e\delta$ is constant in each interval $\left[x_i - \frac{\Delta x}{2}, x_i + \frac{\Delta x}{2}\right]$ and coincides with $(u_e\delta)_i$, they arrived at the formula

$$u_{ej} = 1 + \frac{1}{\pi\Delta x\sqrt{\mathrm{Re}}} \sum_{i=1}^{N}(u_e\delta)_i D(i-j), \qquad (7.3.4)$$

where the grid function

$$D(i-j) = \frac{4}{1-4(i-j)^2}$$

has a pronounced maximum at the point with $i = j$.

Davis and Werle used this formula in both the semi-inverse and the quasisimultaneous methods; in the latter case it was considered together with the relation

$$v - y\frac{\partial v}{\partial y} \to \frac{d}{dx}(u_e\delta)\,, \quad \text{as } y \to \infty\,, \qquad (7.3.5)$$

which determines the asymptotic behavior of the transverse velocity component v at the outer edge of the boundary layer.

If (7.3.4) is rewritten by analogy with (7.3.2) in the form

$$u_{ej} = \alpha(u_e\delta)_j + b \qquad (7.3.6)$$

and the right-hand side in (7.3.5) is represented by the backward difference

$$v - y\frac{\partial v}{\partial y} = \frac{(u_e\delta)_j - (u_e\delta)_{j-1}}{\Delta x}\,,$$

then eliminating $(u_e\delta)_j$ with the help of (7.3.6), we obtain the relation

$$v - y\frac{\partial v}{\partial y} = \frac{1}{\Delta x}\left[\frac{u_{ej}}{\alpha} - \frac{b}{\alpha} - (u_e\delta)_{j-1}\right]\,.$$

This permits the boundary-layer equations to be solved uniquely at the section x_j.

The quasisimultaneous method constructed in this way was used by Davis (1984) to analyze boundary-layer separation in the vicinity of a corner of a body contour and at a parabolic hump on a flat surface, and also by Rothmayer and Davis (1986) and Barnett (1988) to study the transition from flows with a local separation zone to flows with a more pronounced separation zone in the vicinity of the trailing edge of an airfoil in a stream of incompressible fluid.

This method was considerably modified in the work of Edwards and Carter (1986), devoted to the analysis of flows with complicated geometry, when the solution in the outer inviscid flow can no longer be represented by the integral of thin-airfoil theory. Instead, Edwards and Carter used Laplace's equation for the stream function ψ in the inviscid flow.[4] This equation was written in an orthogonal curvilinear coordinate system chosen according to the particular problem considered. It is convenient to align the x axis with the body surface, and then the condition of matching with the solution inside the boundary layer is written as

$$\psi(x,0) = -u_e\delta . \tag{7.3.7}$$

If the derivatives in Laplace's equation are represented by central differences, the following system of equations is obtained for values of the stream function at grid nodes on the line $x = x_j$:

$$a_k\psi_{j,k-1} + b_k\psi_{j,k} + c_k\psi_{j,k+1} + d_k = 0 .$$

The solution is constructed by using the Thomas technique:

$$\psi_{j,k} = R_k\psi_{j,k-1} + Q_k .$$

Upon differentiating ψ with respect to the coordinate normal to the body surface, and using the condition (7.3.7), we find the longitudinal velocity component u_{ej} at the outer edge of the boundary layer as a function of δ_j. Since this function is linear,

$$u_{ej} = \alpha\delta_j + b ,$$

the problem for the boundary layer is reduced to the one considered earlier.

Edwards and Carter applied the method just described to calculate

---

[4] Houwink and Veldman (1984) developed the same approach for flows with transonic velocity at the outer edge of the boundary layer.

boundary-layer separation on the wall of a diverging channel. A second problem that they considered is the separation of a stream of incompressible fluid in a cavity on the surface of a flat plate. Following Carter and Wornom (1975), this problem has been studied by many investigators who have used it as a test to verify numerical methods being developed.

An interesting way of accelerating the convergence of quasisimultaneous methods of solution was proposed in the work of Black and Rothmayer (1991). They wrote the interaction condition in a form valid for flows around bodies of finite thickness. The method of Davis and Werle (1982) was used as the basic method. Acceleration of the iterative process is achieved through a new definition for the value of the displacement thickness δ. Let δ^n be the value of the displacement thickness obtained as a result of the nth iteration. If the iterative process had a geometric convergence rate, the limiting result could be expressed as

$$\delta = \frac{(\delta^{n-1})^2 - \delta^n \delta^{n-2}}{2\delta^{n-1} - \delta^n - \delta^{n-2}} .$$

This value is chosen to be the new approximation to δ. As a result of such additional extrapolation, the number of iterations needed is decreased by almost a factor of three. About 40 iterations are required to obtain convergence with accuracy up to 10^{-5}.

The calculation of three-dimensional flows in the region of interaction of a boundary layer with the outer flow is a much more difficult problem. Investigations of this type are still under development, and therefore only two studies are cited here. The first, carried out by Bodonyi and Duck (1988), is devoted to the application of the quasisimultaneous method to the calculation of three-dimensional separation in the boundary layer. Karas and Kovalev (1989), the authors of the second paper, used the semi-inverse method for this purpose.

Spectral method. Quite different principles for organizing the iteration process underlie the spectral method. Burggraf and Duck (1982) tried to realize those advantages that follow from the application of operational methods in the asymptotic interaction theory. One of the most fruitful directions of the theory is known to be associated with these methods. The first to have applied the Fourier transformation to solve the interaction problem was Stewartson (1970a). He constructed an analytical solution of the linearized interaction problem for incompressible flow past a corner point on a body contour. The analysis of the flow is presented in Chapter 2, Section 3, with the system of equations (2.3.31) describing the corresponding interaction problem. This coincides with the problem (7.1.1)–(7.1.4), (7.1.6) if the function $Y_0(x)$ in (7.1.6) is taken to be

$$Y_0 = \begin{cases} 0 & \text{if } x < 0, \\ \alpha x & \text{if } x > 0. \end{cases}$$

Recall that α is the flow similarity parameter. Since this is multiplied by $\mathrm{Re}^{-1/4}a_0^{1/2}$, where $a_0 = 0.33206$ is the dimensionless skin friction in the Blasius solution, it follows that the tangent to the body contour is turned through the angle

$$\theta = \mathrm{Re}^{-1/4}a_0^{1/2}\alpha$$

in passing the corner point. If the solution to (7.1.1)–(7.1.4), (7.1.6) is represented in the form

$$u = y + u', \quad v = v', \quad p = p', \quad \delta = \delta',$$

then the interaction problem becomes

$$\frac{\partial^2 u'}{\partial y^2} - y\frac{\partial u'}{\partial x} - v' - \frac{dp'}{dx} = R; \quad \frac{\partial u'}{\partial x} + \frac{\partial v'}{\partial y} = 0,$$

$$p' = \frac{1}{\pi}\int_{-\infty}^{\infty}\frac{dY_0/ds + d\delta'/ds}{s - x}ds,$$

$$u' = v' = 0 \quad \text{at } y = 0, \tag{7.3.8}$$

$$u' \to 0 \quad \text{as } x \to \pm\infty,$$

$$u' \to -\delta' \quad \text{as } y \to +\infty.$$

The function R on the right-hand side of the momentum equation is

$$R = u'\frac{\partial u'}{\partial x} + v'\frac{\partial u'}{\partial y}.$$

If $\theta = 0$, the flow in the interaction region remains undisturbed:

$$u = y, \quad v = p = \delta = 0.$$

However, if θ differs from zero, but is small compared with unity, then all functions designated by primes are proportional to θ, and R is a quantity of order θ^2. On these grounds, Stewartson omitted the terms on the right-hand side of the momentum equation and, noting that the coefficients in the resulting linear equation did not depend on x, applied a Fourier transformation to the interaction problem. Instead of the functions u', v', p', and δ' he introduced their Fourier transforms $\overline{u}$, $\overline{v}$, $\overline{p}$, and $\overline{\delta}$ in accordance with the rule

$$\overline{u} = \int_{-\infty}^{\infty} u'(x,y)e^{-i\kappa x}dx.$$

As a result, the equations and boundary conditions for the viscous wall layer are transformed to the form

$$\frac{d^2\overline{u}}{dy^2} - i\kappa y\overline{u} - \overline{v} - i\kappa\overline{p} = 0,$$

$$\frac{d\overline{v}}{dy} + i\kappa\overline{u} = 0, \quad \overline{u} = \overline{v} = 0 \quad \text{at } y = 0, \quad \overline{u} = -\overline{\delta} \quad \text{at } y = \infty,$$

(7.3.9)

and the integral of thin-airfoil theory can be replaced by the simple algebraic relation

$$\overline{p} = -|\kappa|(\overline{\delta} + \overline{Y}_0).$$

After differentiating the first equation in (7.3.9) with respect to y, and eliminating $\overline{v}$ by use of the second equation, we obtain for the derivative $d\overline{u}/dy$ the equation

$$\frac{d^2}{dy^2}\left(\frac{d\overline{u}}{dy}\right) - i\kappa y\frac{d\overline{u}}{dy} = 0,$$

which is reduced to the Airy equation by the substitution of a new independent variable $z = (i\kappa)^{1/3}y$. The Airy equation is known to have two linearly independent solutions. One of them, the Airy function $\mathrm{Ai}(z)$, tends to zero as $z \to \infty$, and the second one, $\mathrm{Bi}(z)$, increases exponentially. Therefore, taking into account the boundedness of $\overline{u}$ as $y \to \infty$, we should take

$$\frac{d\overline{u}}{dz} = c\mathrm{Ai}(z).$$

(7.3.10)

The coefficient c in this formula can be determined by considering the first equation of the system (7.3.9) at $y = 0$,

$$c\mathrm{Ai}'(0) = (i\kappa)^{1/3}\overline{p} = -(i\kappa)^{1/3}|\kappa|(\overline{\delta} + \overline{Y}_0).$$

(7.3.11)

It remains to integrate (7.3.10), taking into account the no-slip condition at the body surface

$$\overline{u} = -\frac{(i\kappa)^{1/3}|\kappa|(\overline{\delta} + \overline{Y}_0)}{\mathrm{Ai}'(0)}\int_0^z \mathrm{Ai}(\xi)d\xi$$

(7.3.12)

and taking the limit as $z \to \infty$ to satisfy the condition at the outer edge of the wall layer. Since the integral of the Airy function over the interval $[0, \infty)$ is equal to $1/3$, we obtain the result

$$\bar{\delta} = -\frac{(i\kappa)^{1/3}|\kappa|\overline{Y}_0}{(i\kappa)^{1/3}|\kappa| - 3\text{Ai}'(0)}.$$

This formula, together with (7.3.11) and (7.3.12), provides the solution of the interaction problem in the Fourier-transform space.

Burggraf and Duck (1982) generalized the above procedure to the case of nonlinear disturbances, when the function R on the right-hand side of the momentum equation in (7.3.8) is not small in comparison with the other terms of this equation. They proposed seeking the solution of the interaction problem with the help of the method of successive approximations, for which the momentum equation is represented as

$$\left(\frac{\partial^2 u'}{\partial y^2} - y\frac{\partial u'}{\partial x} - v' - \frac{dp'}{dx}\right)^{n+1} = R^n, \qquad (7.3.13)$$

and all the remaining relations entering the formulation of the interaction problem (7.3.8) are considered at the new iteration level.

Differentiating equation (7.3.13) with respect to y and applying a Fourier transformation with respect to x, we obtain the following equation for the Fourier transform of the vorticity $\omega' = \partial u'/\partial y$:

$$\frac{d^2\overline{\omega}^{n+1}}{dy^2} - i\kappa y\overline{\omega}^{n+1} = \overline{\left(u'\frac{\partial\omega'}{\partial x} + v'\frac{\partial\omega'}{\partial y}\right)}^n. \qquad (7.3.14)$$

The solution satisfying the condition $\overline{\omega}^{n+1} \to 0$ as $y \to \infty$ can be represented (after substitution of the independent variable $z = (i\kappa)^{1/3}y$) in the form

$$\overline{\omega}^{n+1}(z) = c\text{Ai}(z) + \text{Ai}(z)\int_0^z \frac{d\xi}{\text{Ai}^2(\xi)}\int_\infty^\xi g(s)\text{Ai}(s)ds,$$

where

$$g = (i\kappa)^{-2/3}\overline{\left(u'\frac{\partial\omega'}{\partial x} + v'\frac{\partial\omega'}{\partial y}\right)}^n$$

and then the problem is reduced, as before, to the determination of the coefficient c multiplying the Airy function. However, in this case it is more convenient to represent the second derivative on the left-hand side of equation (7.3.14) by central differences and to use for its solution the Thomas technique, which allows us to express the Fourier transform of the vorticity $\overline{\omega}_k^{n+1}$ at interior grid points y_k in terms of its value $\overline{\omega}_1^{n+1}$ at the body surface

$$\overline{\omega}_k^{n+1} = R_k \overline{\omega}_{k-1}^{n+1} + Q_k, \quad k = 2, \ldots, M. \tag{7.3.15}$$

To determine $\overline{\omega}_1^{n+1}$, it is necessary to carry out the same operations as for the solution of the linear problem, that is, to consider equation (7.3.13) at $y = 0$

$$\left. \frac{d\overline{\omega}^{n+1}}{dy} \right|_{y=0} = \frac{(R_2 - 1)\overline{\omega}_1^{n+1} + Q_2}{\Delta y} = i\kappa \overline{p}^{n+1} \tag{7.3.16}$$

and after constructing a formula analogous to (7.2.13),

$$\overline{u}_k^{n+1} = F\overline{\omega}_1^{n+1} + G,$$

to satisfy the condition at the outer edge of the viscous wall layer,

$$F\overline{\omega}_1^{n+1} + G = -\overline{\delta}^{n+1} = \overline{p}^{n+1}/|\kappa| + \overline{Y}_0. \tag{7.3.17}$$

It remains to solve (7.3.16) and (7.3.17) for $\overline{\omega}_1^{n+1}$ and, using the recurrence formula (7.3.15), to find the distribution of $\overline{\omega}^{n+1}$ across the entire viscous layer. To accelerate the computation process, Burggraf and Duck (1982) applied the algorithm for the fast Fourier transform both for the calculation of the right-hand side of equation (7.3.14) and for the inversion of $\overline{\omega}_k^{n+1}$ at all values of y_k.

In the work of Burggraf and Duck (1982) a spectral method was used for the analysis of boundary-layer separation in the vicinity of an irregularity on the surface of a flat plate – a hump or a cavity. Later, this method was generalized to the case of three-dimensional flow in the interaction region (Duck and Burggraf, 1986). The simplicity in accomplishing the generalization gives the spectral method an advantage over other methods of solving the interaction problem. Duck used this method to analyze unsteady flows within the interaction region. In his 1985 work he applied it in studying wave-packet formation in a boundary layer with a sudden deformation of the body contour, and in his 1988 work he investigated the scattering of acoustic waves by a local irregularity of the surface.

7.4 Direct Methods of Solving Interaction Problems

One may become familiar with the general methodology of constructing direct numerical schemes for solving boundary-value problems with the help of the book by Samarskii and Nikolaev (1978). Here, we repeat briefly the basic points of the approach they proposed, in order to explain in more detail how that approach may be applied to the solution of interaction problems.

For simplicity, suppose that we are concerned with a certain boundary-value problem formulated in terms of a scalar function ψ in the plane of the variables x and y. As usual, the problem consists of a partial differential equation for ψ and boundary conditions prescribed for ψ at the boundary of the region of integration. The region in which the function $\psi(x, y)$ is sought can be covered with a grid of dimensions $N \times M$. Then the finite-difference approximations of the equations for ψ at interior points of the computational grid and of the appropriate conditions at its boundary permit the formulation of a system of $N \times M$ algebraic equations for the values of the unknown function ψ at the grid nodes. This system can be written in the functional form

$$\overrightarrow{B}(\overrightarrow{\psi}) = \overrightarrow{f}, \tag{7.4.1}$$

where

$$\overrightarrow{\psi} = \begin{bmatrix} \psi_1 \\ \vdots \\ \psi_{N \times M} \end{bmatrix}$$

is a vector formed from the unknowns and $\overrightarrow{B}$ is a nonlinear operator acting on $\overrightarrow{\psi}$. Here, $\overrightarrow{B}$ represents a vector function

$$\begin{bmatrix} B_1 \\ \vdots \\ B_{N \times M} \end{bmatrix}$$

of dimensions $N \times M$ in which the elements B_i depend upon $\psi_1, \ldots,$ $\psi_{N \times M}$. The vector $\overrightarrow{f}$ on the right-hand side of equation (7.4.1) is independent of $\overrightarrow{\psi}$.

One of the best-known and most rapidly converging iterative methods for solving equations of the form (7.4.1) is the Newton–Kantorovich method. It makes use of an approximate $\overrightarrow{\psi}^{\,n}$ obtained from a previous iteration to obtain a better approximation $\overrightarrow{\psi}^{\,n+1}$ to the solution of equation (7.4.1):

$$\overrightarrow{B}\left(\overrightarrow{\psi}^{\,n+1}\right) = \overrightarrow{f}. \tag{7.4.2}$$

With accuracy to second order in the small quantities, we expand equation (7.4.2) with respect to $\Delta \vec{\psi} = \vec{\psi}^{\,n+1} - \vec{\psi}^{\,n}$ according to Taylor's theorem. This yields

$$\vec{B}\left(\vec{\psi}^{\,n} + \Delta\vec{\psi}\right) = \vec{B}\left(\vec{\psi}^{\,n}\right) + B'\left(\vec{\psi}^{\,n}\right)\Delta\vec{\psi},$$

that is

$$B'\left(\vec{\psi}^{\,n}\right)\left(\vec{\psi}^{\,n+1} - \vec{\psi}^{\,n}\right) + \vec{B}\left(\vec{\psi}^{\,n}\right) = \vec{f}.$$

The derivative $B'\left(\vec{\psi}^{\,n}\right)$ is given by a Gateaux matrix. Its elements are calculated by the formula

$$B'_{j,k} = \frac{\partial B_j}{\partial \psi_k}, \quad j = 1,\ldots,N; \quad k = 1,\ldots,M.$$

An arbitrary two-stage iterative process for solving problem (7.4.1) may be written in a general form as

$$C^n\left(\vec{\psi}^{\,n}\right)\frac{\vec{\psi}^{\,n+1} - \vec{\psi}^{\,n}}{\tau_n} + \vec{B}\left(\vec{\psi}^{\,n}\right) = \vec{f}.$$

It is known from the theory of iterative methods (Samarskii and Nikolaev, 1978) that if the operator C^n is the Gateaux derivative of the operator $\vec{B}$ at the point $\vec{\psi}^{\,n}$,

$$C^n = B'\left(\vec{\psi}^{\,n}\right),$$

then under certain additional assumptions on the smoothness of the operator $B'\left(\vec{\psi}\right)$, the iteration process

$$B'\left(\vec{\psi}^{\,n}\right)\frac{\vec{\psi}^{\,n+1} - \vec{\psi}^{\,n}}{\tau_n} + \vec{B}\left(\vec{\psi}^{\,n}\right) = \vec{f}$$

has a convergence rate faster than that of a geometric progression, which is typical of the methods of solving interaction problems described in previous paragraphs. With $\tau_n = 1$ the iteration process has a quadratic convergence rate and is called the Newton-Kantorovich method.

A standard inversion of the matrix B' requires $O(M^3 N^3)$ arithmetic operations. However, thanks to the special form of the interaction-theory equations, a considerable number of elements in the matrix B' are zero.

Therefore, to invert the matrix B' a more economical approach is applied, which allows the number of arithmetic operations to be reduced to a quantity of order $N \times (M^3 + N^3)$.

The main ideas of this method, developed by Korolev, will be illustrated using as an example one of the most difficult problems of interaction theory, the problem of incompressible flow separation near the trailing edge of a plate at small angle of attack (see Chapter 3, Section 3).

The problem for the region of interaction at the trailing edge of a plate consists in solving the boundary-layer equations

$$u\frac{\partial u}{\partial x} + v\frac{\partial u}{\partial y} = -\frac{dp}{dx} + \frac{\partial^2 u}{\partial y^2}, \quad \frac{\partial u}{\partial x} + \frac{\partial v}{\partial y} = 0. \qquad (7.4.3)$$

The surface of the plate can sustain a pressure difference; we therefore introduce two functions for the pressure

$$p(x) = p_+(x), \quad y \geq 0,$$
$$p(x) = p_-(x), \quad y \leq 0.$$

Both above $(y > 0)$ and below $(y < 0)$ the plate, on the surface $y = 0$, it is necessary to impose the no-slip conditions

$$u(x, \pm 0) = v(x, \pm 0) = 0 \quad \text{for } x \leq 0. \qquad (7.4.4)$$

Here and in the following, the upper sign in the notation $\pm$ corresponds to values in the upper half-plane and the lower sign to the lower half-plane, while $x = 0$ corresponds to the position of the trailing edge of the plate.

The no-slip conditions no longer apply downstream from the trailing edge, and the first of them, $u(x, 0) = 0$, is replaced by a condition requiring smoothness of the solution. That is, equations (7.4.3) for $x > 0$ should be satisfied not only for $|y| > 0$ but also at $y = 0$. As for the function $v(x, 0)$, this determines the location of the viscous wake behind the plate relative to the x-axis. If the wake were located higher than it should be, the pressure at the upper boundary of the wake would be greater than at the lower boundary. If the wake location were lower than necessary, then the opposite would be true. Therefore, downstream of the trailing edge $v(x, 0)$ is to be found from the condition of equal pressures in the upper and lower half-planes:

$$p_+(x) = p_-(x), \quad \text{for } x \geq 0. \qquad (7.4.5)$$

The conditions of matching with the undisturbed flow ahead of the interaction region have the form

$$u = y + \alpha|x|^{1/6} f_1(\eta) + \alpha^2 f_2(\eta) + \cdots, \quad \eta = y|x|^{-1/3}, \tag{7.4.6}$$

if one considers the flow above the plate, and

$$u = -y - \alpha|x|^{1/6} f_1(|\eta|) + \alpha^2 f_2(|\eta|) + \cdots, \quad \eta = y|x|^{-1/3} \tag{7.4.7}$$

for the flow below it. At large values of $|y|$, the solution of equations (7.4.3) can be written as

$$u = y + \alpha c_1 y^{1/2} + \alpha^2 c_2 \ln y - \delta_+(x) + \cdots \text{ as } y \to \infty, \tag{7.4.8}$$
$$u = -y - \alpha c_1 |y|^{1/2} + \alpha^2 c_2 \ln |y| - \delta_-(x) + \cdots \text{ as } y \to -\infty. \tag{7.4.9}$$

Here, α is a parameter of the problem related to the angle of attack and c_1, c_2, f_1, and f_2 are known constants and functions that are found from matching with the solution for the region lying ahead of the interaction region (Chow and Melnik, 1976). The functions δ_+ and δ_- define the variation of displacement thickness of the viscous sublayer. They are related to the pressure through the interaction condition:

$$\lim_{y \to \infty} \frac{\partial u}{\partial x} = -\delta'_+ = \frac{1}{\pi} \int_{-\infty}^{\infty} \frac{p_+(t) + \alpha|t|^{1/2} H(-t)}{t - x} dt - \alpha x^{1/2} H(x), \tag{7.4.10}$$

$$\lim_{y \to -\infty} \frac{\partial u}{\partial x} = -\delta'_- = \frac{1}{\pi} \int_{-\infty}^{\infty} \frac{p_-(t) - \alpha|t|^{1/2} H(-t)}{t - x} dt + \alpha x^{1/2} H(x), \tag{7.4.11}$$

where $H(x)$ is the Heaviside function. The conditions of matching the pressure upstream and downstream have the following forms (Chow and Melnik, 1976):

$$p_+(x) = -\alpha \left(|x|^{1/2} - \gamma_0 |x|^{-1/2} \right)$$
$$- 2P_a \left(|x|^{-2/3} + 2\gamma_1 |x|^{-5/3} \right) + \cdots, \tag{7.4.12}$$
$$p_-(x) = \alpha \left(|x|^{1/2} - \gamma_0 |x|^{-1/2} \right)$$
$$- 2P_a \left(|x|^{-2/3} + 2\gamma_1 |x|^{-5/3} \right) + \cdots \tag{7.4.13}$$

as $x \to -\infty$, and

$$p_+(x) = p_-(x) = P_a(x^{-2/3} + 2\gamma_1 x^{-5/3}) + \cdots \tag{7.4.14}$$

as $x \to \infty$, with

$$P_a = 0.1717$$

and where γ_0 and γ_1 are constants characterizing the global effect of the displacement thickness of the viscous sublayer in the interaction region, which must be determined from the solution of the problem.

Before undertaking the numerical solution of the problem posed, we apply Prandtl's transposition transformation:

$$z = y, \quad \bar{v} = v(x, y), \quad x \le 0,$$
$$z = y - \theta(x), \quad \bar{v} = v - \theta'u, \quad x > 0,$$

where

$$\theta(x) = \tfrac{1}{2}\left[\delta_+(x) - \delta_-(x) - \delta_+(0) + \delta_-(0)\right]$$

as proposed in the work of Chow and Melnik (1976). In this case, the zero streamline approaches a straight line $z = $ const. asymptotically as $x \to \infty$. The form of the system of equations (7.4.3)–(7.4.7) is unchanged after the transformation, except that in place of the coordinate y one should formally substitute z, and then replace v by $\bar{v} = v - \theta'u$. The interaction conditions (7.4.10), (7.4.11) for $x \le 0$ remain as before, y simply being replaced by z here. For $x > 0$, from (7.4.8) and (7.4.9), changing to the variables x and z, we obtain

$$u(x, z) = z + \alpha c_1 z^{1/2} + \alpha^2 c_2 \ln z - \tfrac{1}{2}[\delta_+(x) + \delta_-(x) + \delta_+(0) - \delta_-(0)] + \cdots \tag{7.4.15}$$

as $z \to \infty$, and

$$u(x, z) = -z - \alpha c_1 |z|^{1/2} + \alpha^2 c_2 \ln|z| - \tfrac{1}{2}[\delta_+(x) + \delta_-(x) - \delta_+(0) + \delta_-(0)] + \cdots \tag{7.4.16}$$

as $z \to -\infty$. Differentiating these expressions with respect to x and using the expressions for δ_+ and δ_- from (7.4.10) and (7.4.11), we obtain the following formulas:

$$\lim_{z \to \infty} \frac{\partial u}{\partial x} = \lim_{z \to -\infty} \frac{\partial u}{\partial x} = \frac{1}{\pi} \int_{-\infty}^{\infty} \frac{p_+(t) + p_-(t)}{2(t - x)} dt, \quad x > 0, \tag{7.4.17}$$

which, together with (7.4.10)–(7.4.11), we will apply later to the numerical solution of the interaction problem.

We introduce a uniform grid (x_j, z_k). The maximum values of j and k are N and M. The values $u(x_j, z_k)$ of the longitudinal velocity component at the nodes of the computational grid are denoted by $u_{j,k}$. The trailing edge of the plate corresponds to $j = NE$, $k = ME$, so that here $x_{NE} = 0$, $z_{ME} = 0$.

Assuming that the left boundary of the computational grid is sufficiently far from the trailing edge of the plate, we write the boundary conditions (7.4.6)–(7.4.7) in the form

$$L_{1,k} = u_{1,k} - |z_k| \mp \alpha |x_1|^{1/6} f_1(\eta_k) - \alpha^2 f_2(\eta_k) = 0,$$
$$\eta_k = |z_k - z_{ME}||x_1|^{-1/3}, \quad k = 1, \dots, M. \tag{7.4.18}$$

If we introduce vectors $\overrightarrow{L}_1$ and $\overrightarrow{U}_1$ of dimension M,

$$\overrightarrow{L}_1 = \begin{bmatrix} L_{1,1} \\ \vdots \\ L_{1,k} \\ \vdots \\ L_{1,M} \end{bmatrix}, \quad \overrightarrow{U}_1 = \begin{bmatrix} u_{1,1} \\ \vdots \\ u_{1,k} \\ \vdots \\ u_{1,M} \end{bmatrix},$$

then the boundary condition (7.4.18) can be written in the more compact form

$$\overrightarrow{L}_1 \left(\overrightarrow{U}_1 \right) = 0. \tag{7.4.19}$$

Further, for all $j = 2, \dots, N$ we use the conditions (7.4.8)–(7.4.9) and (7.4.15)–(7.4.16) at the upper and lower boundaries of the computational grid:

$$L_{j,1} = \frac{4u_{j,2} - u_{j,3} - 3u_{j,1}}{2\Delta z} + 1 - \frac{\alpha c_1}{2}|z_1|^{-1/2} - \alpha^2 c_2 z_1^{-1} = 0, \tag{7.4.20}$$

$$L_{j,M} = \frac{3u_{j,M} - 4u_{j,M-1} + u_{j,M-2}}{2\Delta z} - 1 - \frac{\alpha c_1}{2} z_M^{-1/2} - \alpha^2 c_2 z_M^{-1} = 0. \tag{7.4.21}$$

We will approximate the first of equations (7.4.3) by a second-order finite-difference scheme, namely the Crank–Nicolson scheme, in the region of forward flow ($u_{j,k} \geq 0$) and by a scheme that is of first order with respect to Δx, but remains of second order with respect to Δz, in the region of reverse flow ($u_{j,k} < 0$). The choice of such an approximation scheme is partly justified by the fact that the values of the longitudinal velocity component are usually quite small in the reverse-flow region. Thus, the approximation error that results from representing the term $u \partial u / \partial x$ to first-order accuracy becomes insignificant. We write the finite-difference analog of the momentum equation (7.4.3) as

$$L_{j,k} = \left| \frac{u_{j,k} + u_{j-1,k}}{2} \right| \frac{u_{j,k} - u_{j-s,k}}{\Delta x} + \frac{\overline{v}_{j-0.5,k}}{2} (\lambda_z u_{j,k} + \lambda_z u_{j-1,k})$$
$$+ \frac{p_j - p_{j-1}}{\Delta x} - \frac{1}{2} (\lambda_{zz} u_{j,k} + \lambda_{zz} u_{j-1,k}) = 0. \tag{7.4.22}$$

This equation is valid for $j = 2, \ldots, N$. If $j \leq NE$, then $k = 2, \ldots,$ $ME - 1$, $ME + 1, \ldots, M - 1$. If $j > NE$, then $k = 2, \ldots, M - 1$.

Here, λ_z and λ_{zz} denote operators defined by:

$$\lambda_z u_{j,k} = \frac{u_{j,k+1} - u_{j,k-1}}{2\Delta z}, \quad \lambda_{zz} u_{j,k} = \frac{u_{j,k+1} - 2u_{j,k} + u_{j,k-1}}{\Delta z^2}.$$

Here also, as in the methods described previously, left-hand or right-hand differences are used depending upon the sign of the longitudinal velocity component u. A change in the method of approximating the term $u\, \partial u / \partial x$ in the equation is realized as a result of a change in sign of the index s:

$$s = \text{sign}(u_{j,k} + u_{j-1,k}).$$

To determine the vertical velocity component $\overline{v}_{j-0.5,k}$, the continuity equation is utilized in the form

$$\overline{v}_{j-0.5,k} = \overline{v}_{j-0.5,ME} - \sum_{l=ME+1}^{k-1} (u_{j,l} - u_{j-1,l}) \frac{\Delta z}{\Delta x}$$

$$- (u_{j,ME} - u_{j-1,ME}) \frac{\Delta z}{2\Delta x} - (u_{j,k} - u_{j-1,k}) \frac{\Delta z}{2\Delta x} \quad \text{for } k > ME,$$

$$\overline{v}_{j-0.5,k} = \overline{v}_{j-0.5,ME} + \sum_{l=k+1}^{ME-1} (u_{j,l} - u_{j-1,l}) \frac{\Delta z}{\Delta x}$$

$$+ (u_{j,ME} - u_{j-1,ME}) \frac{\Delta z}{2\Delta x} + (u_{j,k} - u_{j-1,k}) \frac{\Delta z}{2\Delta x} \quad \text{for } k < ME.$$

We note that $\overline{v}_{j-0.5,ME} = 0$ for all $j \leq NE$ and is the unknown function to be determined for $j > NE$. At the plate we employ the no-slip condition

$$L_{j,ME} = u_{j,ME} = 0, \quad 2 \leq j \leq NE.$$

In the flow behind the plate, for $k = ME$ and $j \geq NE + 1, \ldots, N$, we will require that equation (7.4.22) be satisfied.

We introduce the unknown vector functions $\overrightarrow{U}_j$ and $\overrightarrow{P}$ of corresponding dimensions M and $3N - NE + 2$:

$$
\vec{U}_j =
\begin{bmatrix}
u_{j,1} \\
\vdots \\
u_{j,k} \\
\vdots \\
u_{j,M}
\end{bmatrix},
\qquad
\vec{P} =
\begin{bmatrix}
p_{+1} \\
\vdots \\
p_{+N} \\
p_{-1} \\
\vdots \\
p_{-N} \\
\overline{v}_{NE+0.5,ME} \\
\vdots \\
\overline{v}_{N-0.5,ME} \\
\gamma_0 \\
\gamma_1
\end{bmatrix}.
$$

Then for each j the system of equations (7.4.20)–(7.4.22) may be written in the form

$$
\vec{L}_j = \vec{L}_j\left(\vec{U}_{j-1},\ \vec{U}_j,\ \vec{U}_{j+1},\ \vec{P} \right) = 0,
$$

where the first and last elements, $L_{j,1}$ and $L_{j,M}$, of the vector $\vec{L}_j$ coincide with the left-hand sides of equations (7.4.21) and (7.4.20) respectively, while at all intermediate points $k = 2, \ldots, M - 1$ on the jth vertical line $L_{j,k}$ is given by the left-hand side of equation (7.4.22), with the vertical velocity component $\overline{v}_{j-0.5,k}$ having already been expressed in terms of $\vec{U}_{j-1}$, $\vec{U}_j$, and the unknown function $\overline{v}_{j-0.5,ME}$.

We now turn to the interaction condition. The region of integration in equations (7.4.10) and (7.4.11), which are used for $x \leq 0$, and in equation (7.4.17), which is used for $x > 0$, is divided into three intervals, from $-\infty$ to x_1, from x_1 to x_N, and from x_N to ∞. Outside the interval (x_1, x_N) the integrals are calculated with the use of the leading terms in the asymptotic expansions for the pressure, (7.4.12) and (7.4.13) for $x < x_1$, and (7.4.14) for $x > x_N$, and inside the interval (x_1, x_N) they are approximated by finite sums, with the singularity at the point $x_{j-0.5}$ treated separately:

$$
\int_{x_1}^{x_N} \frac{P(t)}{t - x_{j-0.5}}\,dt = \sum_{k=2}^{j-1} \int_{x_{k-1}}^{x_k} \frac{P(t)}{t - x_{j-0.5}}\,dt + \int_{x_{j-1}}^{x_j} \frac{P(t)}{t - x_{j-0.5}}\,dt
$$

$$
+ \sum_{k=j+1}^{N} \int_{x_{k-1}}^{x_k} \frac{P(t)}{t - x_{j-0.5}}\,dt.
$$

Outside the interval (x_{j-1}, x_j) the integrals are estimated in the following way:

$$\int_{x_{k-1}}^{x_k} \frac{P(t)}{t - x_{j-0.5}} dt$$

$$= \int_{x_{k-1}}^{x_k} \frac{\frac{P_k + P_{k-1}}{2} + \frac{P_k - P_{k-1}}{\Delta x}\left(t - \frac{x_k + x_{k-1}}{2}\right)}{t - x_{j-0.5}} dt + O(\Delta x^2)$$

$$= \left[\frac{P_k + P_{k-1}}{2} + (P_k - P_{k-1})\left(x_{j-0.5} - \frac{x_k + x_{k-1}}{2}\right)\right]$$

$$\times \ln\left|\frac{x_k - x_{j-0.5}}{x_{k-1} - x_{j-0.5}}\right| + P_k - P_{k-1} + O(\Delta x^2).$$

Within the interval (x_{j-1}, x_j) we obtain, with second-order accuracy,

$$\int_{x_{j-1}}^{x_j} \frac{P(t)}{t - x_{j-0.5}} = P_j - P_{j-1} + O(\Delta x^2).$$

Then the approximation of the interaction equations takes the following form:

$$0 = Q_j = \frac{u_{j,1} - u_{j-1,1}}{\Delta x}$$

$$- \frac{s_1}{\pi}\left[\left(\sum_{k=2}^{j-1} + \sum_{k=j+1}^{N}\right)(p_{(-)k}d_k + p_{(-)k-1}e_k) + p_{(-)j} - p_{(-)j-1}\right]$$

$$- \frac{r_1}{\pi}\left[\left(\sum_{k=2}^{j-1} + \sum_{k=j+1}^{N}\right)(p_{(+)k}d_k + p_{(+)k-1}e_k) + p_{(+)j} - p_{(+)j-1}\right]$$

$$- G_j(x_j, \alpha, \gamma_0, \gamma_1),$$

$$0 = Q_{j+N} = \frac{u_{j,M} - u_{j-1,M}}{\Delta x} -$$

$$- \frac{s_2}{\pi}\left[\left(\sum_{k=2}^{j-1} + \sum_{k=j+1}^{N}\right)(p_{(-)k}d_k + p_{(-)k-1}e_k) + p_{(-)j} - p_{(-)j-1}\right]$$

$$- \frac{r_2}{\pi}\left[\left(\sum_{k=2}^{j-1} + \sum_{k=j+1}^{N}\right)(p_{(+)k}d_k + p_{(+)k-1}e_k) + p_{(+)j} - p_{(+)j-1}\right]$$

$$- G_{j+N}(x_j, \alpha, \gamma_0, \gamma_1), \quad j = 2, \ldots, N.$$

Here, $s_1 = r_2 = 1$, $r_1 = s_2 = 0$ if $j \leq NE$ and $s_1 = r_1 = s_2 = r_2 = \frac{1}{2}$ if

$j > NE$; d_k and e_k denote coefficients appearing in the approximations of the integrals by finite sums within the interval (x_1, x_N):

$$d_k = \frac{C}{2} + \frac{A+B}{\Delta x}, \quad e_k = \frac{C}{2} - \frac{A+B}{\Delta x},$$

$$A = \left(x_{j-0.5} - \frac{x_k + x_{k-1}}{2}\right) \cdot C,$$

$$B = x_k - x_{k-1}, \quad C = \ln \left| \frac{x_k - x_{j-0.5}}{x_{k-1} - x_{j-0.5}} \right|;$$

and G_j, G_{j+N} take into account the calculations of the integrals outside the interval (x_1, x_N). The boundary conditions (7.4.12)–(7.4.14) and the condition of equal pressures p_+ and p_- behind the trailing edge are written in the form

$$Q_1 = p_{(+)1} + \alpha \left(|x_1|^{1/2} - \gamma_0|x_1|^{-1/2}\right)$$
$$+ 2P_a \left(|x_1|^{-2/3} + 2\gamma_1|x_1|^{-5/3}\right) = 0,$$

$$Q_{N+1} = p_{(-)1} - \alpha \left(|x_1|^{1/2} - \gamma_0|x_1|^{-1/2}\right)$$
$$+ 2P_a \left(|x_1|^{-2/3} + 2\gamma_1|x_1|^{-5/3}\right) = 0,$$

$$Q_{2N+l} = p_{(+)NE+l-1} - p_{(-)NE+l-1} = 0, \quad l = 1, \ldots, N - NE + 1,$$

$$Q_{3N-NE+2} = p_{(+)N} - P_a(x_N^{-2/3} + 2\gamma_1 x_N^{-5/3}) = 0,$$

or, by introducing the vector $\vec{Q}$ of dimension $NP = 3N - NE + 2$, we arrive at the vector equation

$$\vec{Q} = \vec{Q}\left(\vec{U}_1, \ldots, \vec{U}_N, \vec{P}\right) = 0, \quad \vec{Q} = \begin{bmatrix} Q_1 \\ \vdots \\ Q_{NP} \end{bmatrix}.$$

Thus, adding everything together, we obtain the following system of nonlinear algebraic equations, written in vector form as

$$\vec{L}_1\left(\vec{U}_1\right) = 0, \tag{7.4.23}$$

$$\vec{L}_j\left(\vec{U}_{j-1}, \vec{U}_j, \vec{U}_{j+1}, \vec{P}\right) = 0, \quad j = 2, \ldots, N, \tag{7.4.24}$$

$$\vec{Q} = \vec{Q}\left(\vec{U}_1, \ldots, \vec{U}_N, \vec{P}\right) = 0. \tag{7.4.25}$$

This system of nonlinear vector equations is solved by the Newton–Kantorovich method, with the Thomas matrix technique (Samarskii and Nikolaev, 1978) used for inversion of the corresponding matrix at each iteration.

Suppose that we have a certain approximation for the vectors $\vec{U}_j^n$, $\vec{P}^n$ obtained at the previous nth iteration or taken as the zeroth approximation. In order to find a better approximation $\vec{U}_j^{n+1}$, $\vec{P}^{n+1}$ to these vectors we introduce the corresponding correction vectors $\Delta \vec{U}_j = \vec{U}_j^{n+1} - \vec{U}_j^n$, $\Delta \vec{P} = \vec{P}^{n+1} - \vec{P}^n$ and rewrite the system (7.4.23)–(7.4.24) in the form

$$\frac{\partial L_1}{\partial U_1} \Delta \vec{U}_1 + \vec{L}_1 \left(\vec{U}_1^n \right) = 0, \qquad (7.4.26)$$

$$\frac{\partial L_j}{\partial U_{j+1}} \Delta \vec{U}_{j+1} + \frac{\partial L_j}{\partial U_j} \Delta \vec{U}_j + \frac{\partial L_j}{\partial U_{j-1}} \Delta \vec{U}_{j-1} + \frac{\partial L_j}{\partial P} \Delta \vec{P}$$
$$+ \vec{L}_j \left(\vec{U}_{j-1}^n, \vec{U}_j^n, \vec{U}_{j+1}^n, \vec{P}^n \right) = 0, \quad (7.4.27)$$

where the Jacobian matrices multiplied by the correction vectors are evaluated from the formulas

$$\frac{\partial L_j}{\partial U_{j+1}} = \left\{ \frac{\partial L_{j,k}}{\partial u_{j+1,m}} \right\}, \quad \frac{\partial L_j}{\partial U_j} = \left\{ \frac{\partial L_{j,k}}{\partial u_{j,m}} \right\},$$

$$\frac{\partial L_j}{\partial U_{j-1}} = \left\{ \frac{\partial L_{j,k}}{\partial u_{j-1,m}} \right\}, \quad \frac{\partial L_j}{\partial P} = \left\{ \frac{\partial L_{j,k}}{\partial p_l} \right\}.$$

Here, for each j the indices k and m vary over the range from 1 to M and l from 1 to NP.

We seek the solution of the problem (7.4.26)–(7.4.27) in the form

$$\Delta \vec{U}_j = R_j \, \Delta \vec{U}_{j+1} + W_j \, \Delta \vec{P} + \vec{Z}_j \qquad (7.4.28)$$

where R_j and W_j are matrices of dimensions $M \times M$ and $M \times NP$, respectively, and $\vec{Z}_j$ is a vector of dimension M.

From the formula (7.4.28) and the system of equations (7.4.27) for $2 \leq j \leq N$ one finds (as in the case of the standard Thomas technique) recurrence relations for calculating the matrices R_j and W_j and vectors $\vec{Z}_j$:

$$R_j = -B^{-1} \frac{\partial L_j}{\partial U_{j+1}}, \quad B = \frac{\partial L_j}{\partial U_j} + \frac{\partial L_j}{\partial U_{j-1}} R_{j-1},$$

$$W_j = -B^{-1}\left(\frac{\partial L_j}{\partial P} + \frac{\partial L_j}{\partial U_{j-1}}W_{j-1}\right),$$

$$\vec{Z}_j = -B^{-1}\left(\vec{L}_j^n + \frac{\partial L_j}{\partial U_{j-1}}\vec{Z}_{j-1}\right).$$

From (7.4.28) and (7.4.26) with $j = 1$ one finds the initial values

$$R_1 = 0, \quad W_1 = 0, \quad \vec{Z}_1 = -\left(\frac{\partial L_1}{\partial U_1}\right)^{-1}\vec{L}_1^n,$$

which permit starting the calculation with the recurrence relations (7.4.24).

To find the unknown vector $\Delta\vec{P}$ we make use of the system of equations (7.4.25). From these equations we obtain

$$\sum_{l=1}^{N}\frac{\partial Q}{\partial U_l}\Delta\vec{U}_l + \frac{\partial Q}{\partial P}\Delta\vec{P} + \vec{Q}\left(\vec{U}_1^n,\ldots,\vec{U}_N^n,\vec{P}^n\right) = 0, \quad (7.4.29)$$

where the matrices $\partial Q/\partial U_l$ and $\partial Q/\partial P$ are determined by the formulas

$$\frac{\partial Q}{\partial U_l} = \left\{\frac{\partial Q_s}{\partial u_{l,k}}\right\}, \quad \frac{\partial Q}{\partial P} = \left\{\frac{\partial Q_s}{\partial p_m}\right\},$$

and $s = 1,\ldots,NP; k = 1,\ldots,M; m = 1,\ldots,NP$.

In the system of equations (7.4.29) we reduce the number of unknown vectors by successively applying equation (7.4.28). Suppose that at some jth step the system of equations (7.4.29) is represented in the form

$$\sum_{l=j+1}^{N}\frac{\partial Q}{\partial U_l}\Delta\vec{U}_l + T_j\,\Delta\vec{U}_j + H_j\,\Delta\vec{P} + \vec{S}_j = 0, \quad j = 1,\ldots,N. \quad (7.4.30)$$

In particular, for $j = 1$ the system of equations (7.4.30) coincides with the system (7.4.29) if

$$T_1 = \frac{\partial Q}{\partial U_1}, \quad H_1 = \frac{\partial Q}{\partial P}, \quad \vec{S}_1 = \vec{Q}^n. \quad (7.4.31)$$

Now, by substituting $\Delta\vec{U}_j$ from (7.4.28) into (7.4.30), we find successively, for all $j = 1,\ldots,N-1$, recurrence relations for calculation of the matrices T_{j+1}, H_{j+1} and the vectors $\vec{S}_{j+1}$,

$$T_{j+1} = \frac{\partial Q}{\partial U_j} + T_jR_j, \quad H_{j+1} = H_j + T_jW_j, \quad \vec{S}_{j+1} = \vec{S}_j + T_j\vec{Z}_j.$$

We note that the vector $\vec{U}_N$ at the right boundary of the computational domain is determined only at points where $u_{N,k} \geq 0$. The reverse-flow profile $(u_{N,k} < 0)$, if any, should be given at $j = N$ as a boundary condition. Therefore, the system of equations (7.4.24) at $j = N$ does not contain the vector $\vec{U}_{N+1}$, and hence $\partial L_N / \partial U_{N+1} = 0$. Then, from the recurrence relations we obtain $R_N = 0$, and correspondingly from (7.4.28),

$$\Delta \vec{U}_N = W_N \Delta \vec{P} + \vec{Z}_N . \tag{7.4.32}$$

On the other hand, the system (7.4.30) at $j = N$ gives

$$T_N \Delta \vec{U}_N + H_N \Delta \vec{P} + \vec{S}_N = 0 .$$

Substituting $\Delta \vec{U}_N$ from the expression (7.4.32) into this system, we obtain

$$(T_N W_N + H_N) \Delta \vec{P} + \vec{S}_N + T_N \vec{Z}_N = 0 .$$

Evaluating the unknown vector $\Delta \vec{P}$ from this vector equation, we find the vector $\Delta \vec{U}_N$ from (7.4.32) and, further, all values of $\Delta \vec{U}_j$ according to the recurrence formula (7.4.28).

A new approximation for the vectors $\vec{U}_j^{n+1}$ and $\vec{P}^{n+1}$ is taken in the form

$$\vec{U}_j^{n+1} = \vec{U}_j^n + \Delta \vec{U}_j , \quad \vec{P}^{n+1} = \vec{P}^n + \Delta \vec{P} .$$

Then the iteration process is repeated again. Usually it requires about five or six iterations to obtain convergence of the solution with accuracy of order 10^{-6}.

The numerical solution shows the presence of flow regimes with separation at the trailing edge of the plate. Up to the value of $\alpha = 0.48$ the calculation was carried out in such a way that the parameter α was held fixed throughout the iterative process, and after a solution had been obtained, the value of α was increased. However, at large values of α, deterioration of the convergence of the iterative process and singular behavior of γ_0 from (7.4.12) were observed. In particular, the value of the derivative $d\gamma_0 / d\alpha$ began to grow without bound. In this case a continuation method was used. This permitted determining the existence of a critical value of the parameter α and finding a second branch of the solution.

In using this method one takes as the parameter of the problem not the value of the parameter α but rather the arc length of a curve lying in the plane of the variables (α, γ_0). The arc length S of this curve is related to the value of γ_0, which is determined by the interaction, and to the angle of attack α through the following formula:

$$(dS)^2 = (d\alpha)^2 + (d\gamma_0)^2.$$

The parameter S assumes a fixed value, and the values of γ_0 and α are included among the unknowns. The finite-difference analog of such a procedure is written in the form

$$S_l - S_{l-1} = (\alpha_l - \alpha_{l-1}) \left(\frac{\alpha_{l-1} - \alpha_{l-2}}{S_{l-1} - S_{l-2}} \right) + (\gamma_{0l} - \gamma_{0l-1}) \left(\frac{\gamma_{0l-1} - \gamma_{0l-2}}{S_{l-1} - S_{l-2}} \right).$$

Here, S_l denotes the sequence of values of the arc length at the points where the solution of the problem is determined, and l is the number of the corresponding step along this arc. It is assumed that for all values of S_r, $r = l - 1, l - 2, \ldots$, the corresponding values of $\alpha_r = \alpha(S_r)$ and $\gamma_{0r} = \gamma_0(S_r)$ are already known.

The solution is to be sought for a given value of S_l and unknown α_l and γ_{0l}.

As a result of using such a procedure it was found that the local separation zone formed on the upper surface of a plate can exist only under the condition that the angle of attack does not exceed a certain critical value $\alpha = 0.497$ (see Figure 3.19b, Chapter 3). Investigation of the solution with the help of the continuation method in the vicinity of the critical value of the parameter α permitted the discovery of a turning point and the existence of a second branch of the solution. Thus, two flow configurations around a trailing edge become possible.

Use of the Newton–Kantorovich method for solving the equations of interaction theory proved highly successful. Its further development with regard to problems of interaction theory was carried out in the work of Korolev (1987) for calculation of the flow around a surface with small camber and with subsonic and supersonic external flow. With this method, the equations of interaction theory are solved in terms of the vorticity and stream function (7.2.1).

The system of equations (7.1.1)–(7.1.6), written in terms of the vorticity and with the continuity equation and the no-slip condition taken into account, will appear in the following form:

$$\int_0^y \omega dy_1 \frac{\partial \omega}{\partial x} + \left(\int_0^y y_1 \frac{\partial \omega}{\partial x} dy_1 - y \int_0^y \frac{\partial \omega}{\partial x} dy_1 \right) \frac{\partial \omega}{\partial y} = \frac{\partial^2 \omega}{\partial y^2}, \quad (7.4.33)$$

$$\omega = 1 + \cdots \quad \text{as } x \to -\infty, \quad (7.4.34)$$

$$\omega = 1 + \cdots \quad \text{as } y \to \infty, \quad (7.4.35)$$

$$\left.\frac{\partial w}{\partial y}\right|_{y=0} = \frac{dP}{dx},\tag{7.4.36}$$

$$Y_0'(x) - \int_0^\infty \frac{\partial w}{\partial x}dy = \frac{1}{\pi}\int_{-\infty}^\infty \frac{P(t)}{x-t}dt,\tag{7.4.37}$$

$$Y_0'(x) - \int_0^\infty \frac{\partial w}{\partial x}dy = P(x),\tag{7.4.38}$$

where (7.4.37) represents the interaction condition for the case of subsonic external flow, while (7.4.38) refers to supersonic flow.

For the approximation of equation (7.4.33), two finite-difference schemes have been used: a scheme of Crank–Nicolson type (1947) or a three-point difference scheme with respect to x (Carter, 1974). As in the method mentioned above, the unknown vector functions $\overrightarrow{W}_j$ and $\overrightarrow{P}$ are introduced:

$$\overrightarrow{W}_j = \begin{bmatrix} \omega_{j,1} \\ \vdots \\ \omega_{j,N} \end{bmatrix}, \quad \overrightarrow{P} = \begin{bmatrix} P_1 \\ \vdots \\ P_N \end{bmatrix}.$$

After approximating equations (7.4.33) at the interior nodes of the computational grid, and using the conditions (7.4.34)–(7.4.35) at the boundaries of the grid, one can write the resulting system of nonlinear algebraic equations in vector form:

$$\overrightarrow{L}_1\left(\overrightarrow{W}_1\right) = 0,$$

$$\overrightarrow{L}_2\left(\overrightarrow{W}_2, \overrightarrow{W}_1, \overrightarrow{P}\right) = 0,\tag{7.4.39}$$

$$\overrightarrow{L}_j\left(\overrightarrow{W}_{j-1}\overrightarrow{W}_j, \overrightarrow{W}_{j+1}\overrightarrow{P}\right) = 0, \quad j = 3, \ldots, N;$$

where

$$\overrightarrow{L}_j = \begin{bmatrix} L_{j,1} \\ \vdots \\ L_{j,M} \end{bmatrix},$$

for the Crank–Nicolson scheme or

$$\overrightarrow{L}_j\left(\overrightarrow{W}_{j-2}, \overrightarrow{W}_{j-1}, \overrightarrow{W}_j, \overrightarrow{W}_{j+1}, \overrightarrow{W}_{j+2}, \overrightarrow{P}\right), \quad j = 3, \ldots, N \tag{7.4.40}$$

for the three-point difference scheme. Here, $L_{j,k} = 0$ corresponds to the finite-difference equation written for the kth point on the jth vertical line. The interaction condition after approximation of the equations (7.4.37)–(7.4.38) and the required boundary conditions for the pressure are represented in the general case as

$$\vec{Q} = \vec{Q}\left(\vec{W}_1, \ldots, \vec{W}_N, \vec{P}\right) = 0, \quad \vec{Q} = \begin{bmatrix} Q_1 \\ \vdots \\ Q_N \end{bmatrix},$$

where $Q_j = 0$ is the finite-difference equation used in approximating the interaction condition or the boundary conditions for the pressure.

The system of nonlinear algebraic equations (7.4.39)–(7.4.40) is solved by the Newton–Kantorovich method, with the Thomas matrix technique used for inversion of the Jacobian matrix at each iteration. The algorithm in the method is similar in many ways to that described earlier and therefore is not given here.

The greatest efficiency in solving the problem was displayed by the three-point difference scheme. Although it requires a greater number of arithmetic operations than the Crank–Nicolson scheme, it proved to be less sensitive to the zeroth approximation, and the iteration process has a more rapid rate of convergence. As a rule, five or six iterations were required to obtain convergence with accuracy of order 10^{-6}:

$$\max |\Delta \omega_{j,k}, \Delta P_j| < 10^{-6},$$

while the Crank–Nicolson scheme required one or two additional iterations.

With the aid of this method the analysis was successfully carried out for such complex phenomena as the hysteresis of separated flow and the secondary separation occurring at a corner for small deflection angles (Korolev, 1991a, 1992), and the nonuniqueness of the solution of the equations for a laminar boundary layer was also investigated when flow separation takes place (Korolev, 1991c).

Newton's method was also used in the work of Smith and Khorrami (1991) in an investigation of supersonic flow of a viscous gas past a shallow concave corner. To invert the matrix that results from linearization of the system of equations for relatively small increments in the stream function, an iterative method was used rather than a direct one. This somewhat decreases the rate of convergence of the iteration process. The method helped to reveal some interesting features in the solution of the equations. It turned out that with an increase in the deflection angle of the surface, the growth in the length of the separation zone tends to decrease in comparison with the solution for smaller angles. Near the reattachment point a singularity begins to form in the solution of the equations, which is characterized by an abrupt reduction in the magnitude of the friction and a sharp rise in pressure. The numerical results confirm the assumption of Smith (1988a) about the existence of such a singularity. Here also, as for incompressible flows (Korolev, 1991a), the beginning of secondary flow separation was discovered near the corner point of a surface with a concave angle.

We have considered a variety of numerical methods for solving viscous interaction problems. An investigator who intends to use these methods faces the question of evaluating them with respect to the problems of interest to him. Although most of these methods were developed by the authors for the solution of some specific problems, in general, based on the results obtained, it is possible to note the following.

The most effective methods are those that utilize implicit schemes of approximating the interaction condition, such as the quasisimultaneous, spectral, and direct methods. The use of direct methods is recommended for the analysis of the most complicated phenomena connected with flow separation, such as, for example, hysteresis of the separated flow or secondary separation of the boundary layer. When these methods are used, the problem of convergence of the iterations does not arise or, to be more precise, the problem is not as acute as with other methods.

To calculate flows where a solid surface is one of the boundaries of the computational grid, it is better to use methods for solving the equations that are written in terms of the vorticity and the stream function. With the same organization of the iterative process, they require fewer iterations for convergence than methods of solving the equations written in the variables u and v. It is also better to use these methods for solving problems in the vicinity of an airfoil trailing edge when the flow is symmetric about the edge.

We note, however, that the solution of problems with discontinuous boundary conditions and absence of flow symmetry (for example, flow around the trailing edge of an airfoil at angle of attack) is as a rule carried out in the variables u and v, due to the simpler form of programs set up for solving problems in these variables compared with programs for solving problems in the variables ω and ψ.

It is essential to make one more comment. All of the numerical results described above give convergent solutions for problems with steady boundary conditions only within a definite range of the problem parameters. Outside this range the iteration process does not converge. Most commonly, the main cause lies in the appearance of instability of the difference scheme. Application of methods with an implicit scheme for approximating the interaction condition permits expanding this range in some cases. Another possible reason for the failure of the iteration to converge may be the use of a variable stencil in approximating a derivative with respect to the longitudinal coordinate. The work of Napolitano, Werle, and Davis (1979) pointed out this effect in the reattachment region of a separated flow. Moreover, the possibility is not excluded that a steady solution may exist only for a unique value of the parameter as, for example, in the problem of laminar separation from a smooth surface (Chapter 1, Section 7), or only in a limited interval (see the problem of separation from the leading (Chapter 4, Section 5) and trailing edges (Chapter 3, Section 3) of an airfoil). The latter can be related to the existence of a critical value in the solution of the equations and the presence of a second branch of the solution. In such cases the use of a continuation method is recommended.

In conclusion, we note that up to now the mathematical problems of existence and uniqueness of solutions to the equations considered have not been resolved for the differential equations themselves or for their finite-difference counterparts. Therefore, in numerical analysis it is necessary to rely to a greater extent upon the results of computations rather than on rigorous mathematical theory.

References

Achenbach, E. (1968). Distribution of local pressure and skin friction around a circular cylinder in cross-flow up to Re $= 5 \times 10^6$. *J. Fluid Mech.* 34(4), 625–639.

Ackerberg, R. C. (1970). Boundary-layer separation at a free streamline. Part 1. Two-dimensional flow. *J. Fluid Mech.* 44(2), 211–225.

Ackerberg, R. C. (1971). Boundary-layer separation at a free streamline. Part 2. Numerical results. *J. Fluid Mech.* 46(4), 727–736.

Ackerberg, R. C. (1973). Boundary-layer separation at a free streamline. Part 3. Axisymmetric flow and the flow downstream of separation. *J. Fluid Mech.* 59(4), 645–663.

Ackeret, J., F. Feldmann, and N. Rott. (1946). *Investigations of Compression Shocks and Boundary Layers in Gases Moving at High Speed.* NACA TM 1113.

Acrivos, A., D. D. Snowden, A. S. Grove, and E. E. Petersen. (1965). The steady separated flow past a circular cylinder at large Reynolds numbers. *J. Fluid Mech.* 21(4), 737–760.

Acrivos, A., L. G. Leal, D. D. Snowden, and F. Pan. (1968). Further experiments on steady separated flows past bluff objects. *J. Fluid Mech.* 34(1), 25–48.

Adamson, T. C., Jr., and A. F. Messiter. (1980). Analysis of two-dimensional interactions between shock waves and boundary layers. *Ann. Rev. Fluid Mech.* 12, 103–138.

Barnett, M. (1988). Analysis of crossover between local and massive separation on airfoils. *AIAA J.* 26(5), 513–521.

Barnett, M., and J. M. Verdon. (1987). *Viscid/Inviscid Interaction Analysis of Subsonic Turbulent Trailing-Edge Flows.* AIAA Paper 87–0457.

Batchelor, G. K. (1956a). On steady laminar flow with closed streamlines at large Reynolds number. *J. Fluid Mech.* 1(2), 177–190.

Batchelor, G. K. (1956b). A proposal concerning laminar wakes behind bluff bodies at large Reynolds number. *J. Fluid Mech.* 1(4), 388–398.

Betyaev, S. K. (1980). Evolution of vortex sheets. In *Dynamics of a Continuous Medium with Free Surfaces*, pp. 27–38. Cheboksary, Chuvashkii State Univ.

Birkhoff, G. (1962). Helmholtz and Taylor instability. *Proc. Symp. Appl. Math.* 13, 55–76.

Birkhoff, G., and E. H. Zarantonello. (1957). *Jets, Wakes, and Cavities.* Academic Press, New York.

Black, D. W., and A. P. Rothmayer. (1991). Computation of trailing-edge separation using an accelerated, updated bluff-body, interacting boundary-layer algorithm. In V. V. Kozlov and A. V. Dovgal (Eds.), *Separated Flows and Jets,* pp. 77–88. Springer, Berlin.

Blasius, H. (1908). Grenzschichten in Flüssigkeiten mit kleiner Reibung. *Z. Math. Phys.* 56(1), 1–37. (Engl. transl. NACA TM 1256.)

Bodonyi, R. J., and P. W. Duck. (1988). A numerical method for treating strongly interactive three-dimensional viscous-inviscid flows. *Intl. J. Comput. Fluids* 16(3), 279–290.

Bogdanova, E. V., and O. S. Ryzhov. (1982). On perturbations generated by an oscillator in viscous fluid flow at supercritical frequencies. *Zh. Prikl. Mekh. Tekh. Fiz.* (4), 65–72. (Engl. transl. *J. Appl. Mech. Tech. Phys.* 23(4), 523–529.)

Bogdanova, E. V., and O. S. Ryzhov. (1986). On laminar preseparation flow. *Prikl. Mat. Mekh.* 50(3), 394–402. (Engl. transl. *J. Appl. Math. Mech.* 50(3), 297–303.)

Bogdanova, E. V., and O. S. Ryzhov. (1987). Separation of a flow from the corner point of a body. *Prikl. Mat. Mekh.* 51(3), 425–433. (Engl. transl. *J. Appl. Math. Mech.* 51(3), 331–338.)

Bogolepov, V. V. (1974). Calculation of the interaction of a supersonic boundary layer with a thin obstacle. *Uch. zap. TsAGI* 5(6), 30–38.

Bogolepov, V. V. (1975). Flow around a semicylindrical protuberance on the surface of a plate in a shear flow of a viscous fluid. *Mekh. Zhid. Gaza* (6), 30–37. (Engl. transl. *Fluid Dyn.* 10(6), 896–903.)

Bogolepov, V. V. (1985a). Flow structure near the trailing edge of a plate. *Zh. Prikl. Mekh. Tekh. Fiz.* (3), 95–99. (Engl. transl. *J. Appl. Mech. Tech. Phys.* 26(3), 386–390.)

Bogolepov, V. V. (1985b). Solution of the Navier–Stokes equations near the trailing edge of a plate. *Mekh. Zhid. Gaza* (4), 173–176. (Engl. transl. *Fluid Dyn.* 20(4), 644–647.)

Bogolepov, V. V., and I. I. Lipatov. (1982). Similarity solutions for boundary-layer equations with interaction. *Zh. Prikl. Mekh. Tekh. Fiz.* (4), 60–65. (Engl. transl. *J. Appl. Mech. Tech. Phys.* 23(4), 518–522.)

Bogolepov, V. V., and V. Ya. Neiland. (1971). Supersonic flow of a viscous gas over a small irregularity on a body surface. *Tr. TsAGI* 1363.

Bouard, R., and M. Coutanceau. (1980). The early stage of development of the wake behind an impulsively started cylinder for $40 < \mathrm{Re} < 10^4$. *J. Fluid Mech.* 101(3), 583–607.

Brillouin, M. (1911). Les surfaces de glissement d'Helmholtz et la résistance des fluides. *Ann. Chim. Phys.* 8e sér. 23, 145–230.

Brodetsky, S. (1923). Discontinuous fluid motion past circular and elliptic cylinders. *Proc. Roy. Soc. London,* A 102, 542–553.

Brown, S. N., and P. G. Daniels. (1975). On the viscous flow about the trailing edge of a rapidly oscillating plate. *J. Fluid Mech.* 67(4), 743–761.

Brown, S. N., and K. Stewartson. (1969). Laminar separation. *Ann. Rev. Fluid Mech.* 1, 45–72.

Brown, S. N., and K. Stewartson. (1970). Trailing-edge stall. *J. Fluid Mech.* 42(3), 561–584.

Brown, S. N., and K. Stewartson. (1975). Wake curvature and the Kutta condition in laminar flow. *Aero. Q.* 26(4), 275–280.

Brown, S. N., and K. Stewartson. (1983). On an integral equation of marginal separation. *SIAM J. Appl. Math.* 43(5), 1119–1126.

Burggraf, O. R., and P. W. Duck. (1982). Spectral computation of triple-deck flows. In T. Cebeci (Ed.), *Numerical and Physical Aspects of Aerodynamic Flows*, pp. 145–158. Springer, New York.

Carrier, G. F., and C. C. Lin. (1948). On the nature of the boundary layer near the leading edge of a flat plate. *Q. Appl. Math.* 6(1), 63–68.

Carter, J. E. (1974). *Solutions for Laminar Boundary Layers with Separation and Reattachment.* AIAA Paper 74–583.

Carter, J. E. (1978). *A New Boundary-Layer Interaction Technique for Separated Flows.* NASA TM 78690.

Carter, J. E. (1979). *A New Boundary-Layer Inviscid Iteration Technique for Separated Flow.* AIAA Paper 79–1450.

Carter, J. E., and S. F. Wornom. (1975). Solutions for incompressible separated boundary layers including viscous–inviscid interaction. NASA SP-347, part 1, pp. 125–150.

Cassel, K. W., and J. D. A. Walker. (1993). Viscous–inviscid interactions in unsteady boundary layer separation. In *Intl. Workshop on Advances in Analytical Methods in Aerodynamics.* Miedzyzdroje, Poland.

Cassel, K.W., F.T. Smith, and J.D.A. Walker (1996). The Onset of Instability in Unsteady Boundary-Layer Separation. *J. Fluid Mech.* 315, 223–256.

Catherall, D., and K. W. Mangler. (1966). The integration of the two-dimensional laminar boundary-layer equations past the point of vanishing skin friction. *J. Fluid Mech.* 26(1), 163–182.

Cebeci, T., A. A. Khattab, and S. M. Schimke. (1988). Separation and reattachment near the leading edge of a thin oscillating airfoil. *J. Fluid Mech.* 188, 253–274.

Cebeci, T., K. Stewartson, and P. G. Williams. (1981). *Separation and Reattachment Near the Leading Edge of a Thin Airfoil at Incidence.* AGARD CP 291, paper no. 20.

Chaplygin, S. A. (1899). On the problem of jets in an incompressible fluid. *Proc. Phys. Sect. Soc. Natural Scientists* 10(1), 35–40.

Chaplygin, S. A. (1910). *On the Pressure Exerted by a Two-Dimensional Flow on a Body (in the Aeroplane Theory).* Moscow Imperial University, Moscow.

Chapman, D. R., D. M. Kuehn, and H. K. Larson. (1958). *Investigation of Separated Flows in Supersonic and Subsonic Streams with Emphasis on the Effect of Transition.* NACA Rep. 1356.

Chen, Y. (1983). The Lagrangian boundary layer equation and the unsteady separation of the elliptic cylinder at angle of attack. *Acta Mech. Sinica* (6), 571–578.

Cheng, H. K. (1984). *Laminar Separation from Airfoils beyond Trailing-Edge Stall.* AIAA Paper 84–1612.

Cheng, H. K., and C. J. Lee. (1986). Laminar separation studied as an airfoil problem. In T. Cebeci (Ed.), *Numerical and Physical Aspects of Aerodynamic Flows, III*, pp. 102–125. Springer, New York.

Cheng, H. K., and F. T. Smith. (1982). The influence of airfoil thickness and Reynolds number on separation. *Z. Angew. Math. Phys.* 33(2), 151–180.

Chernyi, G. G. (1974). Boundary layer on a moving surface. In *Collected Works Dedicated to the 60th Birthday of V. N. Chelomei: Selected*

Problems in Applied Mechanics, pp. 709–719. VINITI, Moscow.

Chernyshenko, S. I. (1985). On the asymptotics of stationary solutions of the Navier–Stokes equations at large Reynolds numbers. *Dokl. Akad. Nauk SSSR* 285(6), 1353–1355.

Chernyshenko, S. I. (1988). The asymptotic form of the stationary separated flow around a body at high Reynolds numbers. *Prikl. Mat. Mekh.* 52(6), 958–966. (Engl. transl. *J. Appl. Math. Mech.* 52(6), 746–753.)

Chipman, P. D., and P. W. Duck. (1993). On the high-Reynolds-number flow between non-coaxial rotating cylinders. *Q. J. Mech. Appl. Math.* 46(2), 163–191.

Chow, R., and R. E. Melnik. (1976). Numerical solutions of the triple-deck equations for laminar trailing-edge stall. *Lecture Notes in Phys.* 59, 135–144.

Cole, J. D. (1968). *Perturbation Methods in Applied Mathematics*. Blaisdell Publ. Co., Waltham, Mass. (Rev. ed. J. Kevorkian and J. D. Cole (1981), Springer, New York.)

Collins, W. M., and S. C. R. Dennis. (1973). Flow past an impulsively started circular cylinder. *J. Fluid Mech.* 60(1), 105–127.

Cowley, S. J. (1983). Computer extension and analytic continuation of Blasius' expansion for impulsive flow past a circular cylinder. *J. Fluid Mech.* 135, 389–405.

Cowley, S. J., L. L. Van Dommelen, and S. T. Lam. (1990). On the use of lagrangian variables in descriptions of unsteady boundary-layer separation. *Phil. Trans. Roy. Soc. London*, A 333, 343–378.

Crank, J., and P. Nicolson. (1947). A practical method for numerical evaluation of solutions of partial differential equations of the heat-conduction type. *Proc. Camb. Phil. Soc.* 43(1), 50–67.

Crighton, D. G. (1985). The Kutta condition in unsteady flow. *Ann. Rev. Fluid Mech.* 17, 411–445.

Daniels, P. G. (1974a). Numerical and asymptotic solutions for the supersonic flow near the trailing edge of a flat plate. *Q. J. Mech. Appl. Math.* 27(2), 175–191.

Daniels, P. G. (1974b). Numerical and asymptotic solutions for the supersonic flow near the trailing edge of a flat plate at incidence. *J. Fluid Mech.* 63(4), 641–656.

Daniels, P. G. (1977). Viscous mixing at a trailing edge. *Q. J. Mech. Appl. Math.* 30(3), 319–342.

Davis, R. L., and J. E. Carter. (1986). Counterrotating streamline pattern in a transitional separation bubble. *AIAA J.* 24(5), 850–851.

Davis, R. T. (1984). *A Procedure for Solving the Compressible Interacting Boundary-Layer Equations for Subsonic and Supersonic Flows*. AIAA Paper 84-1614.

Davis, R. T., and M. J. Werle. (1982). Progress on interacting boundary-layer computations at high Reynolds number. In T. Cebeci (Ed.), *Numerical and Physical Aspects of Aerodynamic Flows*, pp. 187–210. Springer, New York.

Dennis, S. C. R., and G.-Z. Chang. (1969). Numerical integration of the Navier–Stokes equations for steady two-dimensional flow. *Phys. Fluids* 12(12, Suppl. II), 88–93.

Dennis, S. C. R., and G.-Z. Chang. (1970). Numerical solutions for steady flow past a circular cylinder at Reynolds numbers up to 100. *J. Fluid Mech.*

42(3), 471–489.

Dennis, S. C. R., and J. Dunwoody. (1966). The steady flow of a viscous fluid past a flat plate. *J. Fluid Mech.* 24(3), 577–595.

Diesperov, V. N. (1984). On the existence and uniqueness of self-similar solutions describing the flow in mixing layers. *Dokl. Akad. Nauk SSSR* 275(6), 1341–1346.

Diesperov, V. N. (1985). On the flow in a Chapman mixing layer. *Dokl. Akad. Nauk SSSR* 284(2), 305–309.

Diesperov, V. N. (1986). Investigation of self-similar solutions describing flows in mixing layers. *Prikl. Mat. Mekh.* 50(3), 403–414. (Engl. transl. *J. Appl. Math. Mech.* 50(3), 303–312.)

Dijkstra, D. (1974). *The Solution of the Navier–Stokes Equations near the Trailing Edge of a Flat Plate.* Rijksuniv. Groningen.

Dijkstra, D. (1979). Separating, incompressible, laminar boundary-layer flow over a smooth step of small height. *Lecture Notes in Phys.* 90, 169–176.

Dimopoulos, H. G., and T. J. Hanratty. (1968). Velocity gradients at the wall for flow around a cylinder for Reynolds numbers between 60 and 360. *J. Fluid Mech.* 33(2), 303–319.

Doligalski, T. L., and J. D. A. Walker. (1984). The boundary layer induced by a convected two-dimensional vortex. *J. Fluid Mech.* 139, 1–28.

Duck, P. W. (1985). Laminar flow over unsteady humps: The formation of waves. *J. Fluid Mech.* 160, 465–498.

Duck, P. W. (1988). The effect of small surface perturbations on the pulsatile boundary layer on a semi-infinite flat plate. *J. Fluid Mech.* 197, 259–293.

Duck, P. W., and O. R. Burggraf. (1986). Spectral solutions for three-dimensional triple-deck flow over surface topography. *J. Fluid Mech.* 162, 1–22.

Ece, M. C., J. D. A. Walker, and T. L. Doligalski. (1984). The boundary layer on an impulsively started rotating and translating cylinder. *Phys. Fluids* 27(5), 1077–1089.

Edwards, D. E., and J. E. Carter. (1986). A quasi-simultaneous finite difference approach for strongly interacting flow. In T. Cebeci (Ed.), *Numerical and Physical Aspects of Aerodynamic Flows III*, pp. 126–142. Springer, New York.

Elliott, J. W., and F. T. Smith. (1986). Separated supersonic flow past a trailing edge at incidence. *Intl. J. Comput. Fluids* 14(2), 109–116.

Elliott, J. W., and F. T. Smith. (1987). Dynamic stall due to unsteady marginal separation. *J. Fluid Mech.* 179, 489–512.

Elliott, J. W., F. T. Smith, and S. J. Cowley. (1983). Breakdown of boundary layers: (i) on moving surfaces; (ii) in semi-similar unsteady flow; (iii) in fully unsteady flow. *Geophys. Astrophys. Fluid Dyn.* 25(1,2), 77–138.

Ely, W. L., and R. N. Herring. (1978). *Laminar Leading Edge Stall Prediction for Thin Airfoils.* AIAA Paper 78–1222.

Ermak, Yu. N. (1969). Viscous incompressible flow over the rounded leading edge of a thin airfoil. *Tr. TsAGI* 1141.

Fomina, I. G. (1983). Asymptotic theory of flow around corner points of a solid body. *Uch. zap. TsAGI* 14(5), 31–38.

Föppl, L. (1913). Wirbelbewegung hinter einem Kreiszylinder. *Sitzungsb. Bayer. Akad. Wiss., Math.-phys. Kl.* (1), 1–17.

Fornberg, B. (1980). A numerical study of steady viscous flow past a circular cylinder. *J. Fluid Mech.* 98(4), 819–855.

Fornberg, B. (1985). Steady viscous flow past a circular cylinder up to Reynolds number 600. *J. Comput. Phys.* 61(2), 297–320.

Gabutti, B. (1984). On the equation of similar profiles. *Z. Angew. Math. Phys.* 35(3), 265–281.

Gittler, P. (1992). On similarity solutions occurring in the theory of interactive laminar boundary layers. *J. Fluid Mech.* 244, 131–147.

Gittler, P., and A. Kluwick. (1987). Triple-deck solutions for supersonic flows past flared cylinders. *J. Fluid Mech.* 179, 469–487.

Gogish, L. V., V. Ya. Neiland, and G. Yu. Stepanov. (1975). The theory of two-dimensional separated flows. *Advances in Science and Engineering: Hydromechanics* 8, 5–73.

Goldstein, M. E. (1985). Scattering of acoustic waves into Tollmien–Schlichting waves by small streamwise variations in surface geometry. *J. Fluid Mech.* 154, 509–529.

Goldstein, M. E., and L. S. Hultgren. (1989). Boundary-layer receptivity to long-wave free-stream disturbances. *Ann. Rev. Fluid Mech.* 21, 137–166.

Goldstein, S. (1930). Concerning some solutions of the boundary layer equations in hydrodynamics. *Proc. Camb. Phil. Soc.* 26(1), 1–30.

Goldstein, S. (1948). On laminar boundary-layer flow near a position of separation. *Q. J. Mech. Appl. Math.* 1(1), 43–69.

Grove, A. S., F. H. Shair, E. E. Petersen, and A. Acrivos. (1964). An experimental investigation of the steady separated flow past a circular cylinder. *J. Fluid Mech.* 19(1), 60–80.

Gurevich, M. I. (1979). *The Theory of Jets in an Ideal Fluid.* Nauka, Moscow.

Hackmüller, G., and A. Kluwick. (1989). The effect of a surface mounted obstacle on marginal separation. *Z. Flugwiss. Weltraumforsch.* 13(6), 365–370.

Hartree, D. R. (1939). A solution of the laminar boundary-layer equation for retarded flow. *Aero. Res. Coun. Rep. and Memo.* 2426 (issued 1949).

Helmholtz, H. (1858). Über Integrale der hydrodynamischen Gleichungen, welche den Wirbelbewegungen entsprechen. *J. Reine Angew. Math.* 55(1), 25–55.

Helmholtz, H. (1868). Über discontinuirliche Flüssigkeits – Bewegungen. *Monatsbericht Akad. Wiss. Berlin,* April, 215–228.

Honji, H., and S. Taneda. (1969). Unsteady flow past a circular cylinder. *J. Phys. Soc. Japan* 27(6), 1668–1677.

Houwink, R., and A. E. P. Veldman. (1984). *Steady and Unsteady Separated Flow Computations for Transonic Airfoils.* AIAA Paper 84–1618.

Howarth, L. (1938). On the solution of the laminar boundary layer equations. *Proc. Roy. Soc. London* A 164, 547–579.

Hunt, J. C. R. (1971). A theory for the laminar wake of a two-dimensional body in a boundary layer. *J. Fluid Mech.* 49(1), 159–178.

Imai, I. (1953). Discontinuous potential flow as the limiting form of the viscous flow for vanishing viscosity. *J. Phys. Soc. Japan* 8(3), 399–402.

Imai, I. (1957a). Second approximation to the laminar boundary-layer flow over a flat plate. *J. Aeronaut. Sci.* 24(2), 155–156.

Imai, I. (1957b). *Theory of Bluff Bodies.* Tech. Note BN–104, Inst. Fluid Dyn. and Appl. Math., Univ. Maryland.

Ingham, D. B. (1984). Unsteady separation. *J. Comput. Phys.* 53(1), 90–99.

Jenson, R., O. R. Burggraf, and D. P. Rizzetta. (1975). Asymptotic solution for supersonic viscous flow past a compression corner. *Lecture Notes in*

Phys. 35, 218–224.

Jobe, C. E., and O. R. Burggraf. (1974). The numerical solution of the asymptotic equations of trailing edge flow. *Proc. Roy. Soc. London* A 340, 91–111.

Jones, B. M. (1934). Stalling. *J. Roy. Aero. Soc.* 38, 753–770.

Kachanov, Yu. S., V. V. Kozlov, and V. Ya. Levchenko. (1982). *The Occurrence of Turbulence in a Boundary Layer*. Nauka, Novosibirsk.

Kaden, H. (1931). Aufwicklung einer unstabilen Unstetigkeitsfläche. *Ing.-Arch.* 2(2), 140–168.

Kaplun, S. (1967). In P. A. Lagerstrom, L. N. Howard, and C. S. Liu (Eds.), *Fluid Mechanics and Singular Perturbations*. Academic Press, New York.

Karas, O. V., and V. E. Kovalev. (1989). Application of an inverse method for calculation of a three-dimensional boundary layer for the problem of viscous flow around a wing. *Uch. zap. TsAGI* 20(5), 1–11.

Keldysh, M. V., and L. I. Sedov. (1937). Efficient solution of certain boundary-value problems for harmonic functions. *Dokl. Akad. Nauk SSSR* 16(1), 7–10.

Khrabrov, A. N. (1985). Nonuniqueness of laminar separated flow over an airfoil at angle of attack according to the Kirchhoff model. *Uch. zap. TsAGI* 16(5), 1–9.

Kirchhoff, G. (1869). Zur Theorie freier Flüssigkeitsstrahlen. *J. Reine Angew. Math.* 70(4), 289–298.

Kiya, M., and M. Arie. (1975). Viscous shear flow past small bluff bodies attached to a plane wall. *J. Fluid Mech.* 69(4), 803–823.

Klineberg, J. M., and J. L. Steger. (1974). *On Laminar Boundary-Layer Separation.* AIAA Paper 74–94.

Kluwick, A., P. Gittler, and R. J. Bodonyi, R. J. (1985). Freely interacting axisymmetric boundary layers on bodies of revolution. *Q. J. Mech. Appl. Math.* 38(4), 575–588.

Korolev, G. L. (1980a). Numerical solution of the asymptotic problem of laminar boundary-layer separation from a smooth surface. *Uch. zap. TsAGI* 11(2), 27–36.

Korolev, G. L. (1980b). On the asymptotic theory of flow near the trailing edge of an elliptic profile. *Uch. zap. TsAGI* 11(4), 8–16.

Korolev, G. L. (1987). A method of solving problems of the asymptotic theory of interaction between a boundary layer and the external flow. *Zh. Vych. Mat. Mat. Fiz.* 27(8), 1224–1232. (Engl. transl. *USSR Comput. Math. and Math. Phys.* 27(4), 175–181.)

Korolev, G. L. (1989). On the theory of flow separation at the trailing edge of a thin profile. *Mekh. Zhid. Gaza* (4), 55–59. (Engl. transl. *Fluid Dyn.* 24(4), 534–537.)

Korolev, G. L. (1991a). Asymptotic theory of laminar flow separation at a corner with small turning angle. *Mekh. Zhid. Gaza* (1), 180–182. (Engl. transl. *Fluid Dyn.* 26(1), 150–152.)

Korolev, G. L. (1991b). Interaction theory and non-uniqueness of separated flows around solid bodies. In V. V. Kozlov and A. V. Dovgal (Eds.), *Separated Flows and Jets*, pp. 139–142. Springer, Berlin.

Korolev, G. L. (1991c). Flow separation and the nonuniqueness of the solution of the boundary-layer equations. *Zh. Vych. Mat. Mat. Fiz.* 31(11), 1706–1715. (Engl. transl. *Comput. Math. and Math. Phys.* 31(11), 73–79.)

Korolev, G. L. (1992). On the nonuniqueness of separated flow around corners with small turning angle. *Mekh. Zhid. Gaza* (3) 178–180. (Engl. transl. *Fluid Dyn.* 27(3), 442–444.)

Koromilas, C. A., and D. P. Telionis. (1980). Unsteady laminar separation: an experimental study. *J. Fluid Mech.* 97(2), 347–384.

Kozlov, V. V., and O. S. Ryzhov. (1990). Receptivity of boundary layers: asymptotic theory and experiment. *Proc. Roy. Soc. London* A 429, 341–373.

Kravtsova, M. A. (1993). Numerical solution of the asymptotic problem of incompressible boundary-layer separation ahead of a body corner point. *Zh. Vych. Mat. Mat. Fiz.* 33(3), 439–449. (Engl. transl. *Comput. Math. and Math. Phys.* 33(3), 397–406.)

Kravtsova, M. A., and A. I. Ruban. (1985). On the unsteady boundary layer in the cross-flow over a cylinder performing rotational oscillations. *Uch. zap. TsAGI* 16(6), 99–102.

Kravtsova, M. A., and A. I. Ruban. (1988). Separation of a supersonic boundary layer ahead of the base of a body contour. *Zh. Vych. Mat. Mat. Fiz.* 28(4), 580–590. (Engl. transl. *USSR Comput. Math. and Math. Phys.* 28(2), 177–184.)

Kuo, Y. H. (1953). On the flow of an incompressible viscous fluid past a flat plate at moderate Reynolds numbers. *J. Math. Phys.* 32(2,3), 83–101.

Kuryanov, A. I., G. I. Stolyarov, and R. I. Steinberg. (1979). On the hysteresis of aerodynamic characteristics. *Uch. zap. TsAGI* 10(3), 12–15.

Kutta, W. M. (1902). Auftriebskräfte in strömenden Flüssigkeiten. *Illustr. Aëron. Mitt.* 6. Jahrg. 3, 133–135.

Kwon, O. K., and R. H. Pletcher. (1979). Prediction of incompressible separated boundary layers including viscous-inviscid interaction. *ASME J. Fluids Eng.* 101(4), 466–472.

Lagerstrom, P. A. (1964). Laminar flow theory. In F. K. Moore (Ed.), *High Speed Aerodynamics and Jet Propulsion 4: Theory of Laminar Flows*, pp. 20–285. Princeton Univ. Press.

Lagerstrom, P. A. (1975). Solutions of the Navier–Stokes equation at large Reynolds number. *SIAM J. Appl. Math.* 28(1), 202–214.

Lagerstrom, P. A., and R. G. Casten. (1972). Basic concepts underlying singular perturbation techniques. *SIAM Rev.* 14(1), 63–120.

Lam, S. T. (1988). *On High-Reynolds-Number Laminar Flows through a Curved Pipe, and Past a Rotating Cylinder*. Ph. D. Thesis, Univ. London.

Landau, L. D., and E. M. Lifshitz. (1944). *Mechanics of Continuous Media*. Gostekhizdat, Moscow.

Lavrent'ev, M. A. (1962). *Variational Methods in Boundary-Value Problems for Systems of Elliptic Equations*. Akad. Nauk SSSR, Moscow.

Le Balleur, J. C. (1978). Couplage visqueux–non-visqueux: méthode numérique et applications aux écoulements bidimensionnels transsoniques et supersoniques. *Rech. Aérosp.* (2), 65–76.

Lee, C. J., and H. K. Cheng. (1990). An airfoil theory of bifurcating laminar separation from thin obstacles. *J. Fluid Mech.* 216, 255–284.

Liepmann, H. W. (1946). The interaction between boundary layer and shock waves in transonic flow. *J. Aeronaut. Sci.* 13(12), 623–637.

Lighthill, M. J. (1953). On boundary layers and upstream influence. II. Supersonic flows without separation. *Proc. Roy. Soc. London* A 217, 478–507.

Lock, R. C. (1951). The velocity distribution in the laminar boundary layer between parallel streams. *Q. J. Mech. Appl. Math.* 4(1), 42–63.

Loitsianskii, L. G. (1962). *Laminar Boundary Layers.* Fizmatgiz, Moscow.

Ludwig, G. R. (1964). *An Experimental Investigation of Laminar Separation from a Moving Wall.* AIAA Paper 64–6.

McLachlan, R. I. (1991a). A steady separated viscous corner flow. *J. Fluid Mech.* 231, 1–34.

McLachlan, R. I. (1991b). The boundary layer on a finite flat plate. *Phys. Fluids* A 3(2), 341–348.

McLeod, J. B. (1972). The existence and uniqueness of a similarity solution arising from separation at a free stream line. *Q. J. Math. Oxford*, 2 ser. 23, 63–77.

Merkin, J. H., and F. T. Smith. (1982). Free convection boundary layers near corners and sharp trailing edges. *Z. Angew. Math. Phys.* 33(1), 36–52.

Messiter, A. F. (1970). Boundary-layer flow near the trailing edge of a flat plate. *SIAM J. Appl. Math.* 18(1), 241–257.

Messiter, A. F. (1975). *Laminar Separation – A Local Asymptotic Flow Description for Constant Pressure Downstream.* AGARD CP168, Flow Separation, Paper No. 4.

Messiter, A. F. (1979). Boundary-layer separation. *Proc. 8th U.S. Natl. Congr. Appl. Mech.*, pp. 157–179. Western Periodicals, North Hollywood, CA.

Messiter, A. F. (1983). Boundary-layer interaction theory. *ASME J. Appl. Mech.* 50(4b), 1104–1113.

Messiter, A. F., and R. L. Enlow. (1973). A model for laminar boundary-layer flow near a separation point. *SIAM J. Appl. Math.* 25(3), 655–670.

Messiter, A. F., and J. J. Hu. (1975). Laminar boundary layer at a discontinuity in wall curvature. *Q. Appl. Math.* 33(2), 175–181.

Messiter, A. F., and A. Liñán. (1976). The vertical plate in laminar free convection: Effects of leading and trailing edges and discontinuous temperature. *Z. Angew. Math. Phys.* 27(5), 633–651.

Mikhailov, V. V. (1981). Asymptotic behavior of the neutral curve of the linear stability problem for a laminar boundary layer. *Mekh. Zhid. Gaza* (5), 39–46. (Engl. transl. *Fluid Dyn.* 16(5), 669–674.)

von Mises, R. (1927). Bemerkungen zur Hydrodynamik. *Z. Angew. Math. Mech.* 7(6), 425–431.

von Mises, R., and K. O. Friedrichs. (1971). *Fluid Dynamics.* Springer, New York.

Moore, F. K. (1958). On the separation of the unsteady laminar boundary-layer. In H. Görtler (Ed.), *Boundary Layer Research*, pp. 296–311. Springer, Berlin.

Morkovin, M. V. (1969). *Critical Evaluation of Transition From Laminar to Turbulent Shear Layers with Emphasis on Hypersonically Traveling Bodies.* Air Force Flight Dyn. Lab. Rep. AFFDL-TR-68-149.

Murman, E. M., and J. D. Cole. (1971). Calculation of plane steady transonic flows. *AIAA J.* 9(1), 114–121.

Muskhelishvili, N. I. (1968). *Singular Integral Equations.* Nauka, Moscow.

Nagata, H., K. Minami, and Y. Murata. (1979). Initial flow past an impulsively started circular cylinder. *Bull. JSME* 22(166), 512–520.

Napolitano, M., M. J. Werle, and R. T. Davis. (1979). Numerical technique for the triple-deck problem. *AIAA J.* 17(7), 699–705.

Negoda, V. V., and Vic. V. Sychev. (1987). The boundary layer on a rapidly rotating cylinder. *Mekh. Zhid. Gaza* (5), 36–45. (Engl. transl. *Fluid Dyn.*

22(5), 686–694.)

Neiland, V. Ya. (1969). Theory of laminar boundary-layer separation in supersonic flow. *Mekh. Zhid. Gaza* (4), 53–57. (Engl. transl. *Fluid Dyn.* 4(4), 33–35.)

Neiland, V. Ya. (1971a). Flow behind the boundary-layer separation point in a supersonic stream. *Mekh. Zhid. Gaza* (3), 19–25. (Engl. transl. *Fluid Dyn.* 6(3), 378–384.)

Neiland, V. Ya. (1971b). The asymptotic theory of the interaction of a supersonic flow with a boundary layer. *Mekh. Zhid. Gaza* (4), 41–47. (Engl. transl. *Fluid Dyn.* 6(4), 587–592.)

Neiland, V. Ya. (1974). Asymptotic problems in the theory of viscous supersonic flows. *Tr. TsAGI* 1529.

Neiland, V. Ya. (1981). Asymptotic theory of boundary-layer separation and interaction with supersonic gas flow. *Usp. Mekh.* 4(2), 3–62.

Neiland, V. Ya., and V. V. Sychev. (1966). Asymptotic solutions of the Navier–Stokes equations in regions with large local perturbations. *Mekh. Zhid. Gaza* (4), 43–49. (Engl. transl. *Fluid Dyn.* 1(4), 29–33.)

Nickel, K. (1958). Einige Eigenschaften von Lösungen der Prandtlschen Grenzschicht-Differentialgleichungen. *Arch. Rat. Mech. Anal.* 2(1), 1–31.

Nikolaev, K. V. (1982). Development of boundary-layer separation on a rotating cylinder in a stream of incompressible fluid. *Uch. zap. TsAGI* 13(6), 32–39.

Nikolskii, A. A. (1957a). On the "second" form of the motion of an ideal fluid around a body (study of separated vortex flows). *Dokl. Akad. Nauk SSSR* 116(2), 193–196.

Nikolskii, A. A. (1957b). The influence on force of the "second" form of hydrodynamic motion over two-dimensional bodies (dynamics of two-dimensional separated flows). *Dokl. Akad. Nauk SSSR* 116(3), 365–368.

Nishioka, M., and T. Miyagi. (1978). Measurements of velocity distributions in the laminar wake of a flat plate. *J. Fluid Mech.* 84(4), 705–715.

Oleinik, O. A. (1963). On a system of equations in boundary layer theory. *Zh. Vych. Mat. Mat. Fiz.* 3(3), 489–507. (Engl. transl. *USSR Comput. Math. and Math. Phys.* 3(3), 650–673.)

Peregrine, D. H. (1985). A note on the steady high-Reynolds-number flow about a circular cylinder. *J. Fluid Mech.* 157, 493–500.

Peridier, V. J., F. T. Smith, and J. D. A. Walker. (1991a). Vortex-induced boundary-layer separation. Part 1. The unsteady limit problem Re → ∞. *J. Fluid Mech.* 232, 99–131.

Peridier, V. J., F. T. Smith, and J. D. A. Walker. (1991b). Vortex-induced boundary-layer separation. Part 2. Unsteady interacting boundary-layer theory. *J. Fluid Mech.* 232, 133–165.

Prandtl, L. (1904). Über Flüssigkeitsbewegung bei sehr kleiner Reibung. *Verh. III. Intern. Math. Kongr., Heidelberg, 1904*, S. 484–491. Teubner, Leipzig, 1905. (Engl. transl. NACA TM 452.)

Prandtl, L. (1924). Über die Entstehung von Wirbeln in der idealen Flüssigkeit, mit Anwendung auf die Tragflügeltheorie und andere Aufgaben. *Vorträge Geb. Hydro-u. Aerodynamik* (hrsg. v. Th. v. Kármán u. T. Levi-Civita), S. 18–33. Springer, Berlin.

Prandtl, L. (1927). The generation of vortices in fluids of small viscosity. *J. Roy. Aero. Soc.* 31, 720–741.

Prandtl, L. (1931). *Proc. 3rd Intl. Congr. Appl. Mech., Stockholm 1930*, v. 1, pp. 41–42. Ab. Sveriges Litogr. Tryckerier, Stockholm.

Prandtl, L. (1935). The mechanics of viscous fluids. In W. F. Durand (Ed.), *Aerodynamic Theory*, v. 3, pp. 34–208. Springer, Berlin.

Prandtl, L. (1938). Zur Berechnung der Grenzschichten. *Z. Angew. Math. Mech.* 18(1), 77–82.

Reyhner, T. A., and I. Flügge-Lotz. (1968). The interaction of a shock wave with a laminar boundary layer. *Intl. J. Nonlinear Mech.* 3(2), 173–199.

Riabouchinsky, D. P. (1920). On steady fluid motions with free surfaces. *Proc. London Math. Soc.*, Ser. 2, 19(3), 206–215.

Riley, N. (1975). Unsteady laminar boundary layers. *SIAM Rev.* 17(2), 274–297.

Riley, N., and K. Stewartson. (1969). Trailing edge flows. *J. Fluid Mech.* 39(1), 193–207.

Rizzetta, D. P., O. R. Burggraf, and R. Jenson. (1978). Triple-deck solutions for viscous supersonic and hypersonic flow past corners. *J. Fluid Mech.* 89(3), 535–552.

Rosenhead, L. (Ed.). (1963). *Laminar Boundary Layers*. Clarendon Press, Oxford.

Roshko, A. (1967). A review of concepts in separated flow. *Proc. Canadian Congr. Appl. Mech.* 3, 81–115. Univ. Toronto.

Rothmayer, A. P., and R. T. Davis. (1986). Massive separation and dynamic stall on a cusped trailing-edge airfoil. In T. Cebeci (Ed.), *Numerical and Physical Aspects of Aerodynamic Flows III*, pp. 286–317. Springer, New York.

Rott, N. (1956). Unsteady viscous flow in the vicinity of a stagnation point. *Q. Appl. Math.* 13(4), 444–451.

Ruban, A. I. (1974). On laminar separation from a corner point on a solid surface. *Uch. zap. TsAGI* 5(2), 44–54.

Ruban, A. I. (1976a). A numerical method for solving the free-interaction problem. *Uch. zap. TsAGI* 7(2), 45–51.

Ruban, A. I. (1976b). On the theory of laminar flow separation of a fluid from a corner point on a solid surface. *Uch. zap. TsAGI* 7(4), 18–28.

Ruban, A. I. (1977). On the asymptotic theory of flow near the trailing edge of a thin airfoil. *Uch. zap. TsAGI* 8(1), 6–11.

Ruban, A. I. (1978). Numerical solution of the local asymptotic problem of the unsteady separation of a laminar boundary layer in a supersonic flow. *Zh. Vych. Mat. Mat. Fiz.* 18(5), 1253–1265. (Engl. transl. *USSR Comput. Math. and Math. Phys.* 18(5), 175–187.)

Ruban, A. I. (1981). A singular solution of the boundary-layer equations which can be extended continuously through the point of zero surface friction. *Mekh. Zhid. Gaza* (6), 42–52. (Engl. transl. *Fluid Dyn.* 16(6), 835–843.)

Ruban, A. I. (1982a). Asymptotic theory of short separation regions at the leading edge of a thin airfoil. *Mekh. Zhid. Gaza* (1), 42–51. (Engl. transl. *Fluid Dyn.* 17(1), 33–41.)

Ruban, A. I. (1982b). Stability of the preseparation boundary layer at the leading edge of a thin airfoil. *Mekh. Zhid. Gaza* (6), 55–63. (Engl. transl. *Fluid Dyn.* 17(6), 860–867.)

Ruban, A. I. (1984). On the generation of Tollmien-Schlichting waves by sound. *Mekh. Zhid. Gaza* (5), 44–52. (Engl. transl. *Fluid Dyn.* 19(5), 709–716.)

Ruban, A. I. (1985). On Tollmien-Schlichting wave generation by sound. In V.

V. Kozlov (Ed.), *Laminar-Turbulent Transition*, pp. 313–320. Springer, Berlin.

Ruban, A. I. (1990). Numerical methods in the theory of boundary-layer –inviscid flow interaction. *Uch. zap.TsAGI* 21(5), 3–25.

Ruban, A. I. (1991). Marginal separation theory. In V. V. Kozlov and A. V. Dovgal (Eds.), *Separated Flows and Jets*, pp. 47–54. Springer, Berlin.

Ruban, A. I. (1994). Numerical methods in the theory of boundary-layer inviscid flow interaction. *The TsAGI Journal*, 1(1), 25–54.

Ruban, A. I., and V. V. Sychev. (1979). Asymptotic theory of laminar boundary-layer separation of an incompressible fluid. *Usp. Mekh.* 2(4), 57–95.

Ryzhov, O. S. (1990). The formation of ordered vortex structures from unstable oscillations in the boundary layer. *Zh. Vych. Mat. Mat. Fiz.* 30(12), 1804–1814. (Engl. transl. *USSR Comput. Math. and Math. Phys.* 30(6), 146–154.)

Ryzhov, O. S., and F. T. Smith. (1984). Short-length instabilities, breakdown and initial value problems in dynamic stall. *Mathematika* 31(2), 163–177.

Ryzhov, O. S., and E. D. Terent'ev. (1977). On an unsteady boundary layer with self-induced pressure. *Prikl. Mat. Mekh.* 41(6), 1007–1023. (Engl. transl. *J. Appl. Math. Mech.* 41(6), 1024–1040.)

Ryzhov, O. S., and E. D. Terent'ev. (1986). On the transition regime characterizing the start of a vibrator in a subsonic boundary layer on a plate. *Prikl. Mat. Mekh.* 50(6), 974–986. (Engl. transl. *J. Appl. Math. Mech.* 50(6), 753–762.)

Ryzhov, O. S., and E. D. Terent'ev. (1987). Vortex spots in the boundary layer. *Fluid Dyn. Trans.* 13, 205–234.

Sadovskii, V. S. (1970). A region of constant vorticity in a two-dimensional potential flow. *Uch. zap. TsAGI* 1(4), 1–9.

Sadovskii, V. S. (1971). Vortex regions in a potential flow with a jump in the Bernoulli constant at the boundary. *Prikl. Mat. Mekh.* 35(5), 773–779. (Engl. transl. *J. Appl. Math. Mech.* 35(5), 729–735.)

Sadovskii, V. S. (1973). A study of solutions of the Euler equations containing regions of constant vorticity. *Tr. TsAGI* 1474.

Samarskii, A. A., and E. S. Nikolaev. (1978). *Methods of Solving Finite-Difference Equations*. Nauka, Moscow.

Schlichting, H. (1979). *Boundary-Layer Theory* (seventh ed.). McGraw-Hill, New York.

Schneider, W. (1974). Upstream propagation of unsteady disturbances in supersonic boundary layers. *J. Fluid Mech.* 63(3), 465–485.

Schwabe, M. (1935). Über Druckermittlung in der nichtstationären ebenen Strömung. *Ing.-Arch.* 6(1), 34–50.

Sears, W. R. (1956). Some recent developments in airfoil theory. *J. Aeronaut. Sci.* 23(5), 490–499.

Sears, W. R., and D. P. Telionis. (1975). Boundary-layer separation in unsteady flow. *SIAM J. Appl. Math.* 28(1), 215–235.

Sedov, L. I. (1966). *Two-Dimensional Problems of Hydrodynamics and Aerodynamics*. Nauka, Moscow.

Shen, S. F. (1978). Unsteady separation according to the boundary-layer equation. *Adv. Appl. Mech.* 18, 177–220.

Shidlovskii, V. P. (1977). The structure of the viscous fluid flow near the edge of a rotating disk. *Prikl. Mat. Mekh.* 41(3), 464–472. (Engl. transl. *J.*

References

Appl. Math. Mech. 41(3), 467–476.)

Smith, F. T. (1973). Laminar flow over a small hump on a flat plate. *J. Fluid Mech.* 57(4), 803–824.

Smith, F. T. (1976). Flow through constricted or dilated pipes and channels: Part 1. *Q. J. Mech. Appl. Math.* 29(3), 343–364.

Smith, F. T. (1977). The laminar separation of an incompressible fluid streaming past a smooth surface. *Proc. Roy. Soc. London* A 356, 443–463.

Smith, F. T. (1978). A note on a wall jet negotiating a trailing edge. *Q. J. Mech. Appl. Math.* 31(4), 473–479.

Smith, F. T. (1979a). Laminar flow of an incompressible fluid past a bluff body: the separation, reattachment, eddy properties and drag. *J. Fluid Mech.* 92(1), 171–205.

Smith, F. T. (1979b). On the non-parallel flow stability of the Blasius boundary layer. *Proc. Roy. Soc. London* A 366, 91–109.

Smith, F. T. (1982a). On the high Reynolds number theory of laminar flows. *IMA J. Appl. Math.* 28(3), 207–281.

Smith, F. T. (1982b). Concerning dynamic stall. *Aero. Q.* 33(4), 331–352.

Smith, F. T. (1983). Interacting flow theory and trailing edge separation – no stall. *J. Fluid Mech.* 131, 219–249.

Smith, F. T. (1985). A structure for laminar flow past a bluff body at high Reynolds number. *J. Fluid Mech.* 155, 175–191.

Smith, F. T. (1986). Two-dimensional disturbance travel, growth and spreading in boundary layers. *J. Fluid Mech.* 169, 353–377.

Smith, F. T. (1988a). A reversed-flow singularity in interacting boundary layers. *Proc. Roy. Soc. London* A 420, 21–52.

Smith, F. T. (1988b). Finite-time break-up can occur in any unsteady interacting boundary layer. *Mathematika* 35(2), 256–273.

Smith, F. T. (1991). Break-up in steady and unsteady separation. In V. V. Kozlov and A. V.Dovgal (Eds.), *Separated Flows and Jets*, pp. 25–37. Springer, Berlin.

Smith, F. T., P. W. M. Brighton, P. S. Jackson, and J. C. R. Hunt. (1981). On boundary-layer flow past two-dimensional obstacles. *J. Fluid Mech.* 113, 123–152.

Smith, F. T., and O. R. Burggraf. (1985). On the development of large-sized short-scaled disturbances in boundary layers. *Proc. Roy. Soc. London* A 399, 25–55.

Smith, F. T., and P. G. Daniels. (1981). Removal of Goldstein's singularity at separation, in flow past obstacles in wall layers. *J. Fluid Mech.* 110, 1–37.

Smith, F. T., and P. W. Duck. (1977). Separation of jets or thermal boundary layers from a wall. *Q. J. Mech. Appl. Math.* 30(2), 143–156.

Smith, F. T., and J. W. Elliott. (1985). On the abrupt turbulent reattachment downstream of leading-edge laminar separation. *Proc. Roy. Soc. London* A 401, 1–27.

Smith, F. T., and A. F. Khorrami. (1991). The interactive breakdown in supersonic ramp flow. *J. Fluid Mech.* 224, 197–215.

Smith, F. T., and J. H. Merkin. (1982). Triple-deck solutions for subsonic flow past humps, steps, concave or convex corners and wedged trailing edges. *Intl. J. Comput. Fluids* 10(1), 7–25.

Squire, H. B. (1934). On the laminar flow of a viscous fluid with vanishing viscosity. *Phil. Mag. Ser. 7*, 17, 1150–1160.

Stewartson, K. (1958). On Goldstein's theory of laminar separation. *Q. J. Mech. Appl. Math.* 11(4), 399–410.

Stewartson, K. (1969). On the flow near the trailing edge of a flat plate II. *Mathematika* 16(1), 106–121.

Stewartson, K. (1970a). On laminar boundary layers near corners. *Q. J. Mech. Appl. Math.* 23(2), 137–152.

Stewartson, K. (1970b). Is the singularity at separation removable? *J. Fluid Mech.* 44(2), 347–364.

Stewartson, K. (1971). On laminar boundary layers near corners. Corrections and an addition. *Q. J. Mech. Appl. Math.* 24(3), 387–389.

Stewartson, K. (1974). Multistructured boundary layers on flat plates and related bodies. *Adv. Appl. Mech.* 14, 145–239.

Stewartson, K. (1980). High Reynolds-number flows. *Lecture Notes in Math.* 771, 505–518.

Stewartson, K. (1981). D'Alembert's paradox. *SIAM Rev.* 23(3), 308–343.

Stewartson, K. (1982). Some recent studies in triple-deck theory. In T. Cebeci (Ed.), *Numerical and Physical Aspects of Aerodynamic Flows*, pp. 129–143. Springer, New York.

Stewartson, K., F. T. Smith, and K. Kaups. (1982). Marginal separation. *Stud. Appl. Math.* 67(1), 45–61.

Stewartson, K., and P. G. Williams. (1969). Self-induced separation. *Proc. Roy. Soc. London* A 312, 181–206.

Suh, Y. K., and C. S. Liu. (1990). Study on the flow structure around a flat plate in a stagnation flow field. *J. Fluid Mech.* 214, 469–487.

Sychev, V. V. (1967). On steady laminar fluid flow behind a bluff body at high Reynolds number. *Proc. 8th Symposium on Current Problems of the Mechanics of Liquids and Gases*, Tarda, Poland. (Engl. transl. 1969, Natl. Aero. Lab., Bangalore, India.)

Sychev, V. V. (1972). Laminar separation. *Mekh. Zhid. Gaza* (3), 47–59. (Engl. transl. *Fluid Dyn.* 7(3), 407–417.)

Sychev, V. V. (1978a). Boundary-layer separation from a flat surface. *Uch. zap. TsAGI* 9(3), 20–29.

Sychev, V. V. (1982). Asymptotic theory of separated flows. *Mekh. Zhid. Gaza* (2), 20–30. (Engl. transl. *Fluid Dyn.* 17(2), 179–188.)

Sychev, Vic. V. (1978b). Breakdown of a plane laminar wake. *Uch. zap. TsAGI* 9(6), 9–16.

Sychev, Vic. V. (1979). Asymptotic theory of unsteady separation. *Mekh. Zhid. Gaza* (6), 21–32. (Engl. transl. *Fluid Dyn.* 14(6), 829–838.)

Sychev, Vic. V. (1980). On some singularities in solutions of the boundary-layer equations on a moving surface. *Prikl. Mat. Mekh.* 44(5), 831–838. (Engl. transl. *J. Appl. Math. Mech.* 44(5), 587–591.)

Sychev, Vic. V. (1983). Theory of unsteady boundary-layer separation and wake breakdown. *Usp. Mekh.* 6(1,2), 13–51.

Sychev, Vic. V. (1984). On the asymptotic theory of laminar separation from a moving surface. *Prikl. Mat. Mekh.* 48(2), 247–253. (Engl. transl. *J. Appl. Math. Mech.* 48(2), 171–176.)

Sychev, Vic. V. (1987a). Analytical solution of the problem of flow near a boundary-layer separation point on a moving surface. *Prikl. Mat. Mekh.* 51(3), 519–521. (Engl. transl. *J. Appl. Math. Mech.* 51(3), 405–407.)

Sychev, Vic. V. (1987b). Singular solution of the boundary-layer equations on a moving surface. *Mekh. Zhid. Gaza* (2), 43–52. (Engl. transl. *Fluid Dyn.*

22(2), 203–211.)

Taganov, G. I. (1970). On the limiting flows of a viscous fluid with steady separation zones as Re → ∞. *Uch. zap. TsAGI* 1(3), 1–14.

Tani, I. (1964). Low-speed flows involving bubble separations. *Progr. Aeronaut. Sci.* 5, 70–103.

Telionis, D. P. (1981). *Unsteady Viscous Flows.* Springer, New York.

Telionis, D. P., and D. T. Tsahalis. (1974). Unsteady laminar separation over impulsively moved cylinders. *Acta Astronaut.* 1(11,12), 1487–1505.

Telionis, D. P., D. T. Tsahalis, and M. J. Werle. (1973). Numerical investigation of unsteady boundary-layer separation. *Phys. Fluids* 16(7), 968–973.

Telionis, D. P., and M. J. Werle. (1973). Boundary-layer separation from downstream moving boundaries. *ASME J. Appl. Mech.* 40(2), 369–374.

Terent'ev, E. D. (1981). The linear problem of a vibrator in a subsonic boundary layer. *Prikl. Mat. Mekh.* 45(6), 1049–1055. (Engl. transl. *J. Appl. Math. Mech.* 45(6), 791–795.)

Terent'ev, E. D. (1984). The linear problem of a vibrator performing harmonic oscillations at supercritical frequencies in a subsonic boundary layer. *Prikl. Mat. Mekh.* 48(2), 264–272. (Engl. transl. *J. Appl. Math. Mech.* 48(2), 184–191.)

Terent'ev, E. D. (1985). Linear problem of a vibrator on a flat plate in a subsonic boundary layer. In V. V. Kozlov (Ed.), *Laminar-Turbulent Transition*, pp. 303–311. Springer, Berlin.

Terent'ev, E. D. (1987). On the formation of a wave packet in a boundary layer on a flat plate. *Prikl. Mat. Mekh.* 51(5), 814–819. (Engl. transl. *J. Appl. Math. Mech.* 51(5), 640–644.)

Terrill, R. M. (1960). Laminar boundary-layer flow near separation with and without suction. *Phil. Trans. Roy. Soc. London* A 253, 55–100.

Thoman, D. C., and A. A. Szewczyk. (1969). Time-dependent viscous flow over a circular cylinder. *Phys. Fluids* 12(12, Suppl. II), 76–86.

Timoshin, S. N. (1985). Separated flow around a thin profile with parabolic leading edge at high Reynolds number. *Uch. zap. TsAGI* 16(2), 24–32.

Timoshin, S. N. (1988a). Elimination of marginal separation through the effect of flow pulsations. *Prikl. Mat. Mekh.* 52(1), 77–81. (Engl. transl. *J. Appl. Math. Mech.* 52(1), 59–62.)

Timoshin, S. N. (1988b). Asymptotic analysis of locally disturbed pulsating flows in the boundary layer. *Zh. Prikl. Mekh. Tekh. Fiz.* (2), 23–29. (Engl. transl. *J. Appl. Mech. Tech. Phys.* 29(2), 179–184.)

Timoshin, S. N. (1988c). On the unsteady interaction regime in a pulsating boundary layer. *Zh. Prikl. Mekh. Tekh. Fiz.* (6), 79–85. (Engl. transl. *J. Appl. Mech. Tech. Phys.* 29(6), 835–841.)

van de Vooren, A. I., and D. Dijkstra. (1970). The Navier–Stokes solution for laminar flow past a semi-infinite flat plate. *J. Eng. Math.* 4(1), 9–27.

van Dommelen, L. L., and S. F. Shen. (1980). The spontaneous generation of the singularity in a separating laminar boundary layer. *J. Comput. Phys.* 38(2), 125–140.

van Dommelen, L. L., and S. F. Shen. (1982). The genesis of separation. In T. Cebeci (Ed.), *Numerical and Physical Aspects of Aerodynamic Flows*, pp. 293–311. Springer, New York.

van Dommelen, L. L., and S. F. Shen. (1983a). Boundary layer separation singularities for an upstream moving wall. *Acta Mech.* 49(3,4), 241–254.

van Dommelen, L. L., and S. F. Shen. (1983b). An unsteady interactive separation process. *AIAA J.* 21(3), 358–362.

van Dommelen, L. L., and S. F. Shen. (1984). Interactive separation from a fixed wall. In T. Cebeci (Ed.), *Numerical and Physical Aspects of Aerodynamic Flows II*, pp. 393–402. Springer, New York.

Van Dyke, M. D. (1956). *Second-Order Subsonic Airfoil Theory Including Edge Effects.* NACA Rep. 1274.

Van Dyke, M. D. (1964). *Perturbation Methods in Fluid Mechanics.* Academic Press, New York. (Annotated ed. 1975, Parabolic Press, Stanford, CA.)

Van Dyke, M. D. (1967). *Survey of Higher-Order Boundary-Layer Theory.* SUDAAR No. 326, Dept. of Aeronautics and Astronautics, Stanford Univ.

Van Dyke, M. D. (1969). Higher-order boundary-layer theory. *Ann. Rev. Fluid Mech.* 1, 265–292.

Varty, R. L., and I. G. Currie. (1984). Measurements near a laminar separation point. *J. Fluid Mech.* 138, 1–19.

Veldman, A. E. P. (1976). *Boundary-Layer Flow Past a Finite Flat Plate.* Rijksuniv. Groningen.

Veldman, A. E. P. (1981). New, quasi-simultaneous method to calculate interacting boundary layers. *AIAA J.* 19(1), 79–85.

Veldman, A. E. P., and D. Dijkstra. (1981). A fast method to solve incompressible boundary-layer interaction problems. *Lecture Notes in Phys.* 141, 411–416.

Veldman, A. E. P., and A. I. van de Vooren. (1975). Drag of a finite flat plate. *Lecture Notes in Phys.* 35, 423–430.

Villat, H. (1914). Sur la validité des solutions de certains problèmes d'hydrodynamique. *J. Math. Pures Appl. 6ᵉ sér.*, 10(3), 231–290.

Walker, J. D. A. (1978). The boundary layer due to rectilinear vortex. *Proc. Roy. Soc. London* A 359, 167–188.

Ward, J. W. (1963). The behaviour and effects of laminar separation bubbles on aerofoils in incompressible flow. *J. Roy. Aero. Soc.* 67, 783–790.

Werlé, H. (1971). Sur l'éclatement des tourbillons. *ONERA Note Tech.*, No. 175.

Werle, M. J., and R. T. Davis. (1972). Incompressible laminar boundary layers on a parabola at angle of attack: A study of the separation point. *ASME J. Appl. Mech.* 39(1), 7–12.

Werle, M. J., A. Polak, and S. D. Bertke. (1973). *Supersonic Boundary-Layer Separation and Reattachment—Finite Difference Solutions.* Rep. No. AFL 72-12-1, Dept. Aerospace Eng., Univ. Cincinnati.

Werle, M. J., and V. N. Vatsa. (1974). New method for supersonic boundary-layer separation. *AIAA J.* 12(11), 1491–1497.

Werle, M. J., and J. M. Verdon. (1979). Solutions for supersonic trailing edges including separation. *AIAA Paper* 79-1544.

Weyl, H. (1942). On the differential equations of the simplest boundary-layer problems. *Ann. Math. 2 sér.*, 43(2), 381–407.

Williams, J. C. (1977). Incompressible boundary-layer separation. *Ann. Rev. Fluid Mech.* 9, 113–144.

Williams, J. C. (1982a). Flow development in the vicinity of the sharp trailing edge on bodies impulsively set into motion. *J. Fluid Mech.* 115, 27–37.

Williams, J. C. (1982b). Unsteady development of the boundary layer in the vicinity of a rear stagnation point. In T. Cebeci (Ed.), *Numerical and Physical Aspects of Aerodynamic Flows*, pp. 347–364. Springer, New York.

Williams, J. C., and W. D. Johnson. (1974a). Note on unsteady
  boundary-layer separation. *AIAA J.* 12(10), 1427–1429.
Williams, J. C., and W. D. Johnson. (1974b). Semisimilar solutions to unsteady
  boundary-layer flows including separation. *AIAA J.* 12(10), 1388–1393.
Williams, J. C., and K. Stewartson. (1983). Flow development in the vicinity
  of the sharp trailing edge on bodies impulsively set into motion. Part 2. *J.
  Fluid Mech.* 131, 177–194.
Williams, P. G. (1975). A reverse flow computation in the theory of
  self-induced separation. *Lecture Notes in Phys.* 35, 445–451.
Wu, T. Y. (1972). Cavity and wake flows. *Ann. Rev. Fluid Mech.* 4, 243–284.
Yoshizawa, A. (1970). Laminar viscous flow past a semi-infinite flat plate. *J.
  Phys. Soc. Japan* 28(3), 776–779.
Zametaev, V. B. (1986). Existence and nonuniqueness of local separation zones
  in viscous jets. *Mekh. Zhid. Gaza* (1), 38–45. (Engl. transl. *Fluid Dyn.*
  21(1), 31–38.)
Zametaev, V. B. (1987). Asymptotic analysis of an integro-differential equation
  in the theory of marginal separation. *Uch. zap. TsAGI* 18(3), 120–124.
Zhuk, V. I. (1982). On local recirculation zones in the supersonic boundary
  layer on a moving surface. *Zh. Vych. Mat. Mat. Fiz.* 22(5), 1255–1260.
  (Engl. transl. *USSR Comput. Math. and Math. Phys.* 22(5), 249–255.)
Zhuk, V. I. (1988). On the flow in the free-interaction region near a permeable
  section of a wall. *Zh. Vych. Mat. Mat. Fiz.* 28(6), 941–945. (Engl. transl.
  USSR Comput. Math. and Math. Phys. 28(3), 216–219.)
Zhuk, V. I., and O. S. Ryzhov. (1979). On a boundary layer with self-induced
  pressure on a moving surface. *Dokl. Akad. Nauk SSSR* 248(2), 314–318.
Zhuk, V. I., and O. S. Ryzhov. (1980). Free interaction and stability of the
  boundary layer in an incompressible fluid. *Dokl. Akad. Nauk SSSR* 253(6),
  1326–1329.
Zhuk, V. I., and O. S. Ryzhov. (1982). On locally inviscid perturbations in a
  boundary layer with self-induced pressure. *Dokl. Akad. Nauk SSSR* 263(1),
  56–59.
Zhukovskii, N. E. (1907). On attached vortices. *Proc. Phys. Sect. Soc. Natural
  Scientists* 13(2), 12–25.
Zhukovskii, N. E. (1910). Über die Konturen der Tragflächen der
  Drachenflieger. *Z. Flugtechn. Motorluftschiff.* Jahrg. 1 (22), 281–284.
Zubarev, V. M. (1983). The boundary layer on the moving surface of a cylinder.
  Mekh. Zhid. Gaza (6), 38–42. (Engl. transl. *Fluid Dyn.* 18(6), 862–866.)
Zubarev, V. M. (1984). Laminar boundary layer on the moving surface of a
  Rankine oval. *Mekh. Zhid. Gaza* (3), 171–174. (Engl. transl. *Fluid Dyn.*
  19(3), 490–494.)
Zubtsov, A. V. (1985). Separation of a laminar boundary layer from a plane
  surface. *Uch. zap. TsAGI* 16(5), 108–110.

Index